Multivariate Time Series Analysis in Climate and Environmental Research

Zhihua Zhang

Multivariate Time Series Analysis in Climate and Environmental Research

Springer

Zhihua Zhang
College of Global Change and Earth System Science
Beijing Normal University
Beijing
China

ISBN 978-3-319-67339-4 ISBN 978-3-319-67340-0 (eBook)
https://doi.org/10.1007/978-3-319-67340-0

Library of Congress Control Number: 2017954476

Printed on acid-free paper

This Springer imprint is published by Springer Nature
The registered company is Springer International Publishing AG
The registered company address is: Gewerbestrasse 11, 6330 Cham, Switzerland

Preface

The Earth's climate is a complex, multidimensional multiscale system in which different physical processes act on different temporal and spatial scales. Due to the increasing atmospheric greenhouse gas concentrations, global average temperatures increase with time as a result of interactions among components of the climate system. These interactions and the resulting variations in various climate parameters occur on a variety of timescales ranging from seasonal cycles, yearly cycles to those with times measured in hundreds of years. Climatologists and environmentalists are striking to extract meaningful information from huge amount of observational record and simulation data for the climate system. Classic univariate time series analysis is not capable to handle well these complex multidimensional data. Recently, the techniques and methods of multivariate time series analysis have gained great important in revealing mechanisms of climate change, modeling tempo-spatial evolution of climate change and predicting the trend of future climate change.

This book covers the comprehensive range of theory, models, and algorithms of state-of-the-art multivariate time series analysis which have been widely used in monitoring, modeling, and prediction of climate and environmental change. Each chapter focuses on a specific issue of importance. Chapter 1 discusses artificial neural networks which can make full use of some unknown information hidden in high-dimensional climate data, although these information cannot be extracted directly; Chap. 2 discusses multivariate Harmonic analysis which can determine how the total variance of multivariate time series is distributed in frequency. Main techniques and methods include Fourier transform, fractional Fourier transform, space–frequency representation, sparse approximation, spherical harmonics, and harmonic analysis on graphs; Chap. 3 discusses wavelet representation for multivariate time series with time-dependent dominant cycles. Main techniques and methods include multiresolution analysis and wavelets, discrete wavelet transform, wavelet packet, wavelet variance, significant tests, wavelet shrinkage, and shearlets, bandelets, and curvelets. Chapter 4 focuses on stochastic representation and modeling, including stationarity and trend tests, principal component analysis, factor analysis, cluster analysis, discriminant analysis, canonical correlation analysis, multidimensional scaling, vector ARMA models, Monte Carlo methods, Black–Scholes model, and stochastic optimization; Chap. 5 discusses multivariate spectral analysis and estimation, including periodogram method, Blackman–Tukey method, maximum entropy method, multitaper method, vector ARMA spectrum, and multichannel SSA; Chap. 6 focuses on the development of climate models and related experiments to understand the climate system and climate change; Chap. 7 gives some latest case studies on regional climate change to demonstrate how the methods and tools in Chaps. 1–6 are used; Chap. 8 discusses basic models and key indices on ecosystem and global carbon cycle; Chap. 9 discusses the methods used to reconstruct paleoclimates from proxy data. Chapter 10 introduces three methods to analyze multivariate time series in climate change economics and related latest researches.

Current climate and environmental research is facing the challenge of complex multidimensional data. This book on multivariate time series analysis starts from first principles, always explains various techniques and methods step by step, and shows clearly how to reveal physical meaning from the analysis of observed and stimulated multidimensional data. It has a comprehensive cover and also includes many of the author's unpublished researches. This book is accessible for researchers and advanced students who want to grasp state-of-the-art techniques and methods in multivariate time series analysis. This book builds a cross-disciplinary bridge between various analysis techniques and methods and latest published studies in the wide branches of climatology and environmental science.

Beijing, China Zhihua Zhang

Contents

Chapter 1
Artificial Neural Network

Multivariate time series analysis in climate and environmental research always requires to process huge amount of data. Inspired by human nervous system, the *artificial neural network* methodology is a powerful tool to handle this kind of difficult and challenge problems and has been widely used to investigate mechanism of climate change and predict the climate change trend. The main advantage is that artificial neural networks make full use of some unknown information hidden in climate data although they cannot extract it. In this chapter, we will introduce various neural networks, including linear networks, radial basis function networks, generalized regression networks, Kohonen self-organizing networks, learning vector quantization networks, and Hopfield networks.

1.1 Network Architectures

Artificial neural network is a structure of interconnected units of large number of *neurons*. Each neuron in the network is able to receive input signals, to process them, and to send an output signal. It consists of a set of the weighted synapses, an adder for summing the input data weighted by the respective synaptic strength, and an activation function for limiting the amplitude of the output of the neuron. Network architectures have two fundamentally different classes including multilayer feedforward networks and recurrent networks.

1.1.1 Multilayer Feedforward Networks

Feedforward networks are currently being used in a variety of climate and environment applications with great success. It consists of a number of neurons organized in

Z. Zhang, *Multivariate Time Series Analysis in Climate and Environmental Research*, https://doi.org/10.1007/978-3-319-67340-0_1

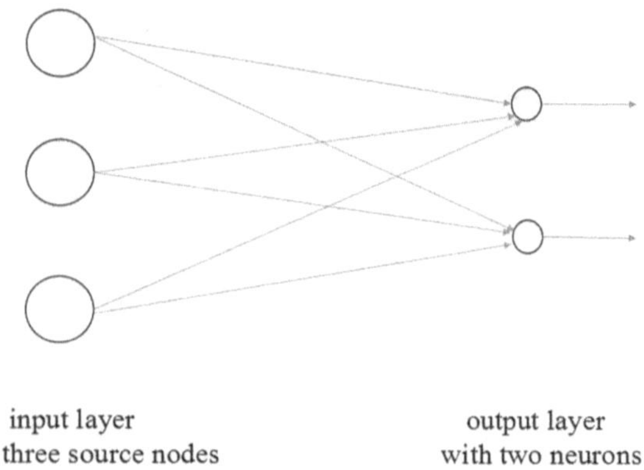

Fig. 1.1 Single-layer feedforward networks

layers. Every neuron in a layer is connected with all the neurons in the previous layer. These connections are not all equal; each connection may have a different strength or weight.

The oldest and simplest artificial neural networks is a single-layer feedforward neural network (see Fig. 1.1). It consists of an input layer of source nodes and an output layer of neurons, where source nodes are projected directly onto the output layer of neurons. The word "single-layer" means that the neural network has only a layer. The layer of source nodes is not counted because no computation is performed.

A multilayer feedforward network consists of an input layer of source nodes, one or more *hidden layers*, and an output layer of neurons (see Fig. 1.2). The hidden layers in the network are not seen directly from either the input or output layer of the network. These hidden layers enable the neural network to extract the higher-order statistical features from its input. Neurons in hidden layers are correspondingly called *hidden neurons*. These hidden neurons have a function to intervene between external input and the network output in some useful manner.

The source nodes in the input layer supply respective elements of the activation pattern to constitute input signals applied to neurons in the first hidden layer. The output signals of neurons in the first hidden layer only are used as input signals to neurons in the second hidden layer. Generally, the output signals of neurons in each hidden layer only are used as input signals to neurons in the adjacent forward hidden layer. There is no connection among neurons in the same layer. Finally, the output signals of neurons in the last hidden layer only are used as input signals to neurons in the output layer. The set of output signals of the neurons in the output layer of the network constitute the overall response of the network to the activation pattern supplied by the source nodes in the input layer of the network.

If every neuron in each layer of the multilayer feedforward network is connected to every neuron in the next layer, then this kind of neural network is called *fully connected*. The simplest fully connected multilayer feedforward network is the network with one hidden layer and one output layer. If such network has m source nodes, h hidden neurons, and n output neurons, for the sake of brevity, this fully connected

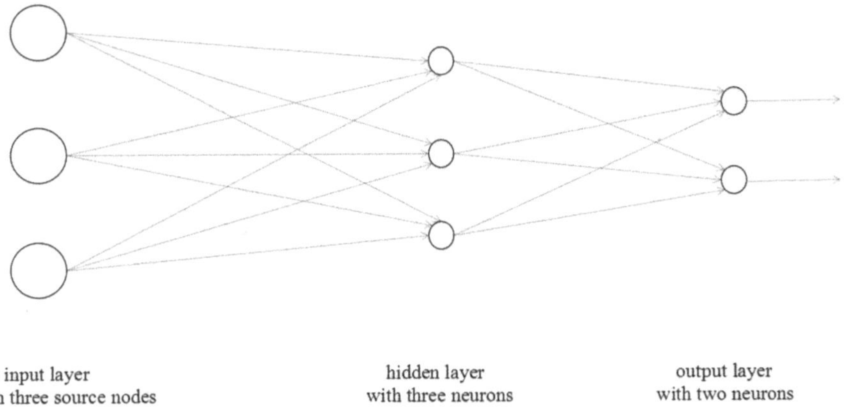

Fig. 1.2 Multilayer feedforward network with a single hidden layer (3-3-2 Network)

multilayer feedforward network is referred to as an $m - h - n$ network. In general, for a fully connected multilayer network with k hidden layers and one output layer, if it has m source nodes in the input layer, h_1 neurons in the first hidden layer, h_2 neurons in the second hidden layer, ..., h_k neurons in the kth hidden layer, and n output neurons, then it is referred to as an $m - h_1 - h_2 - \cdots - h_k - n$ network. If some synaptic connections are missing from the multilayer feedforward network, then the network is called *partially connected.*

1.1.2 Recurrent Networks

Recurrent networks have self-feedback loops or not, the feedback loops involve the use of particular branches composed of unit-delay elements, and recurrent networks also have hidden neurons or not. A recurrent neural network distinguishes from feedforward networks in that it has at least one feedback loop. This offers a lot of flexibility and can approximate arbitrary dynamical systems with arbitrary precision. The network with a single-loop feedback is called a *single-loop feedback network.*

Consider a single-loop feedback network. Denote its input signal by $x(n)$, internal signal by $\widetilde{x}(n)$, and output signal by $y(n)$, where $x(n)$, $\widetilde{x}(n)$, $y(n)$ are dependent on the discrete-time variable n. Assume that the network consists of a forward path $\mathbf{A}$ and a feedback path $\mathbf{B}$, where $\mathbf{A}$ and $\mathbf{B}$ are operators (see Fig. 1.3).

Assume that the output of the forward channel determines in part its own output through the feedback channel. Then, the input and output satisfy the following relationship:

$$y(n) = \mathbf{A}[\widetilde{x}(n)],$$
$$\widetilde{x}(n) = x(n) + \mathbf{B}[y(n)],$$

Fig. 1.3 Single-loop feedback system

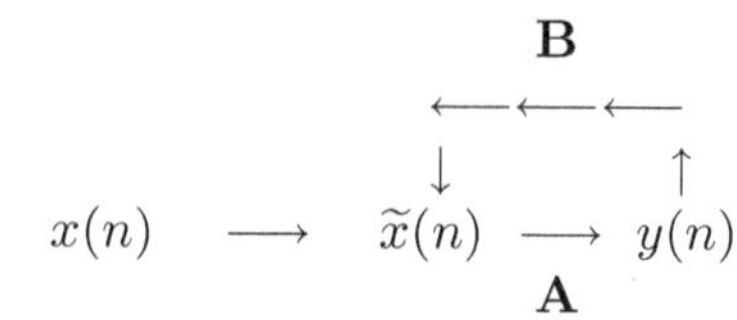

where $\mathbf{A}[\widetilde{x}(n)]$ and $\mathbf{B}[y(n)]$ mean that the operators $\mathbf{A}$ and $\mathbf{B}$ act on $\widetilde{x}(n)$ and $y(n)$, respectively. Eliminating $\widetilde{x}(n)$ from both equations, we get

$$y(n) = \frac{\mathbf{A}}{1 - \mathbf{AB}}[x(n)], \tag{1.1.1}$$

where $\frac{\mathbf{A}}{1-\mathbf{AB}}$ is called a *closed-loop operator* and $\mathbf{AB}$ is called an *open-loop operator*. In general, $\mathbf{AB} \neq \mathbf{BA}$.

If the operator $\mathbf{A}$ is a fixed weight w and the operator $\mathbf{B}$ is a unit-delay operator $\mathbf{z}^{-1}$ whose output is delayed with respect to the input by one time unit, then the closed-loop operator becomes

$$\frac{\mathbf{A}}{1 - \mathbf{AB}} = \frac{w}{1 - w\mathbf{z}^{-1}} = \sum_{l=0}^{\infty} w^{l+1}\mathbf{z}^{-l}.$$

Substituting it into (1.1.1), we get

$$y(n) = \sum_{l=0}^{\infty} w^{l+1}\mathbf{z}^{-l}[x(n)],$$

where $\mathbf{z}^{-l}[x(n)]$ means that the operator $\mathbf{z}^{-l}$ acts on $x(n)$. Since $\mathbf{z}^{-1}$ is a unit-delay operator,

$$\mathbf{z}^{-l}[x(n)] = x(n - l),$$

Furthermore,

$$y(n) = \sum_{l=0}^{\infty} w^{l+1}x(n - l).$$

From this, the dynamic behavior of the single-loop feedback network with the fixed weight w and the unit-delay operator $\mathbf{z}^{-1}$ is determined by the weight w. When $|w| < 1$, the output signal $y(n)$ is *convergent* exponentially. In this case, the system is *stable*. When $|w| \geq 1$, the output signal $y(n)$ is *divergent*. In this case, the system is *unstable*. If $|w| = 1$, the divergence is linear. If $|w| > 1$, the divergence is exponential.

1.2 Perceptrons

The perceptron is a kind of neural networks that can decide whether an input belongs to some specific class. It was the first algorithmically described neural network. Perceptrons can be classified into Rosenblatt's perceptron and multilayer perceptrons.

1.2.1 Rosenblatt's Perceptron

Rosenblatt's perceptron is a network with m input source nodes and a single output neuron, a more general computational model than McCulloch–Pitts model. The perceptron belongs to basically a single-layer neural network and consists of a single neuron with adjustable synaptic weight and bias.

Rosenblatt's perceptron can be described mathematically by the pair of equations:

$$\begin{cases} v = \sum_{i=1}^{m} w_i x_i + b, \\ y = \varphi(v) \end{cases} \tag{1.2.1}$$

or by an equivalent equation:

$$y = \varphi\left(\sum_{i=1}^{m} w_i x_i + b\right),$$

where x_i $(i = 1, \ldots, m)$ are the input signals applied to the perceptron, w_i $(i = 1, \ldots, m)$ are the synaptic weights of the perceptron, b is the externally applied bias, v is called the *induced local field* (or linear combiner), φ is the activation function (or hard limiter), and y is the output signal of the perceptron (see Fig. 1.4). The Rosenblatt's perceptron consists of a linear combiner followed by a hard limiter, and the hard limiter $\varphi(v)$ determines the output of the perceptron in terms of the induced local field v.

Let $x_0 = 1$ and $w_0 = b$ be the input and the weight of a new synapse. An equivalent form of (1.2.1) is

$$\begin{cases} v = \sum_{i=0}^{m} w_i x_i, \\ y = \varphi(v) \end{cases} \quad \text{or} \quad y = \varphi\left(\sum_{i=0}^{m} w_i x_i\right).$$

The activation function in Rosenblatt's perceptron is a threshold function as follows:

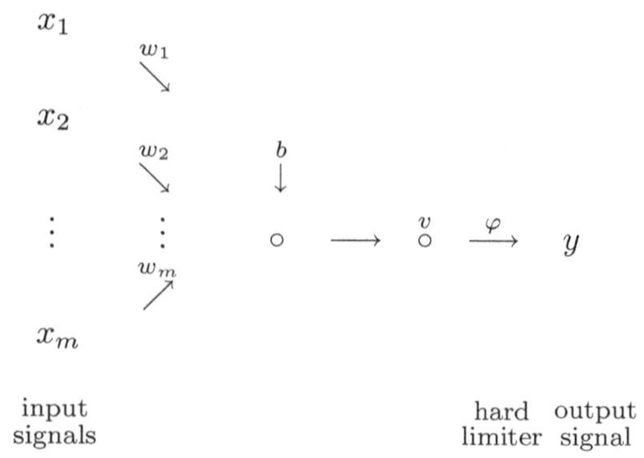

Fig. 1.4 Rosenblatt's perceptron

(a) The Heaviside function is defined as

$$\varphi(v) = \begin{cases} 1 \text{ if } v > 0, \\ 0 \text{ if } v \leq 0. \end{cases}$$

If v is the induced local field and $v = \sum_{i=1}^{m} w_i x_i + b$, then the corresponding output is expressed as

$$y = \begin{cases} 1 \text{ if } v > 0, \\ 0 \text{ if } v \leq 0. \end{cases}$$

(b) The signum function is defined as

$$\varphi(v) = \begin{cases} 1 & \text{if } v > 0, \\ 0 & \text{if } v = 0, \\ -1 & \text{if } v < 0. \end{cases}$$

If v is the induced local field and $v = \sum_{i=1}^{m} w_i x_i + b$, then the corresponding output is expressed as

$$y = \begin{cases} 1 & \text{if } v > 0, \\ 0 & \text{if } v = 0, \\ -1 & \text{if } v < 0. \end{cases}$$

Based on the feature of the harder limiter, Rosenblatt's perceptron can classify correctly the set of externally applied stimuli $x_1, \ldots, x_m$ into one of two classes. The decision rule for the classification is as follows:

- If the input to the hard limiter $v > 0$, when the activation function is the Heaviside function (or the signum function), the points represented by stimuli $x_1, \ldots, x_m$ are assigned to Class 1;
- If the input to the hard limiter $v < 0$, when the activation function is Heaviside function (or the signum function), the points represented by stimuli $x_1, \ldots, x_m$ are assigned to Class 0 (or Class -1).

Here, Class k is the set consisting of points represented by the stimuli $x_1, \ldots, x_m$ having the output $y = k$.

The input to the hard limiter $v = 0$, i.e., the hyperplane

$$\sum_{i=1}^{m} w_i x_i + b = 0$$

is referred to as the *decision boundary* of two classes. The simplest Rosenblatt's perceptron with two source nodes and a single output neuron can classify the set of externally applied stimuli x_1, x_2 into one of two classes, and the decision boundary of these two classes is a straight line:

$$w_1 x_1 + w_2 x_2 + b = 0, \tag{1.2.2}$$

where $w_i (i = 1, 2)$ and b are the synaptic weights and bias of the perceptron, respectively. For example, assume that $w_1 = 1$, $w_2 = 0.5$, $b = -0.5$, and the sampling set of externally applied stimuli x_1, x_2 consists of seven points:

$$\begin{array}{llll} \mathbf{x}^{(1)} = (-1, 2), & \mathbf{x}^{(2)} = (1, 2), & \mathbf{x}^{(3)} = (2, -1), & \\ \mathbf{x}^{(4)} = (1, 1), & \mathbf{x}^{(5)} = (-2, 1), & \mathbf{x}^{(6)} = (-1, -1), & \mathbf{x}^{(7)} = (2, 0). \end{array}$$

By (1.2.2), the decision boundary of two classes is $x_1 + 0.5x_2 - 0.5 = 0$. It is seen that four points $\mathbf{x}^{(2)}, \mathbf{x}^{(3)}, \mathbf{x}^{(4)}, \mathbf{x}^{(7)}$ lie above the straight line and the rest $\mathbf{x}^{(1)}, \mathbf{x}^{(5)}, \mathbf{x}^{(6)}$ lie below the straight line. The decision rule for the classification shows that these seven points can be classified into two classes.

1.2.2 Multilayer Perceptron

The architecture of multilayer perceptrons is very different from the single-layer perceptron. It is a perceptron consisting of an input layer of m source nodes, i hidden layers of h_i neurons, and an output layer of n neurons. Each neuron of the network has a differentiable nonlinear activation function. The architecture of a fully connected multilayer perceptron is that a neuron node in any layer of the network is connected to all neuron nodes in the previous layer and the signal flow through the network progresses in a forward direction from left to right and on a layer-by-layer basis.

The activation function used commonly in the multilayer perceptron is the sigmoid function. The sigmoid function is mathematically convenient and is close to linear near the origin while saturating rather quickly when getting away from the origin. This allows multilayer perceptrons to model well both strongly and mildly nonlinear relations. The logistic function and the hyperbolic tangent function are two popular sigmoid functions. The logistic function is defined as

$$\varphi(v) = \frac{1}{1+e^{-av}} \qquad (a > 0),$$

where a is its slop parameter. Logistic functions of different slopes can be obtained by varying this parameter. The logistic function is differentiable, and its derivative is

$$\varphi'(v) = \frac{ae^{-av}}{(1+e^{-av})^2} = \frac{a}{1+e^{-av}}\left(1 - \frac{1}{1+e^{-av}}\right) = a\varphi(v)(1-\varphi(v)).$$

The hyperbolic tangent function is defined as

$$\varphi(v) = a\tanh(bv) \qquad (a > 0,\ b > 0).$$

The hyperbolic tangent function is also differentiable, and its derivative is

$$\begin{aligned}\varphi'(v) &= ab\,\mathrm{sech}^2(bv) = ab(1-\tanh^2(bv)) = \tfrac{b}{a}(a^2-\varphi^2(v))\\ &= \tfrac{b}{a}(a-\varphi(v))(a+\varphi(v)).\end{aligned}$$

The process of the multilayer perceptron includes forward propagation of function signals and backward propagation of error signals which are identified. *Forward propagation of function signals* is such a process that an input signal comes in at the input end of the network, propagates forward neuron by neuron through the network, and emerges at the output end of the network as an output signal. *Backward propagation of error signals* is such a process that the error signal originates at an output neuron of the network and propagates backward layer-by-layer through the network.

Due to one or more hidden layers, the multilayer perceptron can classify nonlinearly separable patterns, while the single-layer perceptron cannot.

Example 1.2.1 Assume that a sampling set consists of four points:

$$\begin{aligned}\mathbf{x}^{(1)} &= (0,0), \qquad & \mathbf{x}^{(2)} &= (0,1),\\ \mathbf{x}^{(3)} &= (1,0), \qquad & \mathbf{x}^{(4)} &= (1,1).\end{aligned}$$

The XOR problem is to use a perceptron to classify these four points into two classes such that points $\mathbf{x}^{(1)}$, $\mathbf{x}^{(4)}$ are assigned to a class, and points $\mathbf{x}^{(2)}$, $\mathbf{x}^{(3)}$ are assigned to another class. There is no possibility to solve the XOR problem using any single-layer perceptron. In fact, assume that a following single-layer perceptron can solve it:

$$\begin{cases} v = w_1x_1 + w_2x_2 + b,\\ y = \varphi(v),\end{cases}$$

where $w_i (i = 1, 2)$ and b are synaptic weights and bias, respectively, and the activation function φ is the Heaviside function. Since $\mathbf{x}^{(1)} = (0,0)$ and $\mathbf{x}^{(4)} = (1,1)$ are assigned to Class 0, it is clear that $b < 0$ and $w_1 + w_2 + b < 0$. Adding them together gives

$$w_1 + w_2 + 2b < 0. \qquad (1.2.3)$$

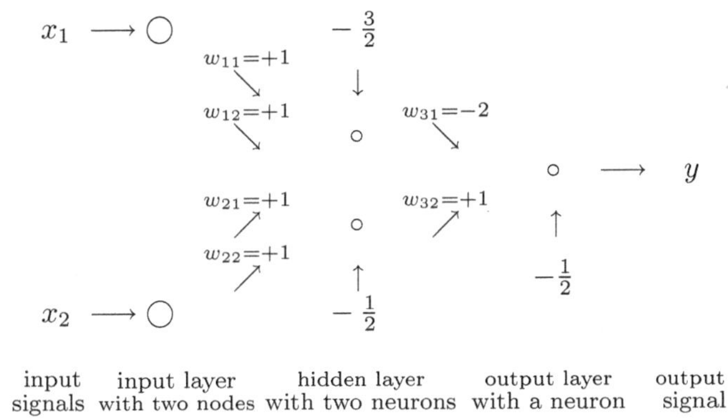

Fig. 1.5 Touretzky–Pomerleau perceptron

Since $\mathbf{x}^{(2)} = (0, 1)$ and $\mathbf{x}^{(3)} = (1, 0)$ are assigned to Class 1, then

$$\begin{aligned} w_2 + b &> 0, \\ w_1 + b &> 0. \end{aligned}$$

Adding them together, we get

$$w_1 + w_2 + 2b > 0.$$

This is contrary to (1.2.3). Thus, there is no possibility to solve the XOR problem using a single-layer perceptron.

Touretzky–Pomerleau perceptron is a multilayer perceptron (see Fig. 1.5), where each "o" represents a neuron and each neuron has Heaviside function as the activation function. Touretzky–Pomerleau perceptron can solve the XOR problem given by Example 1.2.1.

There are three neurons in Touretzky–Pomerleau perceptron, two of them are hidden neurons in the hidden layer and the remainder one is the output neuron in the output layer. For the top hidden neuron, the synaptic weights and bias are, respectively,

$$\begin{aligned} w_{11} &= w_{12} = 1, \\ b_1 &= -\tfrac{3}{2}. \end{aligned}$$

The straight line

$$l_1: \quad x_1 + x_2 - \frac{3}{2} = 0.$$

is the decision boundary formed by the top hidden neuron. It is clear that the point $\mathbf{x}^{(4)}$ lying above the line l_1 is assigned to Class 1, and points $\mathbf{x}^{(1)}, \mathbf{x}^{(2)}, \mathbf{x}^{(3)}$ lying below the line l_1 are assigned to Class 0.

For the bottom hidden neuron, the synaptic weights and bias are, respectively,

$$w_{21} = w_{22} = 1,$$
$$b_2 = -\tfrac{1}{2}.$$

The straight line

$$l_2: \quad x_1 + x_2 - \frac{1}{2} = 0$$

is the decision boundary formed by the bottom hidden neuron. The points $\mathbf{x}^{(2)}$, $\mathbf{x}^{(3)}$, $\mathbf{x}^{(4)}$ lying above the line l_2 are assigned to Class 1, and the point $\mathbf{x}^{(1)}$ lying below the line l_2 is assigned to Class 0.

For the output neuron, the synaptic weights and bias are, respectively,

$$w_{31} = -2, \qquad w_{32} = 1,$$
$$b_3 = -\tfrac{1}{2}.$$

The output neuron constructs a linear combination of decision boundaries formed by two hidden neurons. The decision boundary formed by the output neuron is a straight line:

$$-2y_1 + y_2 - \frac{1}{2} = 0$$

where y_1 is the output from the top hidden neuron and y_2 is the output from the bottom neuron. The decision rule for the classification is that a point that lies above the line l_1 or below the line l_2 is assigned to Class 0 and a point that lies both below the line l_1 and above the line l_2 is assigned to Class 1. Therefore, according to the decision rule for the classification, the points $\mathbf{x}^{(1)}$, $\mathbf{x}^{(4)}$ are assigned to Class 0 and the points $\mathbf{x}^{(2)}$, $\mathbf{x}^{(3)}$ are assigned to Class 1. So the XOR problem for four points $\mathbf{x}^{(i)}$ is solved using the Touretzky–Pomerleau perceptron.

1.3 Linear Network and Bayes Classifier

Linear neural network distinguishes from Rosenblatt's perceptron in that activation function is a linear function. So the output of the linear network may be any value. A linear network of a neuron can be described by a pair of equations:

$$\begin{cases} v = \sum\limits_{i=1}^{m} w_i x_i + b, \\ y = \varphi(v) = v \end{cases}$$

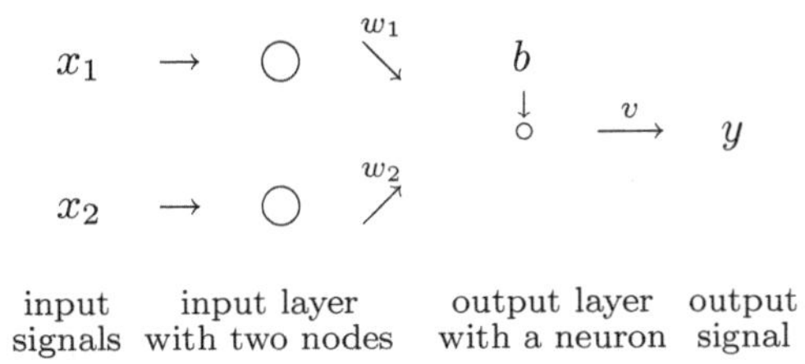

Fig. 1.6 Linear network with double input signals

or simply by an equation:

$$y = \sum_{i=1}^{m} w_i x_i + b = \sum_{i=0}^{m} w_i x_i,$$

where $x_0 = 1$, x_i $(l = 1, \ldots, m)$ are the input signals, w_i $(l = 1, \ldots, m)$ are the synaptic weights, $w_0 = b$ is the bias, and y is the output signal of the neuron. Similar to Rosenblatt's perceptrons, linear neural network is used for classifying linearly separable patterns. For example, a linear network of a neuron with two source nodes is shown as in Fig. 1.6, where x_1, x_2 are two input signals, y is the output signal, w_1, w_2 are two synaptic weights, b is the bias, and $\circ$ represents the neuron. This network can be described by a pair of equations:

$$\begin{cases} v = w_1 x_1 + w_2 x_2 + b, \\ y = v \end{cases}$$

or simply by an equation: $y = w_1 x_1 + w_2 x_2 + b$.
The corresponding decision boundary is a straight line $w_1 x_1 + w_2 x_2 + b = 0$, and the decision rule for classification is that all points that lie above the straight line are assigned to one class and all points that lie below the straight line are assigned to another class.

Below, we discuss *the Bayes classifier*. Consider a two-class problem represented by classes $\mathcal{B}_i$ in the subspace $\mathbb{R}_i$ $(i = 1, 2)$, where $\widetilde{\mathcal{R}} = \mathcal{R}_1 + \mathcal{R}_2$. Denote by $p_{\mathbf{X}}(\mathbf{X}|\mathcal{B}_i)$ $(i = 1, 2)$ the conditional probability density function of a random vector $\mathbf{X}$, given that the observation vector $\mathbf{x}$ is drawn from subspace $\mathbb{R}_i$. In the Bayes classifier, Van Trees defined an *average risk* (AR) of the two-class problem by

$$\begin{aligned} \mathrm{AR} = {} & c_{11} p_1 \int_{\mathcal{R}_1} p_{\mathbf{X}}(\mathbf{X}|\mathcal{B}_1) d\mathbf{X} + c_{22} p_2 \int_{\mathcal{R}_2} p_{\mathbf{X}}(\mathbf{X}|\mathcal{B}_2) d\mathbf{X} \\ & + c_{21} p_1 \int_{\mathcal{R}_2} p_{\mathbf{X}}(\mathbf{X}|\mathcal{B}_1) d\mathbf{X} + c_{12} p_2 \int_{\mathcal{R}_1} p_{\mathbf{X}}(\mathbf{X}|\mathcal{B}_2) d\mathbf{X}, \end{aligned}$$

where c_{ij} is the cost of deciding in favor of class $\mathcal{B}_i$ and p_i is the prior probability of the observation vector $\mathbf{x}$, and $p_1 + p_2 = 1$. Note that

$$\widetilde{\mathcal{R}} = \mathcal{R}_1 + \mathcal{R}_2.$$

The equivalent form of the average risk is

$$\begin{aligned}
\mathrm{AR} &= c_{11}p_1 \int\limits_{\mathcal{R}_1} p_{\mathbf{X}}(\mathbf{X}|\mathcal{B}_1)\mathrm{d}\mathbf{X} + c_{22}p_2 \int\limits_{\tilde{\mathcal{R}}-\mathcal{R}_1} p_{\mathbf{X}}(\mathbf{X}|\mathcal{B}_2)\mathrm{d}\mathbf{X} \\
&\quad + c_{21}p_1 \int\limits_{\tilde{\mathcal{R}}-\mathcal{R}_1} p_{\mathbf{X}}(\mathbf{X}|\mathcal{B}_1)\mathrm{d}\mathbf{X} + c_{12}p_2 \int\limits_{\mathcal{R}_1} p_{\mathbf{X}}(\mathbf{X}|\mathcal{B}_2))\mathrm{d}\mathbf{X} \\
&= c_{11}p_1 \int\limits_{\mathcal{R}_1} p_{\mathbf{X}}(\mathbf{X}|\mathcal{B}_1)\mathrm{d}\mathbf{X} + c_{22}p_2 \int\limits_{\tilde{\mathcal{R}}} p_{\mathbf{X}}(\mathbf{X}|\mathcal{B}_2)\mathrm{d}\mathbf{X} - c_{22}p_2 \int\limits_{\mathcal{R}_1} p_{\mathbf{X}}(\mathbf{X}|\mathcal{B}_2)\mathrm{d}\mathbf{X} \\
&\quad + c_{21}p_1 \int\limits_{\tilde{\mathcal{R}}} p_{\mathbf{X}}(\mathbf{X}|\mathcal{B}_1)\mathrm{d}\mathbf{X} - c_{21}p_1 \int\limits_{\mathcal{R}_1} p_{\mathbf{X}}(\mathbf{X}|\mathcal{B}_1)\mathrm{d}\mathbf{X} + c_{12}p_2 \int\limits_{\mathcal{R}_1} p_{\mathbf{X}}(\mathbf{X}|\mathcal{B}_2))\mathrm{d}\mathbf{X}.
\end{aligned}$$

From this and

$$\begin{aligned}
\textstyle\int_{\tilde{\mathcal{R}}} p_{\mathbf{X}}(\mathbf{X}|\mathcal{B}_1)\mathrm{d}\mathbf{X} = 1, \\
\textstyle\int_{\tilde{\mathcal{R}}} p_{\mathbf{X}}(\mathbf{X}|\mathcal{B}_2)\mathrm{d}\mathbf{X} = 1,
\end{aligned}$$

it follows that the average risk is

$$\mathrm{AR} = c_{21}p_1 + c_{22}p_2 + \int_{\mathcal{R}_1} [p_2(c_{12}-c_{22})p_{\mathbf{X}}(\mathbf{X}|\mathcal{B}_2) - p_1(c_{21}-c_{11})p_{\mathbf{X}}(\mathbf{X}|\mathcal{B}_1)]\mathrm{d}\mathbf{X},$$

where $c_{21}p_1 + c_{22}p_2$ represents a fixed cost. The Bayes classifier requires that the integrand of the last integral is greater than zero, i.e., $p_2(c_{12}-c_{22})p_{\mathbf{X}}(\mathbf{X}|\mathcal{B}_2) - p_1(c_{21}-c_{11})p_{\mathbf{X}}(\mathbf{X}|\mathcal{B}_1) > 0$ or

$$\frac{p_2(c_{12}-c_{22})}{p_1(c_{21}-c_{11})} > \frac{p_{\mathbf{X}}(\mathbf{X}|\mathcal{B}_1)}{p_{\mathbf{X}}(\mathbf{X}|\mathcal{B}_2)}.$$

The quantities

$$\Lambda(\mathbf{X}) = \frac{p_{\mathbf{X}}(\mathbf{X}|\mathcal{B}_1)}{p_{\mathbf{X}}(\mathbf{X}|\mathcal{B}_2)}, \qquad \xi = \frac{p_2(c_{12}-c_{22})}{p_1(c_{21}-c_{11})}.$$

are called *likelihood ratio* and *threshold* of the test, respectively. The likelihood ratio $\Lambda(\mathbf{x})$ and the threshold ξ are both positive. Taking the logarithm of $\Lambda(\mathbf{x})$,

$$\log\Lambda(\mathbf{X}) = \log p_{\mathbf{X}}(\mathbf{X}|\mathcal{B}_1) - \log p_{\mathbf{X}}(\mathbf{X}|\mathcal{B}_2). \tag{1.3.1}$$

The quantity $\log\Lambda(\mathbf{X})$ is called log *−likelihood ratio*. Since the logarithm function is a monotonic function, it is more convenient to compute the log −likelihood ratio than the likelihood ratio.

Consider a special two-class problem:

$$\begin{aligned}
\text{Class } \mathcal{B}_1\text{: } & E[\mathbf{X}] = \mu_1, \\
& E[(\mathbf{X}-\mu_1)(\mathbf{X}-\mu_1)^T] = C; \\
\text{Class } \mathcal{B}_2\text{: } & E[\mathbf{X}] = \mu_2, \\
& E[(\mathbf{X}-\mu_2)(\mathbf{X}-\mu_2)^T] = C,
\end{aligned} \tag{1.3.2}$$

where $\mathbf{X}$ is a random vector. In the two-class problem, the mean values of the random vector are different but their covariance matrix of the random vector is the same. Denote by C^{-1} the inverse matrix of the covariance matrix C. Then, the conditional probability density function may be represented as the multivariate Gaussian distribution:

$$p_{\mathbf{X}}(\mathbf{X}|\mathcal{B}_i) = \frac{1}{(2\pi)^{\frac{m}{2}}(\det(C))^{\frac{1}{2}}} \exp\left(-\frac{1}{2}(\mathbf{x}-\mu_i)^T C^{-1}(\mathbf{x}-\mu_i)\right) \qquad (i=1,2),$$

where m is the dimensionality of the observation vector $\mathbf{x}$ and $\det(C)$ represents the determinant of the matrix C. Assume further in the average risk that

$$p_1 = p_2 = \frac{1}{2}, \quad c_{12} = c_{21}, \quad c_{11} = c_{22} = 0.$$

Note that

$$\begin{aligned}
\log p_{\mathbf{X}}(\mathbf{X}|\mathcal{B}_1) &= -\tfrac{1}{2}(\mathbf{x}-\mu_1)^T C^{-1}(\mathbf{x}-\mu_1) - \log((2\pi)^{\frac{m}{2}}(\det(C))^{\frac{1}{2}}),\\
\log p_{\mathbf{X}}(\mathbf{X}|\mathcal{B}_2) &= -\tfrac{1}{2}(\mathbf{x}-\mu_2)^T C^{-1}(\mathbf{x}-\mu_2) - \log((2\pi)^{\frac{m}{2}}(\det(C))^{\frac{1}{2}}).
\end{aligned}$$

By (1.3.1), the log −likelihood ratio is

$$\begin{aligned}
\log \Lambda(\mathbf{X}) &= \log p_{\mathbf{X}}(\mathbf{X}|\mathcal{B}_1) - \log p_{\mathbf{X}}(\mathbf{X}|\mathcal{B}_2)\\
&= \left(-\tfrac{1}{2}(\mathbf{x}-\mu_1)^T C^{-1}(\mathbf{x}-\mu_1) - \log((2\pi)^{\frac{m}{2}}(\det(C))^{\frac{1}{2}})\right)\\
&\quad - \left(-\tfrac{1}{2}(\mathbf{x}-\mu_2)^T C^{-1}(\mathbf{x}-\mu_2) - \log((2\pi)^{\frac{m}{2}}(\det(C))^{\frac{1}{2}})\right).
\end{aligned}$$

A direct computation shows that

$$\begin{aligned}
\log \Lambda(\mathbf{X}) &= -\tfrac{1}{2}(\mathbf{x}^T - \mu_1^T)C^{-1}(\mathbf{x}-\mu_1) + \tfrac{1}{2}(\mathbf{x}^T - \mu_2^T)C^{-1}(\mathbf{x}-\mu_2)\\
&= -\tfrac{1}{2}\mathbf{x}^T C^{-1}\mathbf{x} + \tfrac{1}{2}\mu_1^T C^{-1}\mathbf{x} + \tfrac{1}{2}\mathbf{x}^T C^{-1}\mu_1 - \tfrac{1}{2}\mu_1^T C^{-1}\mu_1\\
&\quad + \tfrac{1}{2}\mathbf{x}^T C^{-1}\mathbf{x} - \tfrac{1}{2}\mu_2^T C^{-1}\mathbf{x} - \tfrac{1}{2}\mathbf{x}^T C^{-1}\mu_2 + \tfrac{1}{2}\mu_2^T C^{-1}\mu_2\\
&= \tfrac{1}{2}(\mu_1-\mu_2)^T C^{-1}\mathbf{x} + \tfrac{1}{2}\mathbf{x}^T C^{-1}(\mu_1-\mu_2) + \tfrac{1}{2}(\mu_2^T C^{-1}\mu_2 - \mu_1^T C^{-1}\mu_1)\\
&= (\mu_1-\mu_2)^T C^{-1}\mathbf{x} + \tfrac{1}{2}(\mu_2^T C^{-1}\mu_2 - \mu_1^T C^{-1}\mu_1).
\end{aligned} \qquad (1.3.3)$$

By $c_{12} = c_{21}$, $c_{11} = c_{22} = 0$, and $p_1 = p_2 = \frac{1}{2}$, the threshold of the test is

$$\xi = \frac{p_2(c_{12}-c_{22})}{p_1(c_{21}-c_{11})} = \frac{\frac{1}{2}(c_{12}-0)}{\frac{1}{2}(c_{21}-0)} = 1.$$

In (1.3.3), let

$$\begin{aligned}
y &= \log \Lambda(\mathbf{X}),\\
W^T &= (\mu_1-\mu_2)^T C^{-1},\\
b &= \tfrac{1}{2}(\mu_2^T C^{-1}\mu_2 - \mu_1^T C^{-1}\mu_1).
\end{aligned} \qquad (1.3.4)$$

Then, Bayes classifier for multivariate Gaussian distribution, or simply, *Gaussian-distribution classifier*, is

$$y = W^T \mathbf{x} + b. \tag{1.3.5}$$

This equation shows that the Gaussian-distribution classifier is a linear network with the synaptic weight W and the bias b. It is seen from (1.3.5) that the decision boundary of the special two classes $\mathcal{B}_1, \mathcal{B}_2$ is the hyperplane:

$$W^T \mathbf{x} + b = 0.$$

In the Gaussian-distribution classifier, assume that the covariance matrix C is given by $C = aI$, where $a > 0$ is a positive constant and I is the identity matrix. We will find the synaptic-weight vector and bias as well as the representation of the Gaussian-distribution classifier.

Note that $C^{-1} = (aI)^{-1} = \frac{1}{a} I^{-1} = \frac{1}{a} I$. By (1.3.4), the weight vector is equal to

$$W = C^{-1}(\mu_1 - \mu_2) = \frac{1}{a} I (\mu_1 - \mu_2) = \frac{1}{a}(\mu_1 - \mu_2)$$

and the bias is equal to

$$b = \frac{1}{2}(\mu_2^T C^{-1} \mu_2 - \mu_1^T C^{-1} \mu_1) = \frac{1}{2a}(\mu_2^T \mu_2 - \mu_1^T \mu_1).$$

By (1.3.5), the Gaussian-distribution classifier becomes

$$y = W^T \mathbf{x} + b = \frac{1}{a}\left(\mu_1^T - \mu_2^T\right) \mathbf{x} + \frac{1}{2a}(\mu_2^T \mu_2 - \mu_1^T \mu_1).$$

Example 1.3.1 In the one-dimensional case, consider the following two-class problem:

$$\begin{aligned} \text{Class } \mathcal{B}_1\text{: } & E[X] = \mu_1, \\ & E[(X - \mu_1)^2] = C, \\ \text{Class } \mathcal{B}_2\text{: } & E[X] = \mu_2, \\ & E[(X - \mu_2)^2] = C, \end{aligned} \tag{1.3.6}$$

where X is a random variable and μ_1, μ_2, C are all real numbers. So $\mu_1^T = \mu_1$, $\mu_2^T = \mu_2$, and $C^{-1} = \frac{1}{C}$. By (1.3.4), the synaptic weight and the bias are all real numbers and

$$\begin{aligned} W &= C^{-1}(\mu_1 - \mu_2) = \tfrac{\mu_1 - \mu_2}{C}, \\ b &= \tfrac{1}{2}(\mu_2^T C^{-1} \mu_2 - \mu_1^T C^{-1} \mu_1) = \tfrac{1}{2C}(\mu_2^2 - \mu_1^2). \end{aligned}$$

By (1.3.5), the univariate Gaussian-distribution classifier is

$$y = W^T x + b = \frac{\mu_1 - \mu_2}{C} x + \frac{1}{2C}(\mu_2^2 - \mu_1^2)$$

and the decision boundary of the two-class problem (1.3.6) is a point:

$$x^* = -\frac{\frac{1}{2C}(\mu_2^2 - \mu_1^2)}{\frac{\mu_1-\mu_2}{C}} = \frac{\mu_1 + \mu_2}{2}.$$

Assume that $\mu_1 = -10$, $\mu_2 = 10$, and $C = 1$. For the two-class problem (1.3.6), the synaptic weight $W = -20$ and the bias $b = 0$, the univariate Gaussian-distribution classifier is $y = -20x$, and the decision boundary is $x = 0$. The point $x(x > 0)$ and the point $x(x < 0)$ are assigned to Class $\mathcal{B}_1$ and $\mathcal{B}_2$, respectively.

Example 1.3.2 In the two-dimensional case, the means μ_1, μ_2, and the inverse matrix of the covariance matrix are denoted by

$$\mu_1 = (\mu_{11}, \mu_{12})^T, \qquad \mu_2 = (\mu_{21}, \mu_{22})^T,$$
$$C^{-1} = \begin{pmatrix} c_{11} & c_{12} \\ c_{21} & c_{22} \end{pmatrix}.$$

By (1.3.4), the synaptic weight $W = (W_1, W_2)^T$ is

$$W = C^{-1}(\mu_1 - \mu_2) = \begin{pmatrix} c_{11} & c_{12} \\ c_{21} & c_{22} \end{pmatrix} \begin{pmatrix} \mu_{11} - \mu_{21} \\ \mu_{12} - \mu_{22} \end{pmatrix}$$
$$= \begin{pmatrix} c_{11}(\mu_{11} - \mu_{21}) + c_{12}(\mu_{12} - \mu_{22}) \\ c_{21}(\mu_{11} - \mu_{21}) + c_{22}(\mu_{12} - \mu_{22}) \end{pmatrix}.$$

Let $\mathbf{x} = (x_1, x_2)^T$. Then,

$$W^T\mathbf{x} = (c_{11}(\mu_{11} - \mu_{21}) + c_{12}(\mu_{12} - \mu_{22}),\ c_{21}(\mu_{11} - \mu_{21}) + c_{22}(\mu_{12} - \mu_{22})) \begin{pmatrix} x_1 \\ x_2 \end{pmatrix}$$
$$= c_{11}(\mu_{11} - \mu_{21})x_1 + c_{12}(\mu_{12} - \mu_{22})x_1 + c_{21}(\mu_{11} - \mu_{21})x_2 + c_{22}(\mu_{12} - \mu_{22})x_2$$
$$= (\mu_{11} - \mu_{21})(c_{11}x_1 + c_{21}x_2) + (\mu_{12} - \mu_{22})(c_{12}x_1 + c_{22}x_2).$$

By (1.3.4), the bias is

$$b = \frac{1}{2}(\mu_2^T C^{-1}\mu_2 - \mu_1^T C^{-1}\mu_1).$$

Two terms $\mu_2^T C^{-1}\mu_2$ and $\mu_1^T C^{-1}\mu_1$ are computed, respectively, as follows:

$$\mu_2^T C^{-1}\mu_2 = (\mu_{21}, \mu_{22}) \begin{pmatrix} c_{11} & c_{12} \\ c_{21} & c_{22} \end{pmatrix} \begin{pmatrix} \mu_{21} \\ \mu_{22} \end{pmatrix}$$
$$= (\mu_{21}c_{11} + \mu_{22}c_{21},\ \mu_{21}c_{12} + \mu_{22}c_{22}) \begin{pmatrix} \mu_{21} \\ \mu_{22} \end{pmatrix}$$
$$= (\mu_{21}c_{11} + \mu_{22}c_{21})\mu_{21} + (\mu_{21}c_{12} + \mu_{22}c_{22})\mu_{22}$$
$$= \mu_{21}^2 c_{11} + \mu_{21}\mu_{22}c_{21} + \mu_{21}\mu_{22}c_{12} + \mu_{22}^2 c_{22}.$$

$$
\begin{aligned}
\mu_1^T C^{-1} \mu_1 &= (\mu_{11}, \mu_{12}) \begin{pmatrix} c_{11} & c_{12} \\ c_{21} & c_{22} \end{pmatrix} \begin{pmatrix} \mu_{11} \\ \mu_{12} \end{pmatrix} \\
&= (\mu_{11}c_{11} + \mu_{12}c_{21},\ \ \mu_{11}c_{12} + \mu_{12}c_{22}) \begin{pmatrix} \mu_{11} \\ \mu_{12} \end{pmatrix} \\
&= (\mu_{11}c_{11} + \mu_{12}c_{21})\mu_{11} + (\mu_{11}c_{12} + \mu_{12}c_{22})\mu_{12} \\
&= \mu_{11}^2 c_{11} + \mu_{11}\mu_{22}c_{11} + \mu_{11}\mu_{12}c_{12} + \mu_{12}^2 c_{22}.
\end{aligned}
$$

So the bias is

$$
\begin{aligned}
b &= \tfrac{1}{2}(\mu_{21}^2 c_{11} + \mu_{21}\mu_{22}c_{21} + \mu_{21}\mu_{22}c_{12} + \mu_{22}^2 c_{22}) \\
&\quad - \tfrac{1}{2}(\mu_{11}^2 c_{11} + \mu_{11}\mu_{12}c_{21} + \mu_{11}\mu_{12}c_{12} + \mu_{12}^2 c_{22}) \\
&= \tfrac{1}{2}((\mu_{21}^2 - \mu_{11}^2)c_{11} + (\mu_{21}\mu_{22} - \mu_{11}\mu_{12})(c_{21} + c_{12}) + (\mu_{22}^2 - \mu_{12}^2)c_{22})
\end{aligned}
$$

By (1.3.5), the bivariate Gaussian-distribution classifier is

$$
\begin{aligned}
\mathbf{y} &= W^T \mathbf{x} + b \\
&= (\mu_{11} - \mu_{21})(c_{11}x_1 + c_{21}x_2) + (\mu_{12} - \mu_{22})(c_{12}x_1 + c_{22}x_2) \\
&\quad + \tfrac{1}{2}(\mu_{21}^2 - \mu_{11}^2)c_{11} + \tfrac{1}{2}(\mu_{21}\mu_{22} - \mu_{11}\mu_{12})(c_{21} + c_{12}) + \tfrac{1}{2}(\mu_{22}^2 - \mu_{12}^2)c_{22}.
\end{aligned}
$$

1.4 Radial Basis Function Network

Radial basis function network is derived from the theory of function approximation and interpolation. It uses radial basis functions as activation functions.

1.4.1 Radial Basis Function

For a given set of distinct points $\mathbf{x}_1, \ldots, \mathbf{x}_N \in \mathbb{R}^m$, the radial basis function technique is to find a function $F(\mathbf{x})$ that has the form:

$$
F(\mathbf{x}) = \sum_{j=1}^{N} w_j \varphi(\| \mathbf{x} - \mathbf{x}_j \|) = \sum_{j=1}^{N} w_j \varphi(\mathbf{x}, \mathbf{x}_j), \tag{1.4.1}
$$

where $w_j (j = 1, \ldots, N)$ are components of the weight vector $\mathbf{w}$, $\{\varphi(\mathbf{x}, \mathbf{x}_j)\}_{j=1,\ldots,N}$ is a set of *radial basis functions*, and $\mathbf{x}_j$ is the *center* of the radial basis function $\varphi(\mathbf{x}, \mathbf{x}_j) = \varphi(\| \mathbf{x} - \mathbf{x}_j \|)$; here, $\| \cdot \|$ is the Euclidean distance.

The radial basis functions used widely are as follows.

(a) Gaussian function:

$$
\varphi(x) = \exp\left(-\frac{x^2}{2\sigma^2}\right) \quad \text{for some } \sigma > 0 \text{ and } x \in \mathbb{R},
$$

where σ is the width of the Gaussian function.

(b) Multiquadric function: $\varphi(x) = (x^2 + c^2)^{\frac{1}{2}}$ for some $c > 0$ and $x \in \mathbb{R}$.

(c) Inverse multiquadric function: $\varphi(x) = (x^2 + c^2)^{-\frac{1}{2}}$ for some $c > 0$ and $x \in \mathbb{R}$.

1.4.2 Interpolation

The interpolation technique is used for finding the weight vector $\mathbf{w}$.

Given a set of distinct points $\mathbf{x}_1, \ldots, \mathbf{x}_N \in \mathbb{R}^m$ and a corresponding set of real numbers $d_1, \ldots, d_N \in \mathbb{R}$. The interpolation problem is to seek a function $F : \mathbb{R}^m \to \mathbb{R}$ satisfying the interpolation condition:

$$F(\mathbf{x}_i) = d_i \qquad (i = 1, \ldots, N), \tag{1.4.2}$$

where $F(\mathbf{x})$ is called the *interpolation surface* (or the *interpolation function*). The interpolation surface $F(\mathbf{x})$ is constrained to pass through all the training data points $\{\mathbf{x}_i, d_i\}_{i=1,\ldots,N}$, where N is called the *size of the training sample*. The combination of (1.4.1) and (1.4.2) gives

$$d_i = \sum_{j=1}^{N} w_j \varphi(\mathbf{x}_i, \mathbf{x}_j) \qquad (i = 1, \ldots, N).$$

In more detail,

$$\begin{aligned} d_1 &= \sum_{j=1}^{N} w_j \varphi(\mathbf{x}_1, \mathbf{x}_j), \\ d_2 &= \sum_{j=1}^{N} w_j \varphi(\mathbf{x}_2, \mathbf{x}_j), \\ &\vdots \\ d_N &= \sum_{j=1}^{N} w_j \varphi(\mathbf{x}_N, \mathbf{x}_j). \end{aligned} \tag{1.4.3}$$

This is a system of N equations with N unknown weights $w_j (j = 1, \ldots, N)$.

Let $\varphi_{ij} = \varphi(\mathbf{x}_i, \mathbf{x}_j)$ $(i, j = 1, \ldots, N)$. Then, the system (1.4.3) with N unknown w_j can be rewritten in the matrix form

$$\begin{pmatrix} \varphi_{11} & \varphi_{12} & \cdots & \varphi_{1N} \\ \varphi_{21} & \varphi_{22} & \cdots & \varphi_{2N} \\ \vdots & \vdots & \ddots & \vdots \\ \varphi_{N1} & \varphi_{N2} & \cdots & \varphi_{NN} \end{pmatrix} \begin{pmatrix} w_1 \\ w_2 \\ \vdots \\ w_N \end{pmatrix} = \begin{pmatrix} d_1 \\ d_2 \\ \vdots \\ d_N \end{pmatrix},$$

where N is the size of the training sample.

Let

$$\Phi = \begin{pmatrix} \varphi_{11} & \varphi_{12} & \cdots & \varphi_{1N} \\ \varphi_{21} & \varphi_{22} & \cdots & \varphi_{2N} \\ \vdots & \vdots & \ddots & \vdots \\ \varphi_{N1} & \varphi_{N2} & \cdots & \varphi_{NN} \end{pmatrix}, \quad \mathbf{w} = \begin{pmatrix} w_1 \\ w_2 \\ \vdots \\ w_N \end{pmatrix}, \quad \mathbf{d} = \begin{pmatrix} d_1 \\ d_2 \\ \vdots \\ d_N \end{pmatrix}.$$

Then, the system (1.4.3) has the compact form

$$\Phi \mathbf{w} = \mathbf{d}, \tag{1.4.4}$$

where the matrix Φ is called the *interpolation matrix* and the vectors $\mathbf{w}$ and $\mathbf{d}$ are the *linear weight vector* and *desired response vector*, respectively. Micchlli Interpolation Theorem (see Chap. 2) shows that the interpolation matrix Φ is non-singular when $\{\mathbf{x}_i\}_{i=1,\dots,N}$ is a set of distinct points in $\mathbb{R}^{m_0}$. Therefore, its inverse matrix Φ^{-1} exists. So the weight vector $\mathbf{w}$ is given by $\mathbf{w} = \Phi^{-1}\mathbf{d}$.

The *architecture* of radial basis function network is a feedforward network with layered structure. For example, the architecture of a radial basis function network with an input layer, a single hidden layer, and an output layer consisting of a single unit is described as follows.

Given a set of N distinct points $\mathbf{x}_1, \dots, \mathbf{x}_N \in \mathbb{R}^m$, i.e., the size of the input layer is m. The single hidden layer consists of N computation units. Each computational unit is described by a radial basis function:

$$\varphi(\mathbf{x}, \mathbf{x}_j) = \varphi(\| \mathbf{x} - \mathbf{x}_j \|) \qquad (j = 1, \dots, N),$$

where the jth input data point $\mathbf{x}_j$ is the center of the radial basis function and $\mathbf{x}$ is the signal applied to the input layer. The connections between the source nodes and hidden units are direct connections with no weights. The output layer consists of a single computational unit. The size of the output layer is 1. It is characterized by the weight vector $\mathbf{w} = (w_1, \dots, w_N)$, and the approximating function is

$$F(\mathbf{x}) = \sum_{j=1}^{N} w_j \varphi(\mathbf{x}, \mathbf{x}_j).$$

However, in practice, since the training sample $\{\mathbf{x}_i, d_i\}_{i=1,\dots,N}$ is often noisy, it could be wasteful of computational resources to have a hidden layer of the same size as the input layer. The size K of the hidden layer is required to be less than N, and then the corresponding approximating function is the sum of K weighted radial basis functions.

1.4.3 Receptive Field

The *receptive field* of a computational unit (e.g., the hidden unit) in a neural network is that region of the sensory field (e.g., the input layer of source nodes) from which an adequate sensory stimulus (e.g., pattern) will elicit a response. The receptive field of a computational unit is defined as

$$\psi(\mathbf{x}) = \varphi(\mathbf{x}, \mathbf{x}_j) - \alpha,$$

where α is some positive constant and $\varphi(\mathbf{x}, \mathbf{x}_j)$ is a radial basis function with center $\mathbf{x}_j$.

When Gaussian function is used as a radial basis function, Gaussian hidden unit (i.e., each computational unit in the hidden layer) is given by

$$\varphi(\mathbf{x}, \mathbf{x}_j) = \exp\left(-\frac{1}{2\sigma_j^2} \parallel \mathbf{x} - \mathbf{x}_j \parallel^2\right) \qquad (j = 1, \ldots, N),$$

where σ_j is the width of the jth Gaussian function with center $\mathbf{x}_j$.

If all the Gaussian hidden units are assigned a common width σ, then the parameter that distinguishes one hidden unit from another is the center $\mathbf{x}_j$. The receptive field of Gaussian hidden unit is

$$\psi(\mathbf{x}) = \exp\left(-\frac{1}{2\sigma^2} \parallel \mathbf{x} - \mathbf{x}_j \parallel^2\right) - \alpha,$$

where $0 < \alpha < 1$. Since the minimum permissible value of $\psi(\mathbf{x})$ is zero, it gives

$$\parallel \mathbf{x} - \mathbf{x}_j \parallel = \sigma\sqrt{2\log(1/\alpha)}.$$

It is seen that the receptive field of Gaussian hidden unit is a multidimensional surface centered symmetrically around the point $\mathbf{x}_j$.

The one-dimensional receptive field of Gaussian hidden unit is the closed interval:

$$|x - x_j| \le \sigma\sqrt{2\log(1/\alpha)}.$$

where the center and the radius of the closed interval are x_j and $r = \sigma\sqrt{2\log(1/\alpha)}$, respectively.

The two-dimensional receptive field of Gaussian hidden unit is a circular disk:

$$|x_1 - x_{j1}|^2 + |x_2 - x_{j2}|^2 \le 2\sigma^2 \log(1/\alpha),$$

where the center and the radius of the circular disk are $\mathbf{x}_j = (x_{j1}, x_{j2})$ and $r = \sigma\sqrt{2\log(1/\alpha)}$, respectively.

1.5 Generalized Regression Network

Generalized regression networks are based on the nonlinear kernel regression analysis. They belong to radial basis function networks.

The *kernel regression* builds on the notion of density estimation. Let the random vector $\mathbf{x}$ represent a regressor and the random variable y represent an observation. Then, the regression function $f(\mathbf{x})$ equals the conditional mean of the observation y given the regressor $\mathbf{x}$, i.e.,

$$f(\mathbf{x}) = E[Y|\mathbf{X}] = \int_{\mathbb{R}} y p_{Y|\mathbf{X}}(y|\mathbf{x})\mathrm{d}y, \tag{1.5.1}$$

where $p_{Y|\mathbf{X}}(y|\mathbf{x})$ is the conditional probability density function (pdf) of the random variable Y given the random vector $\mathbf{X}$. It is showed by probability theory that

$$p_{Y|\mathbf{X}}(y|\mathbf{x}) = \frac{p_{\mathbf{X},Y}(\mathbf{x}, y)}{p_{\mathbf{X}}(\mathbf{x})}, \tag{1.5.2}$$

where generally $p_{\mathbf{X}}(\mathbf{x})$ is the pdf of the random vector $\mathbf{X}$ and $p_{\mathbf{X},Y}(\mathbf{x}, y)$ is the joint pdf of the random vector $\mathbf{X}$ and the variable Y. The combination of (1.5.1) and (1.5.2) gives

$$f(\mathbf{x}) = \int_{\mathbb{R}} y \frac{p_{\mathbf{X},Y}(\mathbf{x}, y)}{p_{\mathbf{X}}(\mathbf{x})}\mathrm{d}y = \frac{\int_{\mathbb{R}} y p_{\mathbf{X},Y}(\mathbf{x}, y)\mathrm{d}y}{p_{\mathbf{X}}(\mathbf{x})}, \tag{1.5.3}$$

where the joint pdf $p_{\mathbf{X},Y}(\mathbf{x}, y)$ and the pdf $p_{\mathbf{X}}(\mathbf{x})$ are both unknown.

Assume that the training sample $\{\mathbf{x}_i, y_i\}_{i=1,\ldots,N}$ is given, where $\mathbf{x}_i \in \mathbb{R}^m$, $y_i \in \mathbb{R}$, and that $\{\mathbf{x}_i\}_{i=1,\ldots,N}$ is statistical independent and identically distributed. Define the *Parzen–Rosenblatt density estimator* of the pdf $p_{\mathbf{X}}(\mathbf{x})$ as

$$\hat{p}_{\mathbf{X}}(\mathbf{x}) = \frac{1}{Nh^m} \sum_{i=1}^{N} k\left(\frac{\mathbf{x} - \mathbf{x}_i}{h}\right) \qquad (\mathbf{x} \in \mathbb{R}^m), \tag{1.5.4}$$

where the smoothing parameter h, called *bandwidth*, is a positive number and controls the size of the kernel k. If h is a function of N such that $\lim_{N\to\infty} h(N) = 0$, then

$$\lim_{N\to\infty} E[\hat{p}_{\mathbf{X}}(\mathbf{x})] = p_{\mathbf{X}}(\mathbf{x}).$$

Similarly, define the *Parzen–Rosenblatt density estimator* of the joint pdf $p_{\mathbf{X},Y}(\mathbf{x}, y)$ as

$$\hat{p}_{\mathbf{X},Y}(\mathbf{x}, y) = \frac{1}{Nh^{m+1}} \sum_{i=1}^{N} k\left(\frac{\mathbf{x} - \mathbf{x}_i}{h}\right) k\left(\frac{y - y_i}{h}\right) \qquad (\mathbf{x} \in \mathbb{R}^m,\ y \in \mathbb{R}).$$

If h is a function of N such that $\lim_{N\to\infty} h(N) = 0$, then

$$\lim_{N\to\infty} E[\,\hat{p}_{\mathbf{X},Y}(\mathbf{x}, y)\,] = p_{\mathbf{X},Y}(\mathbf{x}, y).$$

Integrating $y\hat{p}_{\mathbf{X},Y}(\mathbf{x}, y)$ with respect to y and using $\int_{\mathbb{R}} k(\xi)\mathrm{d}\xi = 1$ and $\int_{\mathbb{R}} \xi k(\xi)\mathrm{d}\xi = 0$, it follows that

$$\int_{\mathbb{R}} y\hat{p}_{\mathbf{X},Y}(\mathbf{x}, y)\mathrm{d}y = \frac{1}{Nh^{m+1}} \sum_{i=1}^{N} k\left(\frac{\mathbf{x}-\mathbf{x}_i}{h}\right) \int_{\mathbb{R}} yk\left(\frac{y-y_i}{h}\right)\mathrm{d}y.$$

Let $\xi = \frac{y-y_i}{h}$. Then,

$$\begin{aligned}
\int_{\mathbb{R}} y\hat{p}_{\mathbf{X},Y}(\mathbf{x}, y)\mathrm{d}y &= \tfrac{1}{Nh^{m+1}} \sum_{i=1}^{N} k\left(\tfrac{\mathbf{x}-\mathbf{x}_i}{h}\right) \int_{\mathbb{R}}(y_i + h\xi)k(\xi)h\mathrm{d}\xi \\
&= \tfrac{1}{Nh^{m}} \sum_{i=1}^{N} k\left(\tfrac{\mathbf{x}-\mathbf{x}_i}{h}\right) \left(y_i \int_{\mathbb{R}} k(\xi)\mathrm{d}\xi + h\int_{\mathbb{R}} \xi k(\xi)\mathrm{d}\xi\right) \\
&= \tfrac{1}{Nh^{m}} \sum_{i=1}^{N} y_i k\left(\tfrac{\mathbf{x}-\mathbf{x}_i}{h}\right).
\end{aligned} \tag{1.5.5}$$

By (1.5.3), define the *Parzen–Rosenblatt density estimator* of the pdf $f_{\mathbf{X}}(\mathbf{x})$ as

$$\hat{f}(\mathbf{x}) = \frac{\int_{\mathbb{R}} y\hat{p}_{\mathbf{X},Y}(\mathbf{x}, y)\mathrm{d}y}{\hat{p}_{\mathbf{X}}(\mathbf{x})}$$

which is also called *kernel regression estimator*. By (1.5.4) and (1.5.5), the *kernel regression estimator* becomes

$$\hat{f}(\mathbf{x}) = \frac{\frac{1}{Nh^m}\sum_{i=1}^{N} y_i k\left(\frac{\mathbf{x}-\mathbf{x}_i}{h}\right)}{\frac{1}{Nh^m}\sum_{i=1}^{N} k\left(\frac{\mathbf{x}-\mathbf{x}_i}{h}\right)} = \frac{\sum_{i=1}^{N} y_i k\left(\frac{\mathbf{x}-\mathbf{x}_i}{h}\right)}{\sum_{j=1}^{N} k\left(\frac{\mathbf{x}-\mathbf{x}_j}{h}\right)}. \tag{1.5.6}$$

The kernel $k(\mathbf{x})$ is often required to be spherical symmetry, i.e.,

$$k\left(\frac{\mathbf{x}-\mathbf{x}_i}{h}\right) = k\left(\frac{\|\mathbf{x}-\mathbf{x}_i\|}{h}\right) \qquad (i = 1, \ldots, N),$$

where $\|\cdot\|$ is the Euclidean distance, and defined the normalized radial basis function as

$$\psi_N(\mathbf{x}, \mathbf{x}_i) = \frac{k\left(\frac{\|\mathbf{x}-\mathbf{x}_i\|}{h}\right)}{\sum_{j=1}^{N} k\left(\frac{\|\mathbf{x}-\mathbf{x}_j\|}{h}\right)} \qquad (i = 1, \ldots, N)$$

with $0 \leq \psi_i(\mathbf{x}) \leq 1$. The function $\psi_N(\mathbf{x}, \mathbf{x}_i)$ represents the probability of an event described by the input vector $\mathbf{x}$, conditional on $\mathbf{x}_i$. Let $w_i = y_i$ $(i = 1, \ldots, N)$. By (1.5.6), the kernel regression estimator becomes

$$\hat{f}(\mathbf{x}) = \frac{\sum_{i=1}^{N} w_i k\left(\frac{\mathbf{x}-\mathbf{x}_i}{h}\right)}{\sum_{j=1}^{N} k\left(\frac{\mathbf{x}-\mathbf{x}_j}{h}\right)} = \sum_{i=1}^{N} w_i \left(\frac{k\left(\frac{\mathbf{x}-\mathbf{x}_i}{h}\right)}{\sum_{j=1}^{N} k\left(\frac{\mathbf{x}-\mathbf{x}_j}{h}\right)} \right) = \sum_{i=1}^{N} w_i \psi_N(\mathbf{x}, \mathbf{x}_i).$$

This equation represents the input–output mapping of a normalized radial basis function network.

The architecture of generalized regression network is a feedforward network with layered structure. For example, the architecture of a generalized regression network with an input layer, a single hidden regression layer, and output layer consisting of two computational units is described as follows

The input layer consists of m source nodes, where m is the dimensionality of the input vector $\mathbf{x} \in \mathbb{R}^m$. The single hidden regression layer consists of N computational units. Each computational unit is described by the normalized radial basis function:

$$\psi_N(\mathbf{x}, \mathbf{x}_i) = \frac{k\left(\frac{\|\mathbf{x}-\mathbf{x}_i\|}{\sigma}\right)}{\sum_{j=1}^{N} k\left(\frac{\|\mathbf{x}-\mathbf{x}_j\|}{\sigma}\right)} \qquad (i = 1, \ldots, N),$$

where the kernel function k is the Gaussian distribution:

$$k(\mathbf{x}) = \frac{1}{(2\pi)^{\frac{m}{2}}} \exp\left(-\frac{\| \mathbf{x} \|}{2}\right).$$

So the corresponding kernel function with the center $\mathbf{x}_i \in \mathbb{R}^{m_0}$ and a common bandwidth σ is

$$k\left(\frac{\| \mathbf{x} - \mathbf{x}_i \|}{\sigma}\right) = \frac{1}{(2\pi\sigma^2)^{\frac{m}{2}}} \exp\left(-\frac{\| \mathbf{x} - \mathbf{x}_i \|}{2\sigma^2}\right) \qquad (i = 1, \ldots, N),$$

Each computational unit is

$$\psi_N(\mathbf{x}, \mathbf{x}_i) = \frac{\frac{1}{(2\pi\sigma^2)^{\frac{m}{2}}} \exp\left(-\frac{\|\mathbf{x}-\mathbf{x}_i\|}{2\sigma^2}\right)}{\sum_{j=1}^{N} \frac{1}{(2\pi\sigma^2)^{\frac{m}{2}}} \exp\left(-\frac{\|\mathbf{x}-\mathbf{x}_j\|}{2\sigma^2}\right)} = \frac{\exp\left(-\frac{\|\mathbf{x}-\mathbf{x}_i\|}{2\sigma^2}\right)}{\sum_{j=1}^{N} \exp\left(-\frac{\|\mathbf{x}-\mathbf{x}_j\|}{2\sigma^2}\right)} \qquad (i = 1, \ldots, N).$$

The links between source nodes and hidden units are direct connections with no weights. The output layer consists of two computational units:

$$\sum_{i=1}^{N} w_i \exp(-\tfrac{\|\mathbf{x}-\mathbf{x}_i\|}{2\sigma^2}),$$
$$\sum_{j=1}^{N} \exp(-\tfrac{\|\mathbf{x}-\mathbf{x}_j\|}{2\sigma^2}).$$

The size of the output layer is 2. The approximating function is

$$\hat{f}(\mathbf{x}) = \frac{\sum\limits_{i=1}^{N} w_i \exp\left(-\frac{\|\mathbf{x}-\mathbf{x}_i\|}{2\sigma^2}\right)}{\sum\limits_{j=1}^{N} \exp\left(-\frac{\|\mathbf{x}-\mathbf{x}_j\|}{2\sigma^2}\right)}$$

1.6 Self-organizing Network

The aims of the self-organizing network are to simplify and speed up the planning, configuration, management, and optimization of huge amount of data. It consists of an input layer and a competitive layer. Each neuron node in the competitive layer connects all source nodes in the input layer, and the neuron nodes in the competitive layer connect each other. The main advantages of self-organizing networks lie in that they can vary parameters and structure of networks self-organizationally and adaptively and find out the inherent rule of the sample data automatically. The task of the input layer is to receive external information and transfer the input patterns to the competitive layer, while the task of the competitive layer is to find out inherent rules of patterns and then classify them. The self-organizing networks can be divided into the Kohonen self-organizing map network, the learning vector quantization network, and so on.

1.6.1 Kohonen Self-organizing Map Network

The self-organizing map (SOM) network was introduced by Kohonen, so one always calls it Kohonen self-organizing map network. A SOM network consists of an input layer and a competitive layer, where the source nodes of the input layer transfer external information to every neuron in the competitive layer through synaptic-weight vector. When the signals enter the input layer, the network transforms adaptively the incoming signals of arbitrary dimension into a one-dimensional or two-dimensional discrete map in a topologically ordered fashion. The output neurons of the network compete among themselves to be activated. Only one output neuron is activated at any one time. This neuron is called a *winning neuron*. Such competition can be implemented by having lateral inhibition connections (negative feedback) between the neurons. The final result is outputted from the winning neuron of the competitive layer. The SOM network is often viewed as a nonlinear generalization of principal component analysis (see Chap. 4).

Competitive process is an essential process in the self-organizing network. This is based on the best match of the input vector with the synaptic-weight vector as follows:

Let $\mathbf{x} = (x_1, \ldots, x_m)^T$ be an input signal, here m is the dimension of the input space and $\| \mathbf{x} \| = \sqrt{\sum_{j=1}^m x_j^2}$, and let $\mathbf{w}_i = (w_{i1}, \ldots, w_{im})^T$ $(i = 1, \ldots, k)$ be the synaptic-weight vector, where k is the total number of neurons in the network. Correspondingly, the unit vectors are

$$\hat{\mathbf{x}} = \frac{\mathbf{x}}{\| \mathbf{x} \|} = \left(\frac{x_1}{\sqrt{\sum\limits_{j=1}^m x_j^2}}, \ldots, \frac{x_m}{\sqrt{\sum\limits_{j=1}^m x_j^2}} \right)^T =: (\hat{x}_1, \ldots, \hat{x}_m)^T,$$

$$\hat{\mathbf{w}}_i = \frac{\mathbf{w}_i}{\| \mathbf{w}_i \|} = \left(\frac{w_{i1}}{\sqrt{\sum\limits_{j=1}^m w_{ij}^2}}, \ldots, \frac{w_{im}}{\sqrt{\sum\limits_{j=1}^m w_{ij}^2}} \right)^T =: (\hat{w}_{i1}, \ldots, \hat{w}_{im})^T \quad (i = 1, \ldots, k).$$

It is clear that

$$\| \hat{\mathbf{x}} \| = \sum_{i=1}^m \hat{x}_i^2 = \sum_{i=1}^m \left(\frac{x_i}{\sqrt{\sum\limits_{j=1}^m x_j^2}} \right)^2 = \frac{\sum\limits_{i=1}^m x_i^2}{\sum\limits_{j=1}^m x_j^2} = 1,$$

$$\| \hat{\mathbf{w}}_i \| = \sum_{l=1}^m \hat{w}_{il}^2 = \sum_{l=1}^m \left(\frac{w_{il}}{\sqrt{\sum\limits_{j=1}^m w_{ij}^2}} \right)^2 = \frac{\sum\limits_{l=1}^m w_{il}^2}{\sum\limits_{j=1}^m w_{ij}^2} = 1 \qquad (i = 1, \ldots, k).$$

A commonly used criterion is the Euclidean distance between $\hat{\mathbf{x}}$ and $\hat{\mathbf{w}}_i$:

$$\| \hat{\mathbf{x}} - \hat{\mathbf{w}}_i \| = \sqrt{(\hat{\mathbf{x}} - \hat{\mathbf{w}}_i)^T (\hat{\mathbf{x}} - \hat{\mathbf{w}}_i)} \qquad (i = 1, \ldots, k). \tag{1.6.1}$$

The neuron $i_{\mathbf{x}}^*$ satisfying the condition

$$\| \hat{\mathbf{x}} - \hat{\mathbf{w}}_{i_{\mathbf{x}}^*} \| = \min_{i=1,\ldots,k} \| \hat{\mathbf{x}} - \hat{\mathbf{w}}_i \| \tag{1.6.2}$$

is the *winning neuron* for the input vector $\mathbf{x}$. Note that

$$\begin{aligned}
\hat{\mathbf{x}}^T\hat{\mathbf{x}} &= \hat{x}_1^2+\cdots+\hat{x}_m^2 =\|\hat{\mathbf{x}}\|^2=1,\\
\hat{\mathbf{w}}_i^T\hat{\mathbf{w}}_i &= \hat{w}_{i1}^2+\cdots+\hat{w}_{im}^2 =\|\hat{\mathbf{w}}_i\|^2=1\ (i=1,\ldots,k),\\
\hat{\mathbf{x}}^T\hat{\mathbf{w}}_i &= \hat{x}_1\hat{w}_{i1}+\cdots+\hat{x}_m\hat{w}_{im}=\hat{\mathbf{w}}_i^T\hat{\mathbf{x}}\quad (i=1,\ldots,k).
\end{aligned}$$

Then, the product $(\hat{\mathbf{x}}-\hat{\mathbf{w}}_i)^T(\hat{\mathbf{x}}-\hat{\mathbf{w}}_i)$ in (1.6.1) becomes

$$\begin{aligned}
&(\hat{\mathbf{x}}-\hat{\mathbf{w}}_i)^T(\hat{\mathbf{x}}-\hat{\mathbf{w}}_i)=(\hat{\mathbf{x}}^T-\hat{\mathbf{w}}_i^T)(\hat{\mathbf{x}}-\hat{\mathbf{w}}_i)\\
&=\hat{\mathbf{x}}^T\hat{\mathbf{x}}-\hat{\mathbf{w}}_i^T\hat{\mathbf{x}}-\hat{\mathbf{x}}^T\hat{\mathbf{w}}_i+\hat{\mathbf{w}}_i^T\hat{\mathbf{w}}_i=2(1-\hat{\mathbf{w}}_i^T\hat{\mathbf{x}})\qquad (i=1,\ldots,k).
\end{aligned}$$

From this and (1.6.1), the Euclidean distance between $\hat{\mathbf{x}}$ and $\hat{\mathbf{w}}_i$ becomes

$$\|\hat{\mathbf{x}}-\hat{\mathbf{w}}_i\|=\sqrt{2(1-\hat{\mathbf{w}}_i^T\hat{\mathbf{x}})}\qquad (i=1,\ldots,k),$$

where $\hat{\mathbf{w}}_i^T\hat{\mathbf{x}}$ is the inner product of $\hat{\mathbf{w}}_i$ and $\hat{\mathbf{x}}$. From this, it is seen that the minimization of the Euclidean distance $d(\hat{\mathbf{x}},\hat{\mathbf{w}}_i)$ corresponds to the maximization of the inner product $\hat{\mathbf{w}}_i^T\hat{\mathbf{x}}$. Therefore, by (1.6.2), the neuron $i_{\mathbf{x}}^*$ satisfying the condition

$$\hat{\mathbf{w}}_{i_{\mathbf{x}}^*}^T\hat{\mathbf{x}}=\max_{i=1,\ldots,k}\left(\hat{\mathbf{w}}_i^T\hat{\mathbf{x}}\right)\tag{1.6.3}$$

is the winning neuron for the input vector $\mathbf{x}$.

Let ψ_i be the angle between the unit synaptic-weight vector $\hat{\mathbf{w}}_i^T$ and the unit input vector $\hat{\mathbf{x}}$. Note that $\|\hat{\mathbf{w}}_i^T\|=1$ and $\|\hat{\mathbf{x}}\|=1$. So the inner product $\hat{\mathbf{w}}_i^T\hat{\mathbf{x}}$ is represented as

$$\hat{\mathbf{w}}_i^T\hat{\mathbf{x}}=\|\hat{\mathbf{w}}_i^T\|\|\hat{\mathbf{x}}\|\cos\psi_i=\cos\psi_i\qquad (i=1,\ldots,k).$$

Therefore, the maximization of the inner product $\hat{\mathbf{w}}_i^T\hat{\mathbf{x}}$ corresponds to the minimization of the angle ψ_i. Furthermore, by (1.6.3), the neuron $i_{\mathbf{x}}^*$ satisfying the condition

$$\psi_{i_{\mathbf{x}}^*}=\min_{i=1,\ldots,k}\psi_i$$

is the winning neuron for the input vector $\mathbf{x}$.

The topological neighborhood is an important concept in the self-organizing network. When one neuron is activated, its closest neighbors tend to get excited more than those further away. Let neuron j be a typical one of the set of excited neurons, and let $d_{ji_{\mathbf{x}}^*}$ be the lateral distance between the winning neuron $i_{\mathbf{x}}^*$ and the excited neuron j. Main topological neighborhoods $h_{ji_{\mathbf{x}}^*}$ are stated as follows:

(a) Gaussian topological neighborhood is

$$h_{ji_{\mathbf{x}}^*}^{(G)}=\exp\left(-\frac{d_{ji_{\mathbf{x}}^*}^2}{2\sigma^2}\right).$$

Gaussian topological neighborhood is symmetric about 0, and it attains the maximum value 1 when $d_{ji^*_{\mathbf{x}}} = 0$; in other words, it attains the maximum at the winning neuron $i^*_{\mathbf{x}}$ when $d_{ji^*_{\mathbf{x}}} = 0$. The amplitude of Gaussian topological neighborhood decreases monotonically when $d_{ji^*_{\mathbf{x}}}$ increases, and it decays to zero when $d_{ji^*_{\mathbf{x}}}$ tends to infinity. Gaussian topological neighborhood makes the computation of the network converge quick. The parameter σ in a Gaussian topological neighborhood is time varying. As the parameter shrinks with time, the size of the topological neighborhood shrinks with time.

(b) Mexican-hat topological neighborhood is

$$h^{(M)}_{ji^*_{\mathbf{x}}} = \left(1 - \frac{d^2_{ji^*_{\mathbf{x}}}}{2\sigma^2}\right) \exp\left(-\frac{d^2_{ji^*_{\mathbf{x}}}}{2\sigma^2}\right).$$

Mexican-hat topological neighborhood is symmetric about 0 and attains the maximum 1 when $d_{ji^*_{\mathbf{x}}} = 0$. Its amplitude is monotonically positive when $d_{ji^*_{\mathbf{x}}}$ increases from 0 to $\sqrt{2}\sigma$ and is monotonically negative and tends to zero when $d_{ji^*_{\mathbf{x}}}$ increases from $\sqrt{2}\sigma$ to infinity. In application, due to complex computation, one uses often its simple forms, i.e., the big-hat topological neighborhood or the rectangular topological neighborhood.

(c) Big-hat topological neighborhood is

$$h^{(B)}_{ji^*_{\mathbf{x}}} = \begin{cases} -\frac{1}{3}, & -\sqrt{2}\sigma - \frac{1}{3} < d_{ji^*_{\mathbf{x}}} < -\sqrt{2}\sigma, \\ +1, & -\sqrt{2}\sigma < d_{ji^*_{\mathbf{x}}} < \sqrt{2}\sigma, \\ -\frac{1}{3}, & \sqrt{2}\sigma < d_{ji^*_{\mathbf{x}}} < \sqrt{2}\sigma + \frac{1}{3}, \\ 0, & \text{otherwise.} \end{cases}$$

Big-hat topological neighborhood is simpler than the Mexican-hat topological neighborhood. Big-hat topological neighborhood is symmetric about 0 and attains the maximum value 1 when $d_{ji^*_{\mathbf{x}}} = 0$. Its amplitude is 1 when $d_{ji^*_{\mathbf{x}}}$ increases from 0 to $\sqrt{2}\sigma$ and is $-\frac{1}{3}$ when $d_{ji^*_{\mathbf{x}}}$ increases from $\sqrt{2}\sigma$ to $\sqrt{2}\sigma + \frac{1}{3}$.

(d) Rectangular topological neighborhood is

$$h^{(R)}_{ji^*_{\mathbf{x}}} = \begin{cases} +1, & -\sqrt{2}\sigma < d_{ji^*_{\mathbf{x}}} < \sqrt{2}\sigma, \\ 0, & \text{otherwise.} \end{cases}$$

Rectangular topological neighborhood is simpler than the big-hat topological neighborhood or the Mexican-hat topological neighborhood. Rectangular topological neighborhood is symmetric about 0 and attains the maximum value 1 when $d_{ji^*_{\mathbf{x}}} = 0$. Its amplitude is 1 when $d_{ji^*_{\mathbf{x}}}$ increases from 0 to $\sqrt{2}\sigma$.

Finally, the synaptic-weight vectors of all excited neurons are adjusted by the updated formula:

$$\Delta \mathbf{w}_j = \eta h_{j i^*_{\mathbf{x}}} (\mathbf{x} - \mathbf{w}_j),$$

where $i^*_{\mathbf{x}}$ is the winning neuron, j is the excited neuron, η is time varying, and $\Delta \mathbf{w}_j$ is the change of the weight vector of neuron j. Based on these new weights, the new round of competitive process will be done.

1.6.2 Learning Vector Quantization Network

Learning vector quantization (LVQ) network is proposed based on the architecture of the self-organizing competitive network. A single hidden layer, called *Kohonen hidden layer*, is added to the self-organizing network. The LVQ network consists of an input layer, a Kohonen hidden layer, and an output layer. The network is a feedforward network. Source nodes in the input layer and the neuron nodes in Kohonen hidden layer are fully connected. Neuron nodes in the hidden layer are clustered. Neuron nodes in the output layer connect to different clusters of neuron nodes in the hidden layer.

The Kohonen hidden layer uses competitive learning to perform the cluster of patterns. Neurons in the Kohonen hidden layer compete among themselves to yield the only wining neuron. The competitive criterion is the Euclidean distance between $\hat{\mathbf{x}}$ and $\hat{\mathbf{w}}_i$:

$$\| \hat{\mathbf{x}} - \hat{\mathbf{w}}_i \| = \sqrt{\sum_{j=1}^{m} (\hat{x}_j - \hat{w}_{ji})^2} \qquad (i = 1, \ldots, k),$$

where $\hat{\mathbf{x}} = (\hat{x}_1, \ldots, \hat{x}_m)$ is the unit vector of the input sample and $\hat{\mathbf{w}}_i = (\hat{w}_{1i}, \ldots, \hat{w}_{mi})$ is the unit synaptic-weight vector. The winning neuron $i^*_{\mathbf{x}}$ satisfies the condition:

$$\| \hat{\mathbf{x}} - \hat{\mathbf{w}}_{i^*_{\mathbf{x}}} \| = \min_{i=1,\ldots,k} \| \hat{\mathbf{x}} - \hat{\mathbf{w}}_i \| .$$

The output of the winning neuron $i^*_{\mathbf{x}}$ is 1, and the outputs of other neurons are all 0.

The output layer uses vector quantization learning to perform the pattern's classification. Vector quantization is based on an encoding–decoding process as follows:

Let $c(\mathbf{x})$ act as an encoder of the input vector $\mathbf{x}$ and $\mathbf{x}^*(c)$ act as a decoder of $c(\mathbf{x})$. The *expected distortion* D between the input vector $\mathbf{x}$ and its reconstruction vector $\mathbf{x}^* = \mathbf{x}^*(c(\mathbf{x}))$ is defined by

$$D = \frac{1}{2} \int_{-\infty}^{\infty} p_{\mathbf{x}}(\mathbf{x}) d(\mathbf{x}, \mathbf{x}^*) \mathrm{d}\mathbf{x},$$

where $p_{\mathbf{x}}(\mathbf{x})$ is a probability density function and $d(\mathbf{x}, \mathbf{x}^*)$ is a distortion measure. A distortion measure used commonly is the Euclidean distance, i.e.,

$$d(\mathbf{x}, \mathbf{x}^*) = \| \mathbf{x} - \mathbf{x}^* \| .$$

The optimum encoding–decoding scheme is determined by varying $c(\mathbf{x})$ and $\mathbf{x}^*(c)$ so as to minimize the expected distortion. The generalized Lloyd algorithm shows two necessary conditions for minimization of the expected distortion as follows:

(a) Given the input vector $\mathbf{x}$, choose the code $c = c(\mathbf{x})$ to minimize $d(\mathbf{x}, \mathbf{x}^*(c))$;
(b) Given the code c, compute the vector $\mathbf{x}^* = \mathbf{x}^*(c)$ as the centroid of input vectors $\mathbf{x}$ satisfying (a).

To implement vector quantization, the algorithm alternately optimizes the encoder $c(\mathbf{x})$ in accordance with (a) and then optimizes the decoder $\mathbf{x}^*(c)$ in accordance with (b) until the expected distortion D reaches a minimum.

In the output layer, neurons are prescribed which class they belong to. The synaptic weights connecting the hidden nodes to the output nodes are adjusted iteratively in a step-by-step fashion. If the winning neuron is in the prescribed class, the synaptic-weight vectors of all excited neurons are adjusted by the updated formula:

$$\Delta \mathbf{w}_i = +\eta(\mathbf{x} - \mathbf{w}_i) \qquad (i = 1, \ldots, k).$$

If not, they are adjusted by the updated formula:

$$\Delta \mathbf{w}_i = -\eta(\mathbf{x} - \mathbf{w}_i) \qquad (i = 1, \ldots, k),$$

where η is time varying.

1.7 Hopfield Network

Hopfield neural network, or simply *Hopfield network*, is a recurrent network with a single layer of neurons. Its activation functions may be continuous or discrete functions, so Hopfield networks are classified into continuous and discrete Hopfield networks.

1.7.1 Continuous Hopfield Network

A continuous Hopfield network consists of a set of fully connected neurons. The number of its feedback loops is equal to the number of neurons. The synaptic weights w_{ij} between neuron i and neuron j are symmetric, i.e., $w_{ij} = w_{ji}$ for all i, j; the adder of every neuron is constituted by a computing amplifier and a corresponding circuit; the work fashions of neurons are synchronous parallel; and the input and the output are both simulating quantities.

(a) Dynamics

Let $v_i(t)$ be the given induced local field. Assume that the activation function of the continuous Hopfield network is the hyperbolic tangent function:

$$x = \varphi_i(v) = \tanh\left(\frac{a_i v}{2}\right) = \frac{1-\exp(-a_i v)}{1+\exp(-a_i v)} \qquad (a_i \in \mathbb{Z}_+), \tag{1.7.1}$$

where the a_i is referred to as the *gain* of neuron i. It is clear by (1.7.1) that $|x| < 1$ and

$$\left.\frac{d\varphi_i}{dv}\right|_{v=0} = \frac{a_i}{2}.$$

So the slope of the activation function at the origin is $\frac{a_i}{2}$. By (1.7.1), it follows that

$$x + x\exp(-a_i v) = 1 - \exp(-a_i v),$$

So the inverse activation function is

$$v = \varphi_i^{-1}(x) = -\frac{1}{a_i}\log\left(\frac{1-x}{1+x}\right) \qquad (a_i \in \mathbb{Z}_+),$$

Since $|x| < 1$, the derivative of the inverse activation function is

$$\frac{d}{dx}\varphi_i^{-1}(x) = \frac{2}{a_i(1-x^2)} > 0.$$

The inverse activation function $\varphi_i^{-1}(x)$ is a monotonically increasing function of x.

If the continuous Hopfield network is of interconnection of N neurons, the dynamics of this network are defined by the system of differential equations:

$$C_j\frac{d}{dt}v_j(t) + \frac{v_j(t)}{R_j} = \sum_{i=1}^{N} w_{ij}\varphi_i(v_i(t)) + I_j \qquad (j = 1, \ldots, N), \tag{1.7.2}$$

where C_j is the leakage capacitance, R_j is the leakage resistance, w_{ij} represents conductance and is the symmetric synaptic weights: $w_{ij} = w_{ji}$ for all i, j, $v_i(t)$ is the given induced local field, and I_j is the externally applied bias.

(b) Energy Function

The dynamics of continuous Hopfield network are to seek out the minima of its *energy function* which is defined as

$$E = -\frac{1}{2}\sum_{i=1}^{N}\sum_{j=1}^{N} w_{ij}x_i x_j + \sum_{j=1}^{N}\frac{1}{R_j}\int_0^{x_j}\varphi_i^{-1}(x)\mathrm{d}x - \sum_{j=1}^{N} I_j x_j, \tag{1.7.3}$$

where x_j is the output of the jth neuron, $w_{ij} = w_{ji}$, I_j is the externally applied bias, R_j is the leakage resistance, and

$$x_i = \varphi_i(v_i(t)) = \tfrac{1-\exp(-a_i v_i(t))}{1+\exp(-a_i v_i(t))},$$
$$\varphi_i^{-1}(x) = -\tfrac{1}{a_i}\log\left(\tfrac{1-x}{1+x}\right).$$

In order to find the minima of the energy function, differentiating the energy function (1.7.3) with respect to the time t, it follows by (1.7.2) and that

$$\frac{dE}{dt} = -\sum_{j=1}^{N}\left(\sum_{i=1}^{N} w_{ij}x_i - \frac{\varphi_i^{-1}(x_j)}{R_j} + I_j\right)\frac{dx_j}{dt}.$$

Note that $x_j = \varphi_i(v_j(t))$ or $\varphi_i^{-1}(x_j) = v_j$. By (1.7.2), it follows that

$$\frac{dE}{dt} = -\sum_{j=1}^{N}\left(\sum_{i=1}^{N} w_{ij}\varphi_i(v_i) - \frac{v_j}{R_j} + I_j\right)\frac{dx_j}{dt} = -\sum_{j=1}^{N} C_j \frac{d}{dt}v_j(t)\frac{dx_j}{dt}.$$

Note that $v_j(t) = \varphi_j^{-1}(x_j)$. So

$$\frac{d}{dt}v_j(t) = \frac{d}{dx_j}\varphi_j^{-1}(x_j)\frac{dx_j}{dt},$$

and so

$$\frac{dE}{dt} = -\sum_{j=1}^{N} C_j \frac{d}{dx_j}\varphi_j^{-1}(x_j)\left(\frac{dx_j}{dt}\right)^2.$$

Note that

$$\left(\frac{dx_j}{dt}\right)^2 \geq 0, \qquad \frac{d}{dx_j}\varphi_j^{-1}(x_j) \geq 0$$

since $\varphi_j^{-1}(x_j)$ is a monotonically increasing function of x_j. Therefore, for all time t,

$$\frac{dE}{dt} \leq 0. \tag{1.7.4}$$

So the energy function of continuous Hopfield network is a monotonically decreasing function of time. If $\frac{dx_j}{dt} = 0$ for all time t, then $\frac{dE}{dt} = 0$. From this and (1.7.4), $\frac{dE}{dt} < 0$ except at the fixed point which is also called the *attractor*. Therefore, the continuous Hopfield network is asymptotically stable in the Lyapunov sense and the fixed points are the minima of the energy function, and vice versa.

1.7.2 Discrete Hopfield Network

A discrete Hopfield network consists of interconnection of a set of neurons with corresponding unit time delays. The number of feedback loops is equal to the number of neurons. The synaptic weights w_{ij} between neuron i and neuron j are symmetric, i.e., $w_{ij} = w_{ji}$. There is no self-feedback in the network, i.e., $w_{ii} = 0$. This means that the output of each neuron is fed back by a unit-time delay element to each of the other neurons in the network.

(a) Energy Function

For the discrete Hopfield network consisting of interconnection of N neurons, the state of the network is defined by $\mathbf{x} = (x_1, \ldots, x_N)^T$. The synaptic weight matrix $W = (w_{ij})_{i,j=1,\ldots,N}$ is symmetric: $w_{ij} = w_{ji}$ for all i, j and $w_{ii} = 0$ for all i, so the induced local field v_i of neuron i is

$$v_i = \sum_{j=1}^{N} w_{ij}x_j + b_i = \sum_{\substack{j=1 \\ j \neq i}}^{N} w_{ij}x_j + b_i \qquad (i = 1, \ldots, N),$$

where b_i is the externally applied bias to neuron i, and the state x_i of neuron i is modified by

$$x_i = \operatorname{sgn}(v_i) = \begin{cases} 1 & \text{if } v_i > 0 \\ -1 & \text{if } v_i < 0 \end{cases}$$

If $v_i = 0$, then the common convention is the neuron i remains in its previous state. The energy function of the discrete Hopfield network is defined as

$$E = -\frac{1}{2} \sum_{i=1}^{N} \sum_{\substack{j=1 \\ j \neq i}}^{N} w_{ij}x_i x_j - \sum_{i=1}^{N} b_i x_i, \tag{1.7.5}$$

Let $\mathbf{b} = (b_1, \ldots, b_N)^T$. Since

$$W = \begin{pmatrix} 0 & w_{12} & w_{13} & \cdots & w_{1N} \\ w_{21} & 0 & w_{23} & \cdots & w_{2N} \\ w_{31} & w_{32} & 0 & \cdots & w_{3N} \\ \vdots & \vdots & \vdots & \ddots & \vdots \\ w_{N1} & w_{N2} & w_{N3} & \cdots & 0 \end{pmatrix},$$

it is clear that

$$\mathbf{x}^T W\mathbf{x} = (x_1, \ldots, x_N)\begin{pmatrix} 0 & w_{12} & w_{13} & \cdots & w_{1N} \\ w_{21} & 0 & w_{23} & \cdots & w_{2N} \\ w_{31} & w_{32} & 0 & \cdots & w_{3N} \\ \vdots & \vdots & \vdots & \ddots & \vdots \\ w_{N1} & w_{N2} & w_{N3} & \cdots & 0 \end{pmatrix}\begin{pmatrix} x_1 \\ x_2 \\ x_3 \\ \vdots \\ x_N \end{pmatrix}$$

$$= \left(\sum_{\substack{j=1 \\ j\neq 1}}^{N} w_{j1}x_j, \sum_{\substack{j=1 \\ j\neq 2}}^{N} w_{j2}x_j, \ldots, \sum_{\substack{j=1 \\ j\neq N}}^{N} w_{jN}x_j \right)\begin{pmatrix} x_1 \\ x_2 \\ \vdots \\ x_N \end{pmatrix}$$

$$= \sum_{i=1}^{N} \sum_{\substack{j=1 \\ j\neq i}}^{N} w_{ji}x_jx_i$$

and

$$\mathbf{x}^T\mathbf{b} = (x_1, \ldots, x_N)\begin{pmatrix} b_1 \\ \vdots \\ b_N \end{pmatrix} = \sum_{i=1}^{N} b_ix_i$$

From this and (1.7.5), the energy function of the discrete Hopfield network can be rewritten as

$$E = -\frac{1}{2}\mathbf{x}^T W\mathbf{x} - \mathbf{x}^T\mathbf{b}.$$

The energy of the discrete Hopfield network varies with the state of any neuron of the network. When the state of neuron m of the discrete Hopfield network varies from -1 to 1 due to $v_m > 0$, the corresponding energy of the network varies from the original energy E_1 to the latter energy E_2. Since the synaptic weight matrix is symmetric, by (1.7.5), the original energy E_1 is

$$\begin{aligned} E_1 &= -\tfrac{1}{2}\sum_{i=1}^{N} \sum_{\substack{j=1 \\ j\neq i}}^{N} w_{ij}x_ix_j - \sum_{i=1}^{N} b_ix_i \\ &= -\tfrac{1}{2}\sum_{\substack{i=1 \\ i\neq m}}^{N} \sum_{\substack{j=1 \\ j\neq i,m}}^{N} w_{ij}x_ix_j - \sum_{\substack{i=1 \\ i\neq m}}^{N} b_ix_i - \sum_{\substack{j=1 \\ j\neq m}}^{N} w_{mj}x_mx_j - b_mx_m. \end{aligned}$$

Note that the original state of the neuron m is -1, i.e., $x_m = -1$. Then, the original energy is

$$E_1 = -\frac{1}{2}\sum_{\substack{i=1 \\ i\neq m}}^{N} \sum_{\substack{j=1 \\ j\neq i,m}}^{N} w_{ij}x_ix_j - \sum_{\substack{i=1 \\ i\neq m}}^{N} b_ix_i + \sum_{\substack{j=1 \\ j\neq m}}^{N} w_{mj}x_j + b_m \tag{1.7.6}$$

and the latter energy E_2 is

$$E_2 = -\frac{1}{2}\sum_{\substack{i=1\\ i\neq m}}^{N}\sum_{\substack{j=1\\ j\neq i,m}}^{N} w_{ij}x_i x_j - \sum_{\substack{i=1\\ i\neq m}}^{N} b_i x_i - \sum_{\substack{j=1\\ j\neq m}}^{N} w_{mj}x_j - b_m.$$

Combining this with (1.7.6), the difference of the energy is

$$\Delta E = E_2 - E_1 = -2\left(\sum_{\substack{j=1\\ j\neq m}}^{N} w_{mj}x_j + b_m\right) = -2v_m.$$

Due to $v_m > 0$, it is clear that $\Delta E < 0$. It means that when the state of any neuron of the discrete Hopfield network increases from -1 to 1, the energy of the network decreases. Similarly, when the state of any neuron of the discrete Hopfield network decreases from 1 to -1 due to $v_m < 0$, the energy of the network increases.

(b) A Content-Addressable Memory

Discrete Hopfield network is applied as a content-addressable memory. The operation includes two phases: the storage phase and the retrieval phase.

- Storage Phase

Let M vectors $\xi_1, \xi_2, \ldots, \xi_M$ represent a known set of N-dimensional fundamental memories. The synaptic weights from neuron i to neuron j for the discrete Hopfield network are defined as

$$w_{ij} = \begin{cases} \frac{1}{N}\sum\limits_{\mu=1}^{M} \xi_{\mu i}\xi_{\mu j} & (i \neq j), \\ 0 & (i = j), \end{cases}$$

where

$$\xi_\mu = (\xi_{\mu 1}, \ldots, \xi_{\mu N})^T, \qquad \xi_{\mu i} = \pm 1 \qquad (i = 1, \ldots, N).$$

The matrix form of the synaptic weight matrix of the network is

$$W = \frac{1}{N}\sum_{\mu=1}^{M} \xi_\mu \xi_\mu^T - \frac{M}{N} I,$$

where I is the identity matrix and ξ_μ is stated as above, and the synaptic weight matrix is an $N \times N$ symmetric matrix: $W = W^T$.

- Retrieval Phase

Starting from an $N-$dimensional vector termed a probe, the asynchronous updating procedure is performed until the discrete Hopfield network produces a time-invariant state vector $\mathbf{y}$ which satisfies the *stability condition* $\mathbf{y} = \operatorname{sgn}(W\mathbf{y} + \mathbf{b})$ which is also called the *alignment condition*, where W is synaptic weight matrix of discrete Hopfield network and $\mathbf{b}$ is the externally applied bias vector. The state vector $\mathbf{y}$ that satisfies the alignment condition is called a *stable state*. Hence, the discrete Hopfield network always converges to a stable state.

Further Reading

H.Z. Abyaneh, M.B. Varkeshi, G. Golmohammadi, K. Mohammadi, Soil temperature estimation using an artificial neural network and co-active neuro-fuzzy inference system in two differen climates. Arab. J. Geosci. **9**, 377 (2016)

M.A. Amiri, Y. Amerian, M.S. Mesgari, Spatial and temporal monthly precipitation forecasting using wavelet transform and neural networks, Qara-Qum catchment, Iran. Arab. J. Geosci. **9**, 421 (2016)

P.D. Brooks, D. McKnight, K. Elder, Carbon limitation of soil respiration under winter snowpacks: potential feedbacks between growing season and winter carbon fluxes. Glob. Change Biol. **11**, 231–238 (2004)

A. Gersho, R.M. Gray, *Vector Quantization and Signal Compression* (Kluwer, Norwell, MA, 1992)

R.L. Hardy, Multiquadric equations of topography and other irregular surfaces. J. Geophys. Res. **76**, 1905–1915 (1971)

S. Haykin, *Neural Networks and Learning Machines*, 3rd edn. (New York, Pearson Education, 2008)

J.J. Hopfield, Neural networks and physical systems with emergent collective computational abilities. Proc. Natl. Acad. Sci. U.S.A. **79**, 2554–2558 (1982)

J.J. Hopfield, Neurons with graded response have collective computational properties like those of two-state neurons. Proc. Natl. Acad. Sci. U.S.A. **81**, 3088–3092 (1984)

J.J. Hopfield, Neurons, dynamics and computation. Phys. Today **47**, 40–46 (1994)

T. Kohonen, Self-organized formation of topologically correct feature maps. Biol. Cybern. **43**, 59–69 (1982)

T. Kohonen, The self-organizing map. Proc. Inst. Electr. Electron. Eng. **78**, 1464–1480 (1990)

T. Kohonen, *Self-organizing Maps*, 2nd edn. (Springer, Berlin, 1997)

A. Krzyzak, T. Linder, G. Lugosi, Nonparametric estimation and classification using radial basis functions. IEEE Trans. Neural Networks **7**, 475–487 (1996)

S.P. Luttrell, *Self-organization: A Derivation from First Principle of a Class of Learning Algorithms* (IEEE Conference on Neural Networks, Washington, DC, 1989)

C.A. Micchelli, Interpolation of scattered data: distance matrices and conditionally positive definite function. Constr. Approx. **2**, 11–22 (1986)

E.A. Nadaraya, On estimating regression. Theory Probab. Appl. **9**, 141–142 (1964)

K. Obermayer, H. Ritter, K. Schulten, Development and spatial structure of cortical feature maps: a model study, *Advances in Neural Information Proceeding Systems* (Morgan Kaufmann, San Mateo, CA, 1991), pp. 11–17

M. Ozturk, O. Salman, M. Koc, Artificial neural network model for estimating the soil temperature. Can. J. Soil Sci. **91**, 551–562 (2011)

E. Parzen, On estimation of a probability density function and mode. Ann. Math. Stat. **23**, 1065–1076 (1962)

H. Ritter, T. Martinetz, K. Schulten, *Neural Computation and Self-organizing Maps: An Introduction* (Addison-Wesley, Reading, MA, 1992)

M. Rosenblatt, Remarks on some nonparametric estimates of a density function. Ann. Math. Stat. **27**, 832–837 (1956)

M. Rosenblatt, Density estimates and Morkov sequences, in *Nonparametric Techniques in Statistical Inference*, ed. by M. Puri (Cambridge University Press, London, 1970)

A.J. Tenge, F.B.S. Kaihura, R. Lal, B.R. Singh, Diurnal soil temperature fluctuations for different erosional classes of an oxisol at Mlingano, Tanzania. Soil Tillage Res. **49**, 211–217 (1998)

D.S. Tourtzky, D.A. Pomerleau, What is hidden in the hidden layers? Byte **14**, 227–233 (1989)

H.L. Van Trees, *Trees, Detection, Estimation, and Modulation Theory, Part I* (Wiley, New York, 1968)

G.S. Watson, Smooth regression analysis. Sankhyā Indian J. Stat. Ser. A **26**, 359–372 (1964)

L. Xu, A. Krzyzak, A. Yuille, On radial basis function nets and kernel regression: statistical consistency, convergency rates, and receptive field size. Neural Netw. **7**, 609–628 (1994)

Chapter 2
Multivariate Harmonic Analysis

Multivariate harmonic analysis technique has been employed widely in determining cyclic variations of multivariate time series. Although it cannot reveal what causes the cycles, it can reveal the likely frequency, amplitude, and phase of those cycles. Moreover, it can be used to simulate atmospheric and oceanic circulation, rule out significant impacts from anthropogenic factors, and ultimately predict what will happen next under climate change scenarios. In this chapter, we will introduce basic techniques and tools in multivariate harmonic analysis, including Fourier transform, fractional Fourier transform, space–frequency representation, sparse approximation, spherical harmonics, harmonic analysis on graphs.

2.1 Fourier Transform

The multivariate Fourier transform is a systematic way to decompose multivariate time series into a superposition of trigonometric functions. It can determine how the total variance of multivariate time series is distributed in frequency. In this section, we will introduce multivariate Fourier transform, Parseval identity, Poisson summation formula, and Shannon sampling theorem.

The Fourier transform of a $d-$variate complex-valued function $f(t_1, \ldots, t_d)$ is defined as

$$\widehat{f}(\omega_1, \ldots, \omega_d) = \int_{\mathbb{R}} \cdots \int_{\mathbb{R}} f(t_1, \ldots, t_d)\, e^{-2\pi i(\omega_1 t_1 + \cdots + \omega_d t_d)} \mathrm{d}t_1 \cdots \mathrm{d}t_d. \qquad (2.1.1)$$

Let $\mathbf{t} = (t_1, \ldots, t_d)$, $\boldsymbol{\omega} = (\omega_1, \ldots, \omega_d)$, $(\boldsymbol{\omega}, \mathbf{t}) = \sum_{k=1}^{d} \omega_k t_k$, and $\mathbf{dt} = \mathrm{d}t_1 \cdots \mathrm{d}t_d$. Then, (2.1.1) becomes

Z. Zhang, *Multivariate Time Series Analysis in Climate and Environmental Research*, https://doi.org/10.1007/978-3-319-67340-0_2

$$\widehat{f}(\boldsymbol{\omega}) = \int_{\mathbb{R}^d} f(\mathbf{t})\, e^{-2\pi i (\boldsymbol{\omega},\mathbf{t})} \mathbf{dt}$$

If $f(\mathbf{t})$ is integrable on $\mathbb{R}^d$; then its Fourier transform is continuous on $\mathbb{R}^d$. From $|e^{-2\pi i(\boldsymbol{\omega},\mathbf{t})}| = 1$, it follows that

$$|\widehat{f}(\boldsymbol{\omega})| \le \int_{\mathbb{R}^d} |f(\mathbf{t})| \mathbf{dt} \qquad (\boldsymbol{\omega} \in \mathbb{R}^d),$$

i.e., Fourier transform $\widehat{f}(\boldsymbol{\omega})$ is a bounded function on $\mathbb{R}^d$. The Riemann–Lebesgue lemma shows that

$$\widehat{f}(\boldsymbol{\omega}) \to 0 \quad \text{as} \quad \| \boldsymbol{\omega} \| = (\omega_1^2 + \cdots + \omega_d^2)^{\frac{1}{2}} \to \infty.$$

The smoother the function $f(\mathbf{t})$ is, the quicker the Fourier transform $\widehat{f}(\boldsymbol{\omega})$ decays.

Let $f(\mathbf{t})$ be a $d-$dimensional real-valued function. If

$$f(t_1, \ldots, t_{k-1}, -t_k, t_{k+1}, \ldots, t_d) = f(t_1, \ldots, t_{k-1}, t_k, t_{k+1}, \ldots, t_d) \qquad (k = 1, \ldots, d),$$

then $f(\mathbf{t})$ is called an *even function*. Fourier transform of an even, real-valued multivariate function is also an even, real-valued multivariate function. If

$$f(\mathbf{t}) = f_1(t_1) f_2(t_2) \cdots f_d(t_d) \qquad (\mathbf{t} = (t_1, \ldots, t_d)),$$

then $f(\mathbf{t})$ is called a *separable function*. Fourier transform of a separable multivariate function is a product of univariate Fourier transforms, i.e.,

$$\begin{aligned}
\widehat{f}(\boldsymbol{\omega}) &= \textstyle\int_{\mathbb{R}^d} f(\mathbf{t})\, e^{-2\pi i(\boldsymbol{\omega},\mathbf{t})} \mathbf{dt} \\
&= \left(\textstyle\int_{\mathbb{R}^d} f_1(t_1)\, e^{-2\pi i \omega_1 t_1} \mathrm{d}t_1\right) \cdots \left(\textstyle\int_{\mathbb{R}^d} f_d(t_d)\, e^{-2\pi i \omega_d t_d} \mathrm{d}t_d\right) \\
&= \widehat{f_1}(\omega_1) \cdots \widehat{f_d}(\omega_d).
\end{aligned}$$

Below, we give some examples of multivariate Fourier transforms:

Characteristic function of $[-T, T]^d$ is

$$\chi_{[-T,T]^d}(\mathbf{t}) = \begin{cases} 1, \ \mathbf{t} \in [-T, T]^d, \\ 0, \ \mathbf{t} \in \mathbb{R}^d \setminus [-T, T]^d. \end{cases}$$

The corresponding Fourier transform is

$$\widehat{\chi}_{[-T,T]^d}(\boldsymbol{\omega}) = \prod_1^d \frac{\sin(2\pi\omega_k T)}{\pi\omega_k} \qquad (\boldsymbol{\omega} = (\omega_1, \ldots, \omega_d)).$$

Delta function is defined as

$$\delta(\mathbf{t}) = \lim_{T\to 0} \frac{1}{(2T)^d} \chi_{[-T,T]^d}(\mathbf{t}).$$

The corresponding Fourier transform is

$$\widehat{\delta}(\boldsymbol{\omega}) = \lim_{T\to 0} \frac{1}{(2T)^d} \widehat{\chi}_{[-T,T]^d}(\boldsymbol{\omega}) = \prod_1^d \left(\lim_{T\to 0} \frac{\sin(2\pi\omega_k T)}{2\pi\omega_k T} \right) = 1.$$

Gaussian function is defined as $G_M(\mathbf{t}) = e^{-M|\mathbf{t}|^2}$ $(M > 0;\ |\mathbf{t}|^2 = t_1^2 + \cdots + t_d^2)$. The corresponding Fourier transform is

$$\widehat{G}_M(\boldsymbol{\omega}) = \left(\frac{\pi}{M}\right)^{\frac{d}{2}} e^{-\frac{\pi^2}{M}|\boldsymbol{\omega}|^2} \qquad (|\boldsymbol{\omega}|^2 = \omega_1^2 + \cdots + \omega_d^2), \tag{2.1.2}$$

i.e., Fourier transform of Gaussian function is still Gaussian function. It plays an important role in the windowed Fourier transform in Sect. 2.6 and the normal distribution in Sect. 4.1.

Multivariate Fourier transforms have the following properties.

(a) Translation. For $\boldsymbol{\tau} \in \mathbb{R}^d$,

$$(f(\mathbf{t} - \boldsymbol{\tau}))^\wedge(\boldsymbol{\omega}) = e^{-2\pi i(\boldsymbol{\omega},\boldsymbol{\tau})} \widehat{f}(\boldsymbol{\omega}),$$

$$\left(e^{2\pi i(\boldsymbol{\tau},\mathbf{t})} f(\mathbf{t})\right)^\wedge(\boldsymbol{\omega}) = \widehat{f}(\boldsymbol{\omega} - \boldsymbol{\tau}).$$

(b) Dilation. For $a \in \mathbb{R}$ and $a \neq 0$,

$$(f(a\mathbf{t}))^\wedge(\boldsymbol{\omega}) = \frac{1}{|a|^d} \widehat{f}\left(\frac{\boldsymbol{\omega}}{a}\right).$$

(c) Derivation:

$$\frac{\partial \widehat{f}}{\partial \omega_k}(\boldsymbol{\omega}) = (-2\pi i t_k f(\mathbf{t}))^\wedge(\boldsymbol{\omega}),$$

$$\left(\frac{\partial f}{\partial t_k}\right)^\wedge(\boldsymbol{\omega}) = 2\pi i \omega_k \widehat{f}(\boldsymbol{\omega}),$$

where $\mathbf{t} = (t_1, \ldots, t_d)$ and $\boldsymbol{\omega} = (\omega_1, \ldots, \omega_d)$.

(d) Convolution. Let f and g be functions on $\mathbb{R}^d$. Define convolution of f and g as

$$h(\mathbf{t}) = (f * g)(\mathbf{t}) = \int_{\mathbb{R}^d} f(\mathbf{t} - \boldsymbol{\tau}) g(\boldsymbol{\tau}) \mathrm{d}\boldsymbol{\tau}.$$

Fourier transform of convolution is equal to the product of Fourier transforms, i.e., $\widehat{h}(\boldsymbol{\omega}) = \widehat{f}(\boldsymbol{\omega})\widehat{g}(\boldsymbol{\omega})$.

The *inverse Fourier transform* is defined as

$$f(\mathbf{t}) = \int_{\mathbb{R}^d} \widehat{f}(\boldsymbol{\omega})\, e^{2\pi i(\boldsymbol{\omega},\mathbf{t})} \mathrm{d}\boldsymbol{\omega},$$

which can reconstruct the original function $f(\mathbf{t})$ for given a Fourier transform $\widehat{f}(\boldsymbol{\omega})$.

If $f(\mathbf{t})$ is an even function, by the inverse Fourier transform formulas, it follows that

$$f(\mathbf{t}) = f(-\mathbf{t}) = \int_{\mathbb{R}^d} \widehat{f}(\boldsymbol{\omega})\, e^{-2\pi i(\boldsymbol{\omega},\mathbf{t})} \mathrm{d}\mathbf{t},$$

i.e., the even function $f(\mathbf{t})$ is the Fourier transform of $\widehat{f}(\boldsymbol{\omega})$. Again by Fourier transforms of characteristic, delta, and Gaussian functions, it follows that

(a) if $f(\mathbf{t}) = \prod_1^d \frac{\sin(2\pi t_k T)}{\pi t_k}$, then $\widehat{f}(\boldsymbol{\omega}) = \chi_{[-T,T]^d}(\boldsymbol{\omega})$;
(b) if $f(\mathbf{t}) = e^{-\frac{\pi^2}{M}|\mathbf{t}|^2}$, then $\widehat{f}(\boldsymbol{\omega}) = e^{-M|\boldsymbol{\omega}|^2}$;
(c) if $f(\mathbf{t}) = 1$, then $\widehat{f}(\boldsymbol{\omega}) = \delta(\boldsymbol{\omega})$;

by (c) and the translation formula, it follows that

(d) if $f(\mathbf{t}) = e^{2\pi i(\boldsymbol{\tau},\mathbf{t})}$, then $\widehat{f}(\boldsymbol{\omega}) = \delta(\boldsymbol{\omega} - \boldsymbol{\tau})$;

by (d) and Euler's formula, it follows that

(e) if $f(\mathbf{t}) = \cos(2\pi(\boldsymbol{\tau},\mathbf{t}))$, then $\widehat{f}(\boldsymbol{\omega}) = \frac{1}{2}(\delta(\boldsymbol{\omega} - \boldsymbol{\tau}) + \delta(\boldsymbol{\omega} + \boldsymbol{\tau}))$;
(f) if $f(\mathbf{t}) = \sin(2\pi(\boldsymbol{\tau},\mathbf{t}))$, then $\widehat{f}(\boldsymbol{\omega}) = \frac{1}{2i}(\delta(\boldsymbol{\omega} - \boldsymbol{\tau}) - \delta(\boldsymbol{\omega} + \boldsymbol{\tau}))$.

If f and g are square integrable on $\mathbb{R}^d$ (i.e., $\int_{\mathbb{R}^d} |f(\mathbf{t})|^2 \mathrm{d}\mathbf{t} < \infty$ and $\int_{\mathbb{R}^d} |g(\mathbf{t})|^2 \mathrm{d}\mathbf{t} < \infty$), then

Parseval Formula: $\int_{\mathbb{R}^d} f(\mathbf{t})\overline{g(\mathbf{t})}\mathrm{d}\mathbf{t} = \int_{\mathbb{R}^d} \widehat{f}(\boldsymbol{\omega})\overline{\widehat{g}(\boldsymbol{\omega})}\mathrm{d}\boldsymbol{\omega}$;
Plancherel Formula: $\int_{\mathbb{R}^d} |f(\mathbf{t})|^2 \mathrm{d}\mathbf{t} = \int_{\mathbb{R}^d} |\widehat{f}(\boldsymbol{\omega})|^2 \mathrm{d}\boldsymbol{\omega}$.

Suppose that $f(\mathbf{t})$ is square integrable on $[0,1]^d$ and $f(\mathbf{t}+\mathbf{n}) = f(\mathbf{t})$ for all $\mathbf{t} \in \mathbb{R}^d$ and $\mathbf{n} \in \mathbb{Z}^d$. Then, $f(\mathbf{t})$ can be expanded into Fourier series with respect to $\{e^{2\pi i(\mathbf{n},\mathbf{t})}\}_{\mathbf{n}\in\mathbb{Z}^d}$:

$$f(\mathbf{t}) = \sum_{\mathbf{n}\in\mathbb{Z}^d} c_{\mathbf{n}}(f)\, e^{2\pi i(\mathbf{n},\mathbf{t})} \qquad (\mathbf{n} = (n_1, \ldots, n_d),\ \mathbf{t} = (t_1, \ldots, t_d)),$$

where the inner product $(\mathbf{n},\mathbf{t}) = \sum_{k=1}^d n_k t_k$, and Fourier coefficients are

$$c_{\mathbf{n}}(f) = \int_{[0,1]^d} f(\mathbf{t})\, e^{-2\pi i(\mathbf{n},\mathbf{t})} \mathrm{d}\mathbf{t}. \tag{2.1.3}$$

Fourier coefficients have the following important properties:

Riemann–Lebesgue Lemma:

$$c_{\mathbf{n}}(f) \to 0 \text{ as } \| \mathbf{n} \| \to \infty, \text{ where } \| \mathbf{n} \| = (n_1^2 + \cdots + n_d^2)^{\frac{1}{2}}.$$

Parseval Identity: $\sum_{\mathbf{n}\in\mathbb{Z}^d} |c_{\mathbf{n}}(f)|^2 = \int_{[0,1]^d} |f(\mathbf{t})|^2 \mathrm{d}\mathbf{t}.$

Convolution Formula:

$$c_{\mathbf{n}}(f * g) = c_{\mathbf{n}}(f)\, c_{\mathbf{n}}(g), \quad \text{where } f * g = \int_{[0,1]^d} f(\mathbf{t}-\mathbf{s})g(\mathbf{s})\mathrm{d}\mathbf{s}.$$

Poisson Summation Formula: $\sum_{\mathbf{m}\in\mathbb{Z}^d} f(\mathbf{t}+\mathbf{m}) = \sum_{\mathbf{n}\in\mathbb{Z}^d} \widehat{f}(\mathbf{n})\, e^{2\pi i(\mathbf{n},\mathbf{t})}.$

Shannon Sampling Theorem. If the Fourier transform of $f(\mathbf{t})$ satisfies $\widehat{f}(\boldsymbol{\omega}) = 0\ (\boldsymbol{\omega} \notin [-\frac{1}{2}, \frac{1}{2}]^d)$, then the following interpolation formula holds:

$$f(\mathbf{t}) = \sum_{\mathbf{n}\in\mathbb{Z}^d} f(\mathbf{n}) \prod_{k=1}^{d} \frac{\sin \pi(t_k - n_k)}{\pi(t_k - n_k)}.$$

More generally, if $\widehat{f}(\boldsymbol{\omega}) = 0\ (\boldsymbol{\omega} \notin [-\frac{b}{2}, \frac{b}{2}]^d)$ and the sampling interval $\Delta \le \frac{1}{b}$, then the interpolation formula becomes

$$f(\mathbf{t}) = \sum_{\mathbf{n}\in\mathbb{Z}^d} f(\mathbf{n}\Delta) \prod_{k=1}^{d} \frac{\sin \frac{\pi}{\Delta}(t_k - n_k\Delta)}{\frac{\pi}{\Delta}(t_k - n_k\Delta).}$$

If $|\widehat{f}(\boldsymbol{\omega})| < \epsilon\ (\boldsymbol{\omega} \notin [-\frac{b}{2}, \frac{b}{2}]^d)$, then this interpolation formula holds approximately and the error is less than $\frac{\epsilon}{\pi}$.

The Shannon sampling theorem allows the replacement of a continuous band-limited signal by a discrete sequence of its samples without the loss of any information. Moreover, it specifies the lowest sampling rate to reproduce the original signal.

2.2 Discrete Fourier Transform

The discrete Fourier transform (DFT) is a specific kind of Fourier transformation. It requires an input function that is discrete and whose nonzero values have a finite duration. Such inputs are often created by sampling a continuous function.

Start from Fourier coefficient formula (2.1.3), i.e.,

$$c_{\mu_1,\ldots,\mu_d}(f) = \int_{[0,1]^d} f(t_1, \ldots, t_d)\, e^{-2\pi i(\mu_1 t_1 + \cdots + \mu_d t_d)} \mathrm{d}t_1 \cdots \mathrm{d}t_d.$$

Take $(t_1, \ldots, t_d) = (\frac{\nu_1}{N_1}, \ldots, \frac{\nu_d}{N_d})$ $(\nu_l = 0, \ldots, N_l - 1\ (l = 1, \ldots, d))$ as the lattice distribution of $[0, 1]^d$. The numerical calculation of the Fourier coefficients is

$$X_{\mu_1,\ldots,\mu_d} := \frac{1}{N_1 \cdots N_d} \sum_{\nu_1=0}^{N_1-1} \cdots \sum_{\nu_d=0}^{N_d-1} f\left(\frac{\nu_1}{N_1}, \ldots, \frac{\nu_d}{N_d}\right) e^{-2\pi i (\frac{\mu_1\nu_1}{N_1} + \cdots + \frac{\mu_d\nu_d}{N_d})}.$$

Denote $x_{\nu_1,\ldots,\nu_d}$ as the value of $f(t_1, \ldots, t_d)$ at the lattice point $(\frac{\nu_1}{N_1}, \ldots, \frac{\nu_d}{N_d})$. So

$$X_{\mu_1,\ldots,\mu_d} = \frac{1}{N_1 \cdots N_d} \sum_{\nu_1=0}^{N_1-1} \cdots \sum_{\nu_d=0}^{N_d-1} x_{\nu_1,\ldots,\nu_d}\, e^{-2\pi i (\frac{\mu_1\nu_1}{N_1} + \cdots + \frac{\mu_d\nu_d}{N_d})}. \qquad (2.2.1)$$

The above transform from $\{x_{\nu_1,\ldots,\nu_d}\}_{\nu_k=0,\ldots,N_k-1(k=1,\ldots,d)}$ to $\{X_{\mu_1,\ldots,\mu_d}\}_{\mu_k=0,\ldots,N_k-1(k=1,\ldots,d)}$ is called the *$d-$variate discrete Fourier transform* (DFT), denoted by DFT$(x_{\nu_1,\ldots,\nu_d})$.

If $x_{\nu_1,\ldots,\nu_d}$ is separable $x_{\nu_1,\ldots,\nu_d} = x^{(1)}_{\nu_1} x^{(2)}_{\nu_2} \cdots x^{(d)}_{\nu_d}$, then DFT$(x_{\nu_1,\ldots,\nu_d})$ is

$$X_{\mu_1,\ldots,\mu_d} = \left(\frac{1}{N_1} \sum_{\nu_1=0}^{N_1-1} x^{(1)}_{\nu_1}\, e^{-2\pi i \frac{\mu_1\nu_1}{N_1}}\right) \cdots \left(\frac{1}{N_d} \sum_{\nu_d=0}^{N_d-1} x^{(d)}_{\nu_d}\, e^{-2\pi i \frac{\mu_d\nu_d}{N_d}}\right) =: X^{(1)}_{\mu_1} \cdots X^{(d)}_{\mu_d},$$

where $X^{(k)}_{\mu_k}$ is DFT of $x^{(k)}_{\nu_k}$ $(k = 1, \ldots, d)$. This implies DFT$(x_{\nu_1,\ldots,\nu_d})$ is also separable. If $N = N_i (i = 1, \ldots, d)$, Formula (2.2.1) is reduced in the form:

$$X_{\boldsymbol{\mu}} = \frac{1}{N^d} \sum_{\boldsymbol{\nu} \in \Lambda_N} x_{\boldsymbol{\nu}}\, e^{-\frac{2\pi i}{N}(\boldsymbol{\mu},\boldsymbol{\nu})},$$

where $\boldsymbol{\mu} = (\mu_1, \ldots, \mu_d) \in \Lambda_N$ and $\boldsymbol{\nu} = (\nu_1, \ldots, \nu_d)$, and $\Lambda_N = [0, N)^d \bigcap \mathbb{Z}^d$.

The *$d-$variate inverse discrete Fourier transform*(IDFT) is

$$x_{\nu_1,\ldots,\nu_d} = \sum_{\mu_1=0}^{N_1-1} \cdots \sum_{\mu_d=0}^{N_d-1} X_{\mu_1,\ldots,\mu_d}\, e^{2\pi i (\frac{\mu_1\nu_1}{N_1} + \cdots + \frac{\mu_d\nu_d}{N_d})}.$$

For simplicity of notation, we only prove the case $d = 2$.

Exchanging the order of summation gives

$$\sum_{\mu_1=0}^{N_1-1}\sum_{\mu_2=0}^{N_2-1} X_{\mu_1,\mu_2}\, e^{2\pi i(\frac{\mu_1\nu_1}{N_1}+\frac{\mu_2\nu_2}{N_2})}$$

$$= \frac{1}{N_1N_2}\sum_{\mu_1=0}^{N_1-1}\sum_{\mu_2=0}^{N_2-1}\left(\sum_{\alpha_1=0}^{N_1-1}\sum_{\alpha_2=0}^{N_2-1} x_{\alpha_1,\alpha_2}\, e^{-2\pi i\left(\frac{\mu_1\alpha_1}{N_1}+\frac{\mu_2\alpha_2}{N_2}\right)}\right) e^{2\pi i\left(\frac{\mu_1\nu_1}{N_1}+\frac{\mu_2\nu_2}{N_2}\right)} \tag{2.2.2}$$

$$= \frac{1}{N_1N_2}\sum_{\alpha_1=0}^{N_1-1}\sum_{\alpha_2=0}^{N_2-1} x_{\alpha_1,\alpha_2}\left(\sum_{\mu_1=0}^{N_1-1}\sum_{\mu_2=0}^{N_2-1} e^{2\pi i\left(\frac{\nu_1-\alpha_1}{N_1}\mu_1+\frac{\nu_2-\alpha_2}{N_2}\mu_2\right)}\right).$$

Using the known formulas $e^{i(\alpha+\beta)} = e^{i\alpha}e^{i\beta}$ and $\sum_{\lambda=0}^{N-1} e^{\lambda\mu} = \frac{1-e^{N\mu}}{1-e^{\mu}}$ $(\mu \neq 0)$, the term in brackets becomes

$$\sum_{\mu_1=0}^{N_1-1} e^{2\pi i\frac{\nu_1-\alpha_1}{N_1}\mu_1}\sum_{\mu_2=0}^{N_2-1} e^{2\pi i\frac{\nu_2-\alpha_2}{N_2}\mu_2} = \begin{cases} 0, & (\alpha_1,\alpha_2)\neq(\nu_1,\nu_2),\\ N_1N_2, & (\alpha_1,\alpha_2)=(\nu_1,\nu_2).\end{cases}$$

Then, substituting this into (2.2.2) gives

$$\sum_{\mu_1=0}^{N_1-1}\sum_{\mu_2=0}^{N_2-1} X_{\mu_1\mu_2}\, e^{2\pi i\left(\frac{\mu_1\nu_1}{N_1}+\frac{\mu_2\nu_2}{N_2}\right)} = x_{\nu_1\nu_2}.$$

This means that the inverse discrete Fourier transform can reconstruct the original data from its discrete Fourier transform.

Let $\widetilde{x}_{\nu_1,\dots,\nu_d}$ be a $d-$dimensional sequence on $\mathbb{Z}^d$ such that $x_{\nu_1+p_1N_1,\dots,\nu_d+p_dN_d} = x_{\nu_1,\dots,\nu_d}$, where $(N_1,\dots,N_d)\in\mathbb{Z}^d$ and $(p_1,\dots,p_d)\in\mathbb{Z}^d$. Then, $\widetilde{x}_{\nu_1,\dots,\nu_d}$ is called a *recurring sequence with period* $(N_1,\dots,N_d)$.

Discrete Fourier transforms (DFTs) have the following properties:

(a) Linearity. If $x_{\nu_1,\dots,\nu_d}$ and $y_{\nu_1,\dots,\nu_d}$ are two sequences, then

$$\mathrm{DFT}(\alpha x_{\nu_1,\dots,\nu_d}+\beta y_{\nu_1,\dots,\nu_d}) = \alpha\,\mathrm{DFT}(x_{\nu_1,\dots,\nu_d}) + \beta\,\mathrm{DFT}(y_{\nu_1,\dots,\nu_d}) \qquad (\alpha,\beta\in\mathbb{R}).$$

(b) Translation. If $\widetilde{x}_{\nu_1,\dots,\nu_d}$ is a recurring sequence with period $(N_1,\dots,N_d)$, then

$$\mathrm{DFT}\,(\widetilde{x}_{\nu_1-\tau_1,\dots,\nu_d-\tau_d}) = \mathrm{DFT}\,(\widetilde{x}_{\nu_1,\dots,\nu_d})\, e^{-2\pi i\left(\frac{\mu_1\tau_1}{N_1}+\dots+\frac{\mu_d\tau_d}{N_d}\right)} \qquad ((\tau_1,\dots,\tau_d)\in\mathbb{Z}^d).$$

If $\widetilde{X}_{\nu_1,\dots,\nu_d} = \mathrm{DFT}\,(\widetilde{x}_{\nu_1,\dots,\nu_d})$, then

$$\mathrm{IDFT}\,(\widetilde{X}_{\nu_1-\tau_1,\dots,\nu_d-\tau_d}) = \mathrm{IDFT}\,(\widetilde{X}_{\nu_1,\dots,\nu_d})\, e^{2\pi i\left(\frac{\mu_1\tau_1}{N_1}+\dots+\frac{\mu_d\tau_d}{N_d}\right)}.$$

Moreover, if $N = N_i (i = 1,\dots,d)$, then

$$\mathrm{DFT}\,(\widetilde{x}_{\nu-\tau}) = \mathrm{DFT}\,(\widetilde{x}_{\nu})\, e^{-\frac{2\pi i}{N}(\mu,\tau)},$$

$$\text{IDFT}\,(\widetilde{X}_{\boldsymbol{\nu}-\boldsymbol{\tau}}) = \text{IDFT}\,(\widetilde{X}_{\boldsymbol{\nu}})\, e^{\frac{2\pi i}{N}(\boldsymbol{\mu},\boldsymbol{\tau})},$$

where $\boldsymbol{\nu} = (\nu_1, \ldots, \nu_d)$, $\boldsymbol{\tau} = (\tau_1, \ldots, \tau_d)$, and $\boldsymbol{\mu} = (\mu_1, \ldots, \mu_d)$.
(c) Symmetry. Let $X_{\mu_1,\ldots,\mu_d}$ be DFT of $x_{\nu_1,\ldots,\nu_d}$. Then, $\text{DFT}(\overline{x}_{\nu_1,\ldots,\nu_d}) = \overline{X}_{N_1-\mu_1,\ldots,N_d-\mu_d}$.
(d) Parseval Identity. Let $X_{\mu_1,\ldots,\mu_d}$ be DFT of $x_{\nu_1,\ldots,\nu_d}$. Then,

$$\sum_{\mu_1=0}^{N_1-1} \cdots \sum_{\mu_d=0}^{N_d-1} |X_{\mu_1,\ldots,\mu_d}|^2 = N \sum_{\nu_1=0}^{N_1-1} \cdots \sum_{\nu_d=0}^{N_d-1} |x_{\nu_1,\ldots,\nu_d}|^2.$$

(e) Convolutions. Let $\widetilde{x}_{\boldsymbol{\nu}}$, $\widetilde{y}_{\boldsymbol{\nu}}$ ($\boldsymbol{\nu} = (\nu_1, \ldots, \nu_d)$) be both recurring sequences with periods $(N_1, \ldots, N_d)$ and $(M_1, \ldots, M_d)$, respectively. If $N = N_i = M_i$ $(i = 1, \ldots, d)$, the *recurring convolution* of $\widetilde{x}_{\boldsymbol{\nu}}$ and $\widetilde{y}_{\boldsymbol{\nu}}$ is defined as

$$\widetilde{h}_{\mathbf{k}} = (\widetilde{x}_{\boldsymbol{\nu}} * \widetilde{y}_{\boldsymbol{\nu}})_{\mathbf{k}} = \sum_{\boldsymbol{\nu}\in\Lambda_N} \widetilde{x}_{\mathbf{k}-\boldsymbol{\nu}}\widetilde{y}_{\boldsymbol{\nu}} \qquad (\mathbf{k} \in \mathbb{R}^d),$$

where $\Lambda_N = [0, N)^d \bigcap \mathbb{Z}^d$. Noticing that the DFT of $\widetilde{h}_{\mathbf{k}}$ is

$$\begin{aligned}
\widetilde{H}_{\mathbf{n}} &= \sum_{\mathbf{k}\in\Lambda_N} \left(\sum_{\boldsymbol{\nu}\in\Lambda_N} \widetilde{x}_{\mathbf{k}-\boldsymbol{\nu}}\widetilde{y}_{\boldsymbol{\nu}} \right) e^{-\frac{2\pi i(\mathbf{n},\mathbf{k})}{N}} \\
&= \sum_{\mathbf{k}\in\Lambda_N} \left(\sum_{\boldsymbol{\nu}\in\Lambda_N} \widetilde{x}_{\mathbf{k}}\widetilde{y}_{\boldsymbol{\nu}} \right) e^{-\frac{2\pi i(\mathbf{n},\mathbf{k}+\boldsymbol{\nu})}{N}} \\
&= \left(\sum_{\mathbf{k}\in\Lambda_N} \widetilde{x}_{\mathbf{k}}\, e^{-\frac{2\pi i(\mathbf{n},\mathbf{k})}{N}} \right)\left(\sum_{\boldsymbol{\nu}\in\Lambda_N} \widetilde{y}_{\boldsymbol{\nu}}\, e^{-\frac{2\pi i(\mathbf{n},\boldsymbol{\nu})}{N}} \right),
\end{aligned}$$

DFT of the recurring convolution is equal to $\widetilde{X}_{\mathbf{n}}\widetilde{Y}_{\mathbf{n}}$, i.e.,

$$\text{DFT}(\widetilde{x}_{\boldsymbol{\nu}} * \widetilde{y}_{\boldsymbol{\nu}}) = \text{DFT}(\widetilde{x}_{\boldsymbol{\nu}})\text{DFT}(\widetilde{y}_{\boldsymbol{\nu}}).$$

Let $\widetilde{x}_{\nu_1,\ldots,\nu_d}$ $(\nu_l = 0, \ldots, N_l - 1)$ and $\widetilde{y}_{\nu_1,\ldots,\nu_d}$ $(\nu_l = 0, \ldots, M_l - 1)$ be both $d-$dimensional sequences, and

$$P = \max\{N_1, \ldots, N_d, M_1, \ldots, M_d\}.$$

Take $Q = 2P - 1$. Let $\widetilde{x}_{\boldsymbol{\nu}}$ and $\widetilde{y}_{\boldsymbol{\nu}}$ be both recurring sequences with period $\mathbf{Q} = (Q, \ldots, Q)$ and

$$\widetilde{x}_{\nu_1,\ldots,\nu_d} = \begin{cases} x_{\nu_1,\ldots,\nu_d}, & \nu_l = 0, \ldots, N_l - 1, \\ 0, & \nu_l = N_l, \ldots, Q - 1 \quad (l = 1, \ldots, d), \end{cases}$$

$$\widetilde{y}_{\nu_1,\ldots,\nu_d} = \begin{cases} y_{\nu_1,\ldots,\nu_d}, & \nu_l = 0,\ldots,M_l-1, \\ 0, & \nu_l = M_l,\ldots,Q-1 \quad (l=1,\ldots,d). \end{cases}$$

The *linear convolution* of x_ν and y_ν is defined as recurring convolution of $\widetilde{x}_\nu$ and $\widetilde{y}_\nu$, i.e.,

$$\widetilde{h}_{\mathbf{k}} = (x_\nu * y_\nu)_{\mathbf{k}} = (\widetilde{x}_\nu * \widetilde{y}_\nu)_{\mathbf{k}} = \sum_{\nu \in \Lambda_N} \widetilde{x}_{\mathbf{k}-\nu} \widetilde{y}_\nu,$$

where $\mathbf{k} = (k_1,\ldots,k_d)$ and $\boldsymbol{\nu} = (\nu_1,\ldots,\nu_d)$. So the DFT of $h_{\mathbf{k}}$ is $\widetilde{H}_{\mathbf{n}} = \widetilde{X}_{\mathbf{n}} \widetilde{Y}_{\mathbf{n}}$ ($\mathbf{n} \in \Lambda_N$), i.e.,

$$\mathrm{DFT}(x_\nu * y_\nu) = \mathrm{DFT}(\widetilde{x}_\nu)\mathrm{DFT}(\widetilde{y}_\nu).$$

The computation of DFT in (2.2.1) needs $\prod_{k=1}^d N_k$ multiplications and $\prod_{k=1}^d (N_k - 1)$ summations, and the total number of the computation equals approximately $(\prod_{k=1}^d N_k)^2$. The *fast Fourier transform* (FFT) is a fast algorithm to compute DFT through reducing the complexity of computing as follows.

Note that

$$e^{-2\pi i\left(\frac{\mu_1\nu_1}{N_1} + \frac{\mu_2\nu_2}{N_2} + \cdots + \frac{\mu_d\nu_d}{N_d}\right)} = e^{-2\pi i \frac{\mu_1\nu_1}{N_1}} e^{-2\pi i \frac{\mu_2\nu_2}{N_2}} \cdots e^{-2\pi i \frac{\mu_d\nu_d}{N_d}}.$$

The (2.2.1) is rewritten in the form:

$$X_{\mu_1,\ldots,\mu_d} = \frac{1}{N_1} \sum_{\nu_1=0}^{N_1-1} e^{-2\pi i \frac{\mu_1\nu_1}{N_1}} \left(\frac{1}{N_2} \sum_{\nu_2=0}^{N_2-1} e^{-2\pi i \frac{\mu_2\nu_2}{N_2}} \cdots \left(\frac{1}{N_d} \sum_{\nu_d=0}^{N_d-1} x_{\nu_1,\ldots,\nu_d} e^{-2\pi i \frac{\mu_d\nu_d}{N_d}} \right) \cdots \right). \tag{2.2.3}$$

Denote by $R_1(\nu_1,\ldots,\nu_{d-1},\mu_d)$ the last sum, i.e.,

$$R_1(\nu_1,\ldots,\nu_{d-1},\mu_d) = \frac{1}{N_d} \sum_{\nu_d=0}^{N_d-1} x_{\nu_1,\ldots,\nu_d} e^{-2\pi i \frac{\mu_d\nu_d}{N_d}},$$

where $\nu_l = 0,\ldots,N_l-1$ ($l = 1,\ldots,d-1$). Regard $R_1(\nu_1,\ldots,\nu_{d-1},\mu_d)$ as a function of μ_d ($\mu_d = 0,\ldots,N_d-1$). Denote $P_{\mu_d} = R_1(\nu_1,\ldots,\nu_{d-1},\mu_d)$, $p_{\nu_d} = x_{\nu_1,\ldots,\nu_d}$, and $w_{N_d} = e^{-2\pi i \frac{1}{N_d}}$. Then,

$$P_{\mu_d} = \frac{1}{N_d} \sum_{\nu_d=0}^{N_d-1} p_{\nu_d} w_{N_d}^{\mu_d\nu_d}.$$

Without loss of generality, assume $N_d = 2^M$. Otherwise, let $2^{M-1} \le N_d < 2^M$. We extend the original sequence $p_0,\ldots,p_{N_{d-1}}$ into a new sequence $p_0,\ldots,p_{N_d-1}$, $0,\ldots,0$ by adding $2^M - N_d$ zeros behind $p_0,\ldots,p_{N_d-1}$.

First, we decompose the 2^M−point sequences $\{p_{\nu_d}\}$ into two 2^{M-1}−point sequences:

$$y = (p_0, p_2, \dots, p_{2^M-2}) =: (y_0, y_1, \dots, y_{2^{M-1}-1}),$$

$$z = (p_1, p_3, \dots, p_{2^M-1}) =: (z_0, z_1, \dots, z_{2^{M-1}-1}).$$

The first half P_{μ_d} ($\mu_d = 0, 1, \dots, 2^{M-1} - 1$) is computed as follows:

$$P_{\mu_d} = \frac{1}{2^M} \sum_{\nu=0}^{2^{M-1}-1} y_\nu w_{2^M}^{-2\nu\mu_d} + \frac{1}{2^M} \sum_{\nu=0}^{2^{M-1}-1} z_\nu w_{2^M}^{-(2\nu+1)\mu_d}$$

$$= \frac{1}{2}(Y_{\mu_d} + w_{2^M}^{\mu_d} Z_{\mu_d}) \qquad (\mu_d = 0, \dots, 2^{M-1} - 1);$$

the second half $P_{\mu_d}(\mu_d = 2^{M-1}, \dots, 2^M)$ is similarly computed as follows:

$$P_{\mu_d+2^{M-1}} = \frac{1}{2}(Y_{\mu_d} - w_{2^M}^{\mu_d} Z_{\mu_d}) \qquad (\mu_d = 0, \dots, 2^{M-1} - 1),$$

where

$$Y_{\mu_d} = \frac{1}{2^{M-1}} \sum_{\nu=0}^{2^{M-1}-1} y_\nu w_{2^{M-1}}^{\nu\mu_d},$$

$$Z_{\mu_d} = \frac{1}{2^{M-1}} \sum_{\nu=0}^{2^{M-1}-1} z_\nu w_{2^{M-1}}^{\nu\mu_d}$$

are the DFT of y and z, respectively. It means that the computation of P_{μ_d} or R_1 can be reduced to that of Y_{μ_d} and Z_{μ_d}.

Next, we rewrite $Y_{\mu_d}(\mu_d = 0, \dots, 2^{M-1} - 1)$ and Z_{μ_d} ($\mu_d = 0, \dots, 2^{M-1} - 1$), respectively, as follows:

$$Y_{\mu_d} = \frac{1}{2}(Y'_{\mu_d} + w_{2^{M-1}}^{\mu_d} Y''_{\mu_d}),$$

$$Y_{\mu_d+2^{M-2}} = \frac{1}{2}(Y'_{\mu_d} - w_{2^{M-1}}^{\mu_d} Y''_{\mu_d}) \qquad (\mu_d = 0, \dots, 2^{M-2} - 1)$$

and

$$Z_{\mu_d} = \frac{1}{2}(Z'_{\mu_d} + w_{2^{M-1}}^{\mu_d} Z''_{\mu_d}),$$

$$Z_{\mu_d+2^{M-2}} = \frac{1}{2}(Z'_{\mu_d} - w_{2^{M-1}}^{\mu_d} Z''_{\mu_d}) \qquad (\mu_d = 0, \dots, 2^{M-2} - 1),$$

where Y'_{μ_d} and Y''_{μ_d} are DFTs of 2^{M-2}−even samples and 2^{M-2}−odd samples of $\{Y_{\mu_d}\}$, respectively, and Z'_{μ_d} and Z''_{μ_d} are DFTs of 2^{M-2}−even samples and 2^{M-2}−odd samples of $\{Z_{\mu_d}\}$, respectively. It means that the computation of P_{μ_d} or R_1 can be reduced to that of Y'_{μ_d}, Y''_{μ_d}, Z'_{μ_d}, and Z''_{μ_d}.

Finally, by repeating the above procedure again and again, it terminates at the computation of DFT of one sample. This procedure gives a fast algorithm to compute $R_1(\nu_1, \ldots, \nu_{d-1}, \mu_d)$.

Let

$$R_2(\nu_1, \ldots, \nu_{d-2}, \mu_{d-1}, \mu_d) := \frac{1}{N_{d-1}} \sum_{\nu_{d-1}=0}^{N_{d-1}-1} R_1(\nu_1, \ldots, \nu_{d-1}, \mu_d)\, e^{-2\pi i \frac{\mu_{d-1}\nu_{d-1}}{N_{d-1}}}.$$

Using the same algorithm as in R_1, the $R_2(\nu_1, \ldots, \nu_{d-2}, \mu_{d-1}, \mu_d)$ can be computed fast. Continuing this procedure, we finally fast compute

$$X_{\mu_1,\ldots,\mu_d} = \frac{1}{N_1} \sum_{\nu_1=0}^{N_1-1} R_{d-1}(\nu_1, \mu_2, \ldots, \mu_d)\, e^{-2\pi i \frac{\mu_1\nu_1}{N_1}}.$$

The total number of operations in the whole procedure is equal approximately to

$$\left(\prod_{k=1}^{d} N_k\right) \log_2\left(\prod_{k=1}^{d} N_k\right).$$

2.3 Discrete Cosine/Sine Transform

The DFT is a linear transform from a sequence $x_0, \ldots, x_{N-1}$ of real numbers to a sequence $X_0, \ldots, X_{N-1}$ of complex numbers. In order to reduce boundary effects, one often takes DFTs after even extension/odd extension of the original sequence $x_0, \ldots, x_{N-1}$ around end points. As a result, various discrete cosine/sine transforms (DCT/DST) are introduced. The DCT/DST is similar to the DFT: They are also the transform connecting between the time/spatial domain to the frequency domain.

2.3.1 *Four Forms of DCTs*

DCTs transform real numbers $x_0, \ldots, x_{N-1}$ into real numbers $X_0, \ldots, X_{N-1}$.

Define DCT-1 as

$$X_k = \sum_{n=0}^{N-1} \beta_n x_n \cos \frac{\pi n k}{N-1} \qquad (k = 0, \ldots, N-1), \tag{2.3.1}$$

where $\beta_0 = \beta_{N-1} = \frac{1}{2}$ and $\beta_n = 1$ $(n = 1, \ldots, N-2)$. Its inverse transform is

$$x_k = \frac{2}{N-1}\sum_{n=0}^{N-1} \beta_n X_n \cos\frac{\pi n k}{N-1} \qquad (k=0,\ldots,N-1).$$

In fact, substituting this into the right-hand side of (2.3.1), by Euler's formula, we get

$$\begin{aligned} J_k := &\sum_{n=0}^{N-1} \beta_n x_n \cos\tfrac{\pi nk}{N-1} \\ = &\tfrac{2}{N-1}\sum_{n=0}^{N-1}\sum_{l=0}^{N-1} \beta_n\beta_l X_l \cos\tfrac{\pi ln}{N-1}\cos\tfrac{\pi nk}{N-1} \\ = &\tfrac{1}{N-1}\sum_{l=0}^{N-1}\beta_l X_l \,\mathrm{Re}\left\{\sum_{n=0}^{N-1}\beta_n\left(e^{i\frac{\pi n(l+k)}{N-1}}+e^{i\frac{\pi n(l-k)}{N-1}}\right)\right\}. \end{aligned}$$

By the summation formula of geometric series,

$$\mathrm{Re}\left\{\sum_{n=0}^{N-1}\beta_n\left(e^{i\frac{\pi n(l+k)}{N-1}}+e^{i\frac{\pi n(l-k)}{N-1}}\right)\right\} = \begin{cases} 0, & l\neq k\ (k\neq 0,\ N-1), \\ N-1, & l=k\ (k\neq 0,\ N-1), \\ 2(N-1), & l=k\ (k=0,\ N-1). \end{cases}$$

So $J_k = X_k$.

The even extension of the sequence $x_0,\ldots,x_{N-1}$ around x_{N-1} is $x_0,\ldots,x_{N-2}, x_{N-1}, x_{N-2},\ldots,x_1$, and then the DCT-1 of $x_0,\ldots,x_{N-1}$ is equivalent to the DFT of its even extension around x_{N-1}.

In fact, let

$$y_n = \begin{cases} x_n, & n=0,\ldots,N-1, \\ x_{2N-2-n}, & n=N,\ldots,2N-3. \end{cases}$$

The DFT of $y_n (n=0,\ldots,2N-3)$ is equal to

$$\begin{aligned} Y_k = &\tfrac{1}{2N-2}\sum_{n=0}^{2N-3} y_n\, e^{-i\frac{2\pi nk}{2N-2}} \\ = &\tfrac{1}{2N-2}\left(y_0+(-1)^k y_{N-1}+\sum_{n=1}^{N-2} y_n\left(e^{-i\frac{\pi nk}{N-1}}+e^{i\frac{\pi nk}{N-1}}\right)\right) \\ = &\tfrac{1}{N-1}\left(\tfrac{1}{2}(x_0+(-1)^k x_{N-1})+\sum_{n=1}^{N-2} x_n \cos\tfrac{\pi nk}{N-1}\right) = \tfrac{1}{N-1}X_k \quad (k=0,\ldots,N-1). \end{aligned}$$

The *high-dimensional form of DCT-1* is

$$X_{k_1,\ldots,k_d} = \sum_{n_1=0}^{N_1-1} \cdots \sum_{n_d=0}^{N_d-1} x_{n_1,\ldots,n_d} \prod_{j=1}^{d} \beta_{n_j} \cos \frac{\pi n_j k_j}{N_j - 1} \qquad (k_l = 1, \ldots, N_l - 1).$$

Its inverse transform is

$$x_{k_1,\ldots,k_d} = \frac{2^d}{(N_1 - 1)\cdots(N_d - 1)} \sum_{n_1=0}^{N_1-1} \cdots \sum_{n_d=0}^{N_d-1} X_{n_1,\ldots,n_d} \prod_{j=1}^{d} \beta_{n_j} \cos \frac{\pi n_j k_j}{N_j - 1},$$

where $\beta_{n_j} = \frac{1}{2}$ $(n_j = 0, N_j - 1)$ and $\beta_{n_j} = 1$ $(n_j = 1, \ldots, N_j - 2)$.

DCT-2 is the most commonly used form of discrete cosine transform. Define DCT-2 as

$$X_k = \sum_{n=0}^{N-1} x_n \cos \frac{\pi(n + \frac{1}{2})k}{N} \qquad (k = 0, \ldots, N - 1). \tag{2.3.2}$$

The inverse formula is

$$x_k = \frac{2}{N} \sum_{n=0}^{N-1} \alpha_n X_n \cos \frac{\pi n(k + \frac{1}{2})}{N} \qquad (k = 0, \ldots, N - 1), \tag{2.3.3}$$

where $\alpha_0 = \frac{1}{2}$, $\alpha_n = 1$ $(n = 1, \ldots, N - 1)$.

In fact, substituting (2.3.3) into the right-hand side of (2.3.2),

$$\frac{1}{N} \sum_{l=0}^{N-1} \alpha_l X_l \sum_{n=0}^{N-1} 2 \cos \frac{\pi(n + \frac{1}{2})l}{N} \cos \frac{\pi(n + \frac{1}{2})k}{N} =: J_k.$$

Note that

$$\sum_{n=0}^{N-1} 2 \cos \frac{\pi(n + \frac{1}{2})l}{N} \cos \frac{\pi(n + \frac{1}{2})k}{N} = \begin{cases} 0, & k \neq l, \\ N, & k = l \neq 0, \\ 2N, & k = l = 0 \end{cases}$$

So $J_k = X_k$ $(k = 0, \ldots, N - 1)$, i.e., (2.3.3) holds.

Given a sequence $x_0, \ldots, x_{N-1}$, one constructs a new sequence $y_0, \ldots, y_{4N-1}$ satisfying the following:

$$y_{2n} = 0 \qquad (n = 0, \ldots, N - 1),$$

$$y_{2n+1} = x_n \qquad (n = 0, \ldots, N - 1),$$

$$y_{4N-n} = y_n \qquad (n = 1, \ldots, 2N - 1).$$

The DFT of the sequence $y_0, \ldots, y_{4N-1}$ is

$$\begin{aligned} Y_k &= \sum_{n=0}^{4N-1} y_n\, e^{-i\frac{2\pi nk}{4N}} \\ &= \sum_{n=0}^{2N-1} y_{2n+1}\, e^{-i\frac{2\pi(2n+1)k}{4N}} \\ &= 2\sum_{n=0}^{N-1} x_n \cos\frac{\pi(n+\frac{1}{2})k}{N} = 2X_k. \end{aligned}$$

So the DCT-2 of a sequence of N real numbers $x_0, \ldots, x_{N-1}$ is equivalent to the DFT of the sequence of $4N$ real numbers $y_0, \ldots, y_{4N-1}$ of even symmetry whose even-indexed elements are zero.

The *high-dimensional form of DCT-2* of a sequence $\{x_{n_1,\ldots,n_d}\}_{n_i=0,\ldots,N_i-1(i=1,\ldots,d)}$ is

$$X_{k_1,\ldots,k_d} = \sum_{n_1=0}^{N_1-1} \cdots \sum_{n_d=0}^{N_d-1} x_{n_1,\ldots,n_d} \prod_{l=1}^{d} \cos\frac{\pi(n_l+\frac{1}{2})k_l}{N_l}.$$

Its inverse formula is

$$x_{k_1,\ldots,k_d} = \frac{2^d}{N_1\cdots N_d} \sum_{n_1=0}^{N_1-1} \cdots \sum_{n_d=0}^{N_d-1} X_{n_1,\ldots,n_d} \prod_{l=1}^{d} \alpha_{n_l} \cos\frac{\pi n_l(k_l+\frac{1}{2})}{N_l},$$

where $\alpha_{n_l} = \frac{1}{2}$ $(n_l = 0)$ and $\alpha_{n_l} = 1$ $(n_l = 1, \ldots, N_l - 1)$ for $l = 1, \ldots, d$.

The DCT-3 is defined as the inverse transform of the DCT-2.

Define DCT-4 as

$$X_k = \sum_{n=0}^{N-1} x_n \cos\frac{\pi(n+\frac{1}{2})(k+\frac{1}{2})}{N} \qquad (k = 0, \ldots, N-1). \tag{2.3.4}$$

Its inverse formula is

$$x_k = \frac{2}{N}\left(\sum_{n=0}^{N-1} X_n \cos\frac{\pi(n+\frac{1}{2})(k+\frac{1}{2})}{N}\right) \qquad (k = 0, \ldots, N-1).$$

In fact, substituting this into the right-hand side of (2.3.4),

$$
\begin{aligned}
J_k &:= \sum_{n=0}^{N-1} x_n \cos\frac{\pi(n+\frac{1}{2})(k+\frac{1}{2})}{N} \\
&= \frac{2}{N}\sum_{n=0}^{N-1}\sum_{l=0}^{N-1} X_l \cos\frac{\pi(l+\frac{1}{2})(n+\frac{1}{2})}{N}\cos\frac{\pi(n+\frac{1}{2})(k+\frac{1}{2})}{N} \qquad (2.3.5) \\
&= \frac{1}{N}\sum_{l=0}^{N-1} X_l \mathrm{Re}\left(e^{i\frac{\pi(l+k+1)}{2N}}\sum_{n=0}^{N-1} e^{i\frac{\pi n(l+k+1)}{N}} + e^{i\frac{\pi(l-k)}{2N}}\sum_{n=0}^{N-1} e^{i\frac{\pi n(l-k)}{N}}\right).
\end{aligned}
$$

Since

$$
e^{i\frac{\pi(l+k+1)}{2N}}\sum_{n=0}^{N-1} e^{i\frac{\pi n(l+k+1)}{N}} = i\frac{1-(-1)^{l+k+1}}{2\sin\frac{\pi}{2N}(l+k+1)},
$$

$$
e^{i\frac{\pi(l-k)}{2N}}\sum_{n=0}^{N-1} e^{i\frac{\pi n(l-k)}{N}} = \begin{cases} i\frac{1-(-1)^{l-k}}{2\sin\frac{\pi}{2N}(l-k)}, & l \neq k, \\ N, & l = k, \end{cases}
$$

the representative in large parentheses in (2.3.5) is equal to a pure imaginary number for $l \neq k$ and to N for $l = k$. So $J_k = \frac{1}{N} X_k$ $(k = 0, \ldots, N-1)$.

The *high-dimensional form of DCT-4* of a real sequence $\{x_{n_1,\ldots,n_d}\}_{n_i=0,\ldots,N_i-1(i=1,\ldots,d)}$ is defined as

$$
X_{k_1,\ldots,k_d} = \sum_{n_1=0}^{N_1-1}\cdots\sum_{n_d=0}^{N_d-1} x_{n_1,\ldots,n_d}\prod_{j=1}^{d}\cos\frac{\pi(n_j+\frac{1}{2})(k_j+\frac{1}{2})}{N_j} \quad (k_l = 0,\ldots,N_l-1). \tag{2.3.6}
$$

Using the one-dimensional DCT successively, the high-dimensional form of DCT-4 becomes

$$
X_{k_1,\ldots,k_d} = \sum_{n_1=0}^{N_1-1}\cdots\left(\sum_{n_d=0}^{N_d-1} x_{n_1,\ldots,n_d}\cos\frac{\pi(n_d+\frac{1}{2})(k_d+\frac{1}{2})}{N_d}\right)\cdots\cos\frac{\pi(n_1+\frac{1}{2})(k_1+\frac{1}{2})}{N_1}.
$$

Its inverse transform is

$$
x_{k_1,\ldots,k_d} = \frac{2^d}{N_1\cdots N_d}\sum_{n_1=0}^{N_1-1}\cdots\sum_{n_d=0}^{N_d-1} X_{n_1,\ldots,n_d}\prod_{j=1}^{d}\cos\frac{\pi(n_j+\frac{1}{2})(k_j+\frac{1}{2})}{N_j}.
$$

2.3.2 Four Forms of DSTs

Similar to DCTs, there are four forms of DSTs. The DST-1 of a sequence $x_0, \ldots, x_{N-1}$ is defined as

$$X_k = \sum_{n=0}^{N-1} x_n \sin \frac{\pi(n+1)(k+1)}{N+1} \qquad (k = 0, \ldots, N-1). \tag{2.3.7}$$

Its inverse formula is

$$x_k = \frac{2}{N+1} \sum_{n=0}^{N-1} X_n \sin \frac{\pi(n+1)(k+1)}{N+1} \qquad (k = 0, \ldots, N-1). \tag{2.3.8}$$

In fact, substituting (2.3.8) into the right-hand side of (2.3.7),

$$\begin{aligned} J_k := & \frac{2}{N+1} \sum_{n=0}^{N-1} \sum_{j=0}^{N-1} X_l \sin \frac{\pi(l+1)(n+1)}{N+1} \sin \frac{\pi(n+1)(k+1)}{N+1} \\ = & \frac{1}{N+1} \sum_{l=0}^{N-1} X_l (A_{kl} - B_{kl}), \end{aligned}$$

where

$$A_{kl} = \mathrm{Re}\left(\sum_{n=0}^{N-1} e^{i\frac{\pi(n+1)(l-k)}{N+1}} \right) = \begin{cases} N, & l = k, \\ -1, & l \neq k, \ l-k \text{ is even}, \\ 0, & l-k \text{ is odd}, \end{cases}$$

$$B_{kl} = \mathrm{Re}\left(\sum_{n=0}^{N-1} e^{i\frac{\pi(n+1)(l+k+2)}{N+1}} \right) = \begin{cases} -1, & l = k, \\ -1, & l \neq k, \ l-k \text{ is even}, \\ 0, & l-k \text{ is odd}. \end{cases}$$

So $J_k = X_k$.

The *high-dimensional form of DST-1* is

$$X_{k_1,\ldots,k_d} = \sum_{n_1=0}^{N_1-1} \cdots \sum_{n_d=0}^{N_d-1} x_{n_1,\ldots,n_d} \prod_{l=1}^{d} \sin\left(\frac{\pi}{N}(n_l+1)(k_l+1)\right).$$

The *high-dimensional form of DST-2* is

$$X_{k_1,\ldots,k_d} = \sum_{n_1=0}^{N_1-1} \cdots \sum_{n_d=0}^{N_d-1} x_{n_1,\ldots,n_d} \prod_{l=1}^{d} \sin \frac{\pi(n_l+\frac{1}{2})(k_l+1)}{N_l}.$$

The *high-dimensional form of DST-3* is

$$X_{k_1,\ldots,k_d} = \left(\prod_{l=1}^{d} \frac{2}{N_l+1}\right)\left(\sum_{n_1=0}^{N_1-1} \cdots \sum_{n_d=0}^{N_d-1} x_{n_1,\ldots,n_d} \prod_{l=1}^{d} \gamma_{n_l} \sin \frac{\pi(n_l+1)(k_l+\frac{1}{2})}{N_l}\right),$$

where $\gamma_{n_l} = 1$ $(n_l = 0, \ldots, N_l - 2)$ and $\gamma_{n_l} = \frac{1}{2}$ $(n_l = N_l - 1)$.

The *high-dimensional form of DST-4* is

$$X_{k_1,\ldots,k_d} = \sum_{n_1=0}^{N_1-1} \cdots \sum_{n_d=0}^{N_d-1} x_{n_1,\ldots,n_d} \prod_{l=1}^{d} \sin \frac{\pi(n_l+\frac{1}{2})(k_l+\frac{1}{2})}{N_l}.$$

2.4 Filtering

Climatologists and environmentalists might be interested in some phenomena that would be cyclic over a certain period, or exhibit slow changes, so filtering becomes a key preprocessing step of climate and environmental data. The design of filters is closely related to the Fourier transform of the $d-$dimensional sequence which is defined as

$$F(x_{n_1,\ldots,n_d}) = X(e^{i\omega_1}, \ldots, e^{i\omega_d}) = \sum_{n_1\in\mathbb{Z}} \cdots \sum_{n_d\in\mathbb{Z}} x_{n_1,\ldots,n_d}\, e^{-in_1\omega_1} \cdots e^{-in_d\omega_d},$$

where $X(e^{i\omega_1}, \ldots, e^{i\omega_d})$ is called the *frequency characteristic* of $h_{n_1,\ldots,n_d}$. Its inverse transform is defined as

$$\begin{aligned} &F^{-1}(X(e^{i\omega_1}, \ldots, e^{i\omega_d})) \\ &= \frac{1}{(2\pi)^d} \int_{-\pi}^{\pi} \cdots \int_{-\pi}^{\pi} X(e^{i\omega_1}, \ldots, e^{i\omega_d})\, e^{in_1\omega_1} \cdots e^{in_d\omega_d} d\omega_1 \cdots d\omega_d. \end{aligned}$$

With the help of Poisson summation formula in Sect. 2.1, it is clear that the Fourier transform of the $d-$dimensional sequence is a transform from time/spatial domain to frequency domain.

(i) Low-Pass Filter

For a filter $h_{n_1,\ldots,n_d}$, if its frequency characteristic $H(e^{i\theta_1}, \ldots, e^{i\theta_d})$ satisfies the condition:

$$H(e^{i\theta_1}, \ldots, e^{i\theta_d}) = \begin{cases} 1, \ \sqrt{\theta_1^2 + \cdots + \theta_d^2} \le R < \pi, \\ \\ 0, \ \text{otherwise}, \end{cases}$$

the filter $h_{n_1,\ldots,n_d}$ is called a *spherically symmetric low-pass filter*.

In the case $d = 1$. If h_n is a low-pass filter and its frequency characteristic is

$$H(e^{i\theta}) = \begin{cases} 1, \ |\theta| \le R < \pi, \\ 0, \ \text{otherwise}, \end{cases}$$

then the low-pass filter is

$$h_n = \frac{1}{2\pi}\int_{-\pi}^{\pi} H(e^{i\theta})\, e^{in\theta}\mathrm{d}\theta = \frac{1}{2\pi}\int_{-R}^{R} e^{in\theta}\mathrm{d}\theta = \frac{1}{2\pi}\int_{-R}^{R} e^{i\theta}\mathrm{d}\theta = \begin{cases} \frac{\sin(nR)}{\pi n}, & n \ne 0, \\ \frac{R}{\pi}, & n = 0, \end{cases}$$

In the case $d = 2$. If h_{n_1,n_2} is a low-pass filter and its frequency characteristic is

$$H(e^{i\theta_1}, e^{i\theta_2}) = \begin{cases} 1, \ \sqrt{\theta_1^2 + \theta_2^2} \le R < \pi, \\ 0, \ \text{otherwise}, \end{cases}$$

then the low-pass filter is

$$\begin{aligned} h_{n_1,n_2} &= \tfrac{1}{(2\pi)^2} \int\int_{\sqrt{\theta_1^2+\theta_2^2} \le R} e^{in_1\theta_1}\, e^{in_2\theta_2}\, \mathrm{d}\theta_1 \mathrm{d}\theta_2 \\ &= \tfrac{1}{4\pi^2} \int_0^R r \left(\int_0^{2\pi} e^{ir(n_1 \cos\theta + n_2 \sin\theta)} \mathrm{d}\theta \right) \mathrm{d}r, \end{aligned} \tag{2.4.1}$$

where $\theta_1 = r\cos\theta$ and $\theta_2 = r\sin\theta$. Note that

$$n_1 \cos\theta + n_2 \sin\theta = \sqrt{n_1^2 + n_2^2}(\sin\varphi\cos\theta + \cos\varphi\sin\theta) = \sqrt{n_1^2 + n_2^2}\sin(\theta + \varphi),$$

where

$$\sin\varphi = \frac{n_1}{\sqrt{n_1^2 + n_2^2}}, \qquad \cos\varphi = \frac{n_2}{\sqrt{n_1^2 + n_2^2}},$$

and the integral representation of Bessel function of degree 0 is

$$J_0(t) = \frac{1}{2\pi}\int_0^{2\pi} e^{it\sin\theta}\mathrm{d}\theta.$$

By the periodicity of sine function, the inner integral of (2.4.1) is

$$\begin{aligned} \int_0^{2\pi} e^{ir(n_1\cos\theta + n_2\sin\theta)}\mathrm{d}\theta =* \int_0^{2\pi} e^{ir\sqrt{n_1^2+n_2^2}\sin(\theta+\varphi)}\mathrm{d}\theta = \int_0^{2\pi} e^{ir(\sqrt{n_1^2+n_2^2})\sin\theta}\mathrm{d}\theta \\ = J_0\left(r\sqrt{n_1^2 + n_2^2} \right). \end{aligned}$$

By (2.4.1), this follows that the low-pass filter is

$$h_{n_1,n_2} = \frac{1}{4\pi^2} \int_0^R r J_0 \left(r\sqrt{n_1^2 + n_2^2} \right) \mathrm{d}r,$$

where J_0 is the Bessel function of degree 0.

(ii) Band-Pass Filter
For a filter $h_{n_1,\ldots,n_d}$, if its frequency characteristic $H(e^{i\theta_1}, \ldots, e^{i\theta_d})$ satisfies the condition:

$$H(e^{i\theta_1}, \ldots, e^{i\theta_d}) = \begin{cases} 1, \; 0 < R_2 \le \sqrt{\theta_1^2 + \cdots + \theta_d^2} \le R_1 < \pi, \\ 0, \; \text{otherwise}, \end{cases}$$

then the filter $h_{n_1,\ldots,n_d}$ is called a *spherically symmetric band-pass filter.*

In the case $d = 1$. The band-pass filter is

$$h_n = \begin{cases} \frac{1}{\pi n}(\sin(nR_1) - \sin(nR_2)), & n \ne 0, \\ \frac{1}{\pi}(R_1 - R_2), & n = 0. \end{cases}$$

In the case $d = 2$. The band-pass filter is

$$h_{n_1,n_2} = \frac{1}{4\pi^2} \int_{R_2}^{R_1} r J_0 \left(r\sqrt{n_1^2 + n_2^2} \right) \mathrm{d}r,$$

where J_0 is the Bessel function of degree 0.

(iii) High-Pass Filter
For a filter $h_{n_1,\ldots,n_d}$, if its frequency characteristic $H(e^{i\theta_1}, \ldots, e^{i\theta_d})$ satisfies the condition:

$$H(e^{i\theta_1}, \ldots, e^{i\theta_d}) = \begin{cases} 1, \; 0 < R \le \sqrt{\theta_1^2 + \cdots + \theta_d^2} < \pi, \\ 0, \; \text{otherwise}, \end{cases}$$

the filter $h_{n_1,\ldots,n_d}$ is called a *spherically symmetric high-pass filter.*

In the case $d = 1$. The high-pass filter is

$$h_n = \begin{cases} -\frac{\sin(nR)}{\pi n}, & n \ne 0 \\ 1 - \frac{R}{\pi}, & n = 0. \end{cases}$$

In the case $d = 2$. The high-pass filter is

$$h_{n_1,n_2} = \begin{cases} -\frac{1}{4\pi^2}\int_0^R r J_0\left(r\sqrt{n_1^2+n_2^2}\right) dr, & n_1 \neq 0 \text{ or } n_2 \neq 0, \\ 1 - \frac{R^2}{4\pi}, & n_1 = n_2 = 0, \end{cases}$$

where J_0 is the Bessel function of degree 0.

2.5 Fractional Fourier Transform

Fractional Fourier transform (FRFT) is an extension of Fourier transform. It is thought of as Fourier transform to αth power. FRFT can transform a time series into the domain between time and frequency, so it demonstrates obvious advantages over Fourier transform.

2.5.1 *Continuous FRFT*

Starting from successive Fourier transform operator F, we study its nth iterated F^n given by

$$F^n(f) = F(f^{n-1}(f)), \qquad F^{-n} = (F^{-1})^n,$$

where n is a nonnegative integer and $F^0(f) = f$. From Fourier transform operator F and its inverse operator F^{-1}, it follows that

$$F^2(f)(t) = F \circ F(f)(t) = f(-t),$$

$$F^3(f)(t) = F^{-1}(f)(t),$$

$$F^4(f))(t) = f(t).$$

FRFT provides a family of linear transforms that further extend Fourier transform to handle non-integer power $n = \frac{2\alpha}{\pi}$ of the Fourier transform. For a $\alpha \in \mathbb{R}$, the $\alpha-$angle FRFT of a function f is defined as

$$F_\alpha(f)(\omega) = \int_{\mathbb{R}} K_\alpha(t,\omega) f(t) dt$$

and $K_\alpha(t,\omega)$ is called *kernel function* and

$$K_\alpha(t,\omega) = \sqrt{1 - i\cot\alpha}\, e^{\pi i(\omega^2+t^2)\cot\alpha - 2\pi i\omega t\csc\alpha}. \tag{2.5.1}$$

It is clear that $F_{\frac{\pi}{2}}(f)(\omega)$ is the Fourier transform and $F_{-\frac{\pi}{2}}(f)(\omega)$ is the inverse Fourier transform. When α is an integer multiple of π,

$$(F_\alpha f)(\omega) = \begin{cases} f(\omega), & \text{if } \alpha \text{ is a multiple of } 2\pi, \\ f(-\omega), & \text{if } \alpha + \pi \text{ is a multiple of } 2\pi. \end{cases}$$

In FRFT, the variable ω is neither a time nor a frequency, and it is an interpolation between time and frequency. Note that $F_0 f$ is f itself, $F_{\frac{\pi}{2}} f$ is Fourier transform of f, $F_\pi f$ results in an inversion of the time axis of f, and $F_{2\pi} f$ is f itself. So $F_\alpha f$ is regarded as a counterclockwise rotation of the axis by an angle α of f in the time–frequency domain.

(i) FRFT and Eigenfunctions of Fourier Transform

Hermite–Gaussian functions are defined as

$$\psi_k(t) = \frac{2^{\frac{1}{4}}}{\sqrt{2^k k!}} H_k(\sqrt{2\pi}\, t)\, e^{-\pi t^2} \qquad (k = 0, 1, \ldots), \tag{2.5.2}$$

where H_k is the kth Hermite polynomial:

$$H_k(t) = e^{t^2} \frac{d^n (e^{-t^2})}{dt^n}.$$

Hermite–Gaussian functions are the unique finite energy eigensolutions of the Hermite–Gaussian equation:

$$\frac{d^2 f(t)}{dt^2} - 4\pi^2 t^2 f(t) = \lambda f(t). \tag{2.5.3}$$

Let $D = \frac{d}{dt}$, F denote the Fourier transform operator, and $S = D^2 + F D^2 F^{-1}$. Then, the Eq. (2.5.3) can be rewritten in the form $Sf(t) = \lambda f(t)$, and $\psi_k (k = 0, 1, \ldots)$ are the eigenfunctions of the operator S. Since the operators S and F are exchangeable and two exchangeable operators must have a common eigenvector set, the Fourier transform operator F has eigenfunctions $\psi_k(t)(k = 0, 1, \ldots)$, and the eigenvalue corresponding to $\psi_k(t)$ is $e^{-i\frac{\pi k}{2}}$, i.e.,

$$F(\psi_k)(\omega) = e^{-i\frac{\pi k}{2}} \psi_k(\omega). \tag{2.5.4}$$

Since $\{\psi_k(t)\}_{k=0,1,\ldots}$ can form a normal orthogonal basis for $L^2(\mathbb{R})$, any $f \in L^2(\mathbb{R})$ can be expanded into a series

$$f(t) = \sum_{k=0}^{\infty} c_k \psi_k(t),$$

where the coefficients $c_k = \int_{\mathbb{R}} f(t)\psi_k(t)\mathrm{d}t$. Taking Fourier transform on both sides gives

$$F(f)(\omega)=\sum_{k=0}^{\infty}c_kF(\psi_k)(\omega)=\sum_{k=0}^{\infty}c_ke^{-i\frac{\pi k}{2}}\psi_k(\omega)=\int_{\mathbb{R}}K(t,\omega)f(t)\mathrm{d}t, \quad (2.5.5)$$

where $K(t,\omega)=\sum_{k=0}^{\infty}\psi_k(\omega)\psi_k(t)\,e^{-i\frac{\pi k}{2}}$. For FRFT, its kernel function $K_\alpha(t,\omega)$ can be expanded into

$$K_\alpha(t,\omega)=\sum_{k=0}^{\infty}\psi_k(\omega)\psi_k(t)\,e^{-i\alpha\frac{\pi k}{2}}. \quad (2.5.6)$$

This means that FRFT can be derived by the eigenfunctions of Fourier transform.

(ii) Index Additivity

Successive applications of FRFT are equivalent to a single transform whose order is the sum of individual orders, i.e., $F_\alpha\circ F_\beta=F_\alpha(F_\beta)=F_{\alpha+\beta}$.

In fact,

$$F_\alpha\circ F_\beta(u)=\int_{\mathbb{R}}\left(\int_{\mathbb{R}}K_\alpha(u,\omega)K_\beta(t,\omega)\mathrm{d}\omega\right)f(t)\mathrm{d}t.$$

Since $\int_{\mathbb{R}}\psi_k(\omega)\psi_l(\omega)\mathrm{d}\omega=\delta_{k,l}$, by the expansion (2.5.6) of the kernel function $K_\alpha(t,\omega)$, it follows that

$$\int_{\mathbb{R}}K_\alpha(u,\omega)K_\beta(t,\omega)\mathrm{d}\omega=\sum_{k=0}^{\infty}e^{-i\frac{\pi k}{2}(\alpha+\beta)}\psi_k(u)\psi_k(t)=K_{\alpha+\beta}(u,t).$$

So

$$F_\alpha\circ F_\beta(u)=\int_{\mathbb{R}}K_{\alpha+\beta}(u,t)f(t)\mathrm{d}t=F_{\alpha+\beta}(u).$$

According to the index additivity, FRFT has the following properties:

(a) Inverse. $f(t)=\int_{\mathbb{R}}F_\alpha(f)(\omega)K_{-\alpha}(\omega,t)\mathrm{d}\omega$.

(b) Parseval Identity. $\int_{\mathbb{R}}f(t)\overline{g(t)}\mathrm{d}t=\int_{\mathbb{R}}F_\alpha(f)(\omega)\overline{F_\alpha(g)(\omega)}\mathrm{d}\omega$.

(c) Shift. Let $g(t)=f(t-\tau)$. Then,

$$F_\alpha(g)(\omega)=e^{i\pi\tau^2\sin\alpha\cos\alpha}\,e^{2\pi i\omega\tau\sin\alpha}F_\alpha(f)(\omega-\tau\cos\alpha).$$

(d) Modulation. Let $g(t)=f(t)\,e^{i\eta t}$. Then,

$$F_\alpha(g)(\omega)=e^{-i\pi\eta^2\sin\alpha\cos\alpha}e^{2\pi i\omega\eta\cos\alpha}F_\alpha(f)(\omega-\eta\sin\alpha).$$

(e) Dilation. Let $g(t)=f(ct)$. Then,

$$F_\alpha(g)(\omega)=\sqrt{\frac{1-i\cot\alpha}{c^2-i\cot\alpha}}\,e^{i\pi\cot\alpha\left(1-\frac{\cos^2\beta}{\cos^2\alpha}\right)}F_\beta(f)\left(\frac{\omega\sin\beta}{c\sin\alpha}\right),$$

where $\cot\beta=c^{-2}\cot\alpha$.

2.5.2 Discrete FRFT

The discrete FRFT is a discrete version of the continuous FRFT. It can also be thought as DFT to αth power. The algorithm of discrete FRFT is based on a special set of eigenvectors of DFT matrix.

A normalized discrete Fourier transform

$$X_\mu = N^{-\frac{1}{2}} \sum_{\nu=0}^{N-1} x_\nu \, e^{-2\pi i \frac{\mu\nu}{N}} \qquad (\mu = 0, 1, \ldots, N-1) \tag{2.5.7}$$

can be rewritten into the matrix form $\mathbf{X} = F\mathbf{x}$, where

$$\mathbf{X} = (X_0, \ldots, X_{N-1})^T, \quad \mathbf{x} = (x_0, \ldots, x_{N-1})^T, \quad F = \left(N^{-\frac{1}{2}} \, e^{-2\pi i \frac{\mu\nu}{N}}\right)_{\mu,\nu=0,\ldots,N-1}.$$

Let $\{\lambda_l\}_{l=0,\ldots,N-1}$ and $\{\mathbf{p}_l\}_{l=0,\ldots,N-1}$ be the eigenvalues and eigenvectors of the DFT matrix F, respectively, where the eigenvectors $\{\mathbf{p}_l\}_{l=0,\ldots,N-1}$ form a normal orthogonal basis for $\mathbb{R}^N$. For any $\mathbf{x} \in \mathbb{R}^N$,

$$\mathbf{x} = \sum_{l=0}^{N-1} c_l \mathbf{p}_l,$$

where $c_l = (\mathbf{p}_l, \mathbf{x}) = \mathbf{p}_l^T \mathbf{x}$. Taking DFT on both sides,

$$F\mathbf{x} = \sum_{l=0}^{N-1} c_l F\mathbf{p}_l = \sum_{l=0}^{N-1} c_l \lambda_l \mathbf{p}_l = \sum_{l=0}^{N-1} \mathbf{p}_l \lambda_l \mathbf{p}_l^T \mathbf{x}.$$

Therefore, the *spectral decomposition* of DFT matrix F is

$$F = \sum_{l=0}^{N-1} \mathbf{p}_l \lambda_l \mathbf{p}_l^T \qquad \text{or} \qquad F(\mu, \nu) = \sum_{l=0}^{N-1} \mathbf{p}_l(\mu) \lambda_l \mathbf{p}_l(\nu),$$

where $F = (F(\mu, \nu))_{\mu,\nu=0,\ldots,N-1}$ and $\mathbf{p}_l = (p_l(0), \ldots, p_l(N-1))^T$.

The *discrete FRTF matrix* F^α is defined as

$$F^\alpha = \sum_{l=0}^{N-1} \mathbf{p}_l \lambda_l^\alpha \mathbf{p}_l^T \quad \text{or} \quad F^\alpha(\mu, \nu) = \sum_{l=0}^{N-1} \mathbf{p}_l(\mu) \lambda_l^\alpha \mathbf{p}_l(\nu) \quad (\alpha \in \mathbb{R}), \tag{2.5.8}$$

where $\lambda_l^\alpha = e^{-i\frac{\pi l \alpha}{2}}$ and $F^\alpha = (F^\alpha(\mu, \nu))_{\mu,\nu=0,\ldots,N-1}$ and $\mathbf{p}_l = (p_l(0), \ldots, p_l(N-1))^T$. It is clear that F^1 is the DFT matrix F. The *discrete FRFT of order* α is defined as

$$X^\alpha(\mu)=\sum_{\nu=0}^{N-1}x_\nu F^\alpha(\mu,\nu).$$

(i) Index Additivity
For the discrete FRFT, the index additivity is $F^\alpha\circ F^\beta(f)=F^{\alpha+\beta}(f)$ for any $f\in\mathbb{R}^N$.

In fact, by (2.5.8), since $\{p_k\}_{k=0,\dots,N-1}$ are a normal orthogonal basis for $\mathbb{R}^N$, we get

$$\begin{aligned}(F^\alpha\circ F^\beta)(\mu,\nu)&=\sum_{n=0}^{N-1}F^\alpha(\mu,n)F^\beta(n,\nu)\\&=\sum_{l=0}^{N-1}\sum_{s=0}^{N-1}\mathbf{p}_l(\mu)\lambda_l^\alpha\lambda_s^\beta\mathbf{p}_s(\nu)\left(\sum_{n=0}^{N-1}\mathbf{p}_l(n)\mathbf{p}_s(n)\right)\\&=\sum_{l=0}^{N-1}\mathbf{p}_l(\mu)\lambda_l^{\alpha+\beta}\mathbf{p}_l(\nu)=F^{\alpha+\beta}(\mu,\nu).\end{aligned}$$

So $F^{\alpha+\beta}f=F^\alpha\circ F^\beta(f)$ for any $f\in\mathbb{R}^N$. Since $(p_l(\mu))_{l,\mu}$ is an orthogonal matrix, by (2.5.8), $F^0(\mu,\nu)=\delta_{\mu,\nu}$, and so $F^0=I$.

According to the index additivity, the inversion of F^α is equal to $F^{-\alpha}$, i.e., $(F^\alpha)^{-1}=F^{-\alpha}$.

(ii) Discrete Hermite–Gaussian Functions
Note that the set of eigenvectors for DFT matrix F is not unique. In order to make the discrete FRFT completely analogous to continuous FRFT, we give the discrete form of the Hermite–Gaussian equation (2.5.3).

The first term of (2.5.3) is equal approximately to

$$\frac{d^2f}{dt^2}\approx\widetilde{D}^2f(t)=\frac{f(t+h)-2f(t)+f(t-h)}{h^2}.$$

The second term of (2.5.3) is equal approximately to $(-4\pi^2t^2)f(t)\approx(F\widetilde{D}^2F^{-1})f(t)$. So the discrete form of the Hermite–Gaussian equation is

$$(\widetilde{D}^2+F\widetilde{D}^2F^{-1})f(t)=\lambda f(t).\tag{2.5.9}$$

Using the Taylor formula,

$$\widetilde{D}^2f(t)=\frac{f(t+h)-2f(t)+f(t-h)}{h^2}=\frac{2}{h^2}\sum_{n=1}^{\infty}\frac{d^{2n}f(t)}{dt^{2n}}\frac{h^{2n}}{(2n)!}.\tag{2.5.10}$$

Substituting $F^{-1}f$ into f in (2.5.10) and then taking Fourier transform on both sides, we get

$$(F\widetilde{D}^2F^{-1})f(t) = F\left(\frac{2}{h^2}\sum_{n=1}^{\infty}\frac{d^{2n}(F^{-1}f)(t)}{dt^{2n}}\frac{h^{2n}}{(2n)!}\right) = \frac{2}{h^2}\sum_{n=1}^{\infty}F\left(\frac{d^{2n}(F^{-1}f)(t)}{dt^{2n}}\right)\frac{h^{2n}}{(2n)!}.$$

Using the derivative formula of Fourier transform:

$$F\left(\frac{d^{2n}(F^{-1}f)(t)}{dt^{2n}}\right) = (2\pi it)^{2n}F(F^{-1}f)(t) = (2\pi it)^{2n}f(t)$$

and the known formula: $\cos(2\pi ht) = 1 + \sum_{n=1}^{\infty}(-1)^n(2\pi t)^{2n}\frac{h^{2n}}{(2n)!}$, we get

$$(F\widetilde{D}^2F^{-1})f(t) = \frac{2}{h^2}\left(\sum_{n=1}^{\infty}(-1)^n(2\pi t)^{2n}\frac{h^{2n}}{(2n)!}\right)f(t) = \frac{2}{h^2}(\cos(2\pi ht) - 1)f(t).$$

So (2.5.9) is equivalent to

$$f(t+h) - 2f(t) + f(t-h) + 2(\cos(2\pi ht) - 1)f(t) = h^2\lambda f(t). \qquad (2.5.11)$$

Now, we find the solution of (2.5.11).
Let $t = nh$, $g(n) = f(nh)$, and $h = \frac{1}{\sqrt{N}}$. Then, (2.5.11) becomes:

$$g(n+1) + g(n-1) + 2\left(\cos\frac{2\pi n}{N} - 2\right)g(n) = \lambda g(n). \qquad (2.5.12)$$

Since coefficients of (2.5.12) are periodic, its solutions are periodic. Let $n = 0, \ldots, N-1$. Then, for any $\mathbf{g} \in \mathbb{R}^N$,

$$M\mathbf{g} = \lambda\mathbf{g}, \qquad (2.5.13)$$

where $\mathbf{g} = (g(0), \ldots, g(N-1))^T$ and $M = (\gamma_{kl})_{k,l=0,\ldots,N-1}$, and

$$\begin{cases} \gamma_{kk} = 2\cos\frac{2\pi k}{N} - 4 & (k = 0, \ldots, N-1), \\ \gamma_{k,k+1} = \gamma_{k+1,k} = 1 & (k = 0, \ldots, N-1), \\ \gamma_{0,N-1} = \gamma_{N-1,0} = 1, \\ \gamma_{k,l} = 0 & (\text{otherwise}). \end{cases}$$

Namely,

$$M=\begin{pmatrix}
2\cos\frac{2\pi 0}{N}-4 & 1 & 0 & \cdots & \cdots & 0 & 1\\
1 & 2\cos\frac{2\pi}{N}-4 & 1 & 0 & \cdots & \cdots & 0\\
0 & 1 & 2\cos\frac{4\pi}{N}-4 & 1 & 0 & \cdots & 0\\
 & \vdots & \cdots & \cdots & \cdots & \cdots & \vdots\\
\vdots & \vdots & \vdots & \cdots & \cdots & 1 & 0\\
0 & \vdots & \vdots & 0 & 1 & 2\cos\frac{2(N-2)\pi}{N}-4 & 1\\
1 & 0 & \cdots & \cdots & 0 & 1 & 2\cos\frac{2(N-1)\pi}{N}-4
\end{pmatrix}$$

It is easy to prove that the matrices M and F are exchangeable, i.e., $MF = FM$. So the matrices M and F have a common eigenvector set $\{u_k(n)\}_{k=0,\ldots,N-1}$ which is called the *discrete Hermite–Gaussians*. This eigenvector set forms a normal orthogonal basis for $\mathbb{R}^N$. The corresponding discrete FRFT becomes

$$F^\alpha(\mu,\nu)=\sum_{k=0}^{N-1}u_k(\mu)\,e^{-i\pi\frac{k\alpha}{2}}\,u_k(\nu).$$

(iii) Algorithm

We introduce a matrix P that maps the even part of $\{g(n)\}_{n=0,\ldots,N-1}$ in (2.5.13) to the first $\left[\frac{N}{2}\right]+1$ components and its odd part to the remaining components. Since the matrix P is orthogonal symmetric, the similarity transform of the matrix M in (2.5.13) has a block diagonal form as follows:

$$PMP^{-1}=PMP=\begin{pmatrix}E_v & 0\\ 0 & O_d\end{pmatrix}$$

The eigenvectors of PMP^{-1} can be determined separately from E_v and O_d matrices. Sort the eigenvectors of the matrices E_v and O_d in the descending order of eigenvalues. Denote by $\mathbf{e}_k$ ($k=0,\ldots,\left[\frac{N}{2}\right]$) and $\mathbf{o}_k$ ($k=\left[\frac{N}{2}\right]+1,\ldots,N-1$) the corresponding eigenvectors, respectively. Then, the even and odd eigenvectors of M can be formed, respectively, by

$$\begin{aligned}
u_{2k} &= P(\mathbf{e}_k^T,0,\ldots,0)^T,\\
u_{2k+1} &= P(0,\ldots,0,\mathbf{o}_k^T)^T.
\end{aligned}\tag{2.5.14}$$

Finally, the discrete FRFT matrix is given by

$$F^{\alpha}(\mu,\nu)=\begin{cases}\sum_{k=0}^{N-1} u_k(\mu)\,e^{-i\frac{\pi}{2}k\alpha}u_k(\nu) & (N \text{ is odd}),\\ \left(\sum_{k=0}^{N-2} u_k(\mu)\,e^{-i\frac{\pi}{2}k\alpha}u_k(\nu)\right)+u_N(\mu)\,e^{-i\frac{\pi}{2}N\alpha}u_N(\nu) & (N \text{ is even}).\end{cases} \tag{2.5.15}$$

The algorithm of discrete FRFT:
Step 1. For given N, write the matrices M and P [see (2.5.13) and (2.5.14)].
Step 2. Calculate the matrices E_v and O_d defined by

$$PMP=\begin{pmatrix} E_v & 0\\ 0 & O_d\end{pmatrix}$$

and sort the eigenvectors of the matrices E_v and O_d in the descending order of eigenvalues and denote the sorted eigenvectors by $\mathbf{e}_k$ and $\mathbf{o}_k$, respectively.
Step 3. Calculate the discrete Hermite–Gaussian functions as u_k by (2.5.14).
Step 4. The $\alpha-$order discrete FRFT is

$$(F^{\alpha}\mathbf{f})(\mu)=\sum_{\nu=0}^{N-1} F^{\alpha}(\mu,\nu)f(\nu),$$

where $F^{\alpha}(\mu,\nu)$ is stated in (2.5.15) and $\mathbf{f}=(f(0),\ldots,f(N-1))^T$.

2.5.3 Multivariate FRFT

For $\boldsymbol{\alpha}=(\alpha_1,\ldots,\alpha_d)\in\mathbb{R}^d$, the *multivariate FRFT* of a $d-$variate function $f(\mathbf{t})$ is defined as

$$F_{\boldsymbol{\alpha}}(\boldsymbol{\omega})=\int_{\mathbb{R}^d} f(\mathbf{t})K_{\boldsymbol{\alpha}}(\mathbf{t},\boldsymbol{\omega})\mathrm{d}\mathbf{t} \qquad (\mathbf{t}=(t_1,\ldots,t_d),\ \boldsymbol{\omega}=(\omega_1,\ldots,\omega_d)),$$

where the kernel is

$$K_{\boldsymbol{\alpha}}(\mathbf{t},\boldsymbol{\omega})=\prod_{k=1}^{d}\left(\sqrt{1-i\cot\alpha_k}\,e^{\pi i(\omega_k^2+t_k^2)\cot\alpha_k-2\pi i\omega_k t_k\csc\alpha_k}\right).$$

Its inverse formula is $f(\mathbf{t})=\int_{\mathbb{R}^d} F_{\boldsymbol{\alpha}}(\boldsymbol{\omega})\,K_{-\boldsymbol{\alpha}}(\boldsymbol{\omega},\mathbf{t})\mathrm{d}\boldsymbol{\omega}$.

For $\boldsymbol{\alpha}=(\alpha_1,\ldots,\alpha_d)\in\mathbb{R}^d$, the multivariate DFRFT of a $d-$dimensional sequence $\{x_\nu\}_\nu$ is defined as

$$X^{\alpha_1,\ldots,\alpha_d}(\mu_1,\ldots,\mu_d)=\sum_{\nu_d=0}^{N-1}\cdots\sum_{\nu_1=0}^{N-1}x_{\nu_1,\ldots,\nu_d}\prod_{j=1}^{d}F^{\alpha_j}(\mu_j,\nu_j),$$

where $F^{\alpha_j}(\mu_j,\nu_j)$ is stated in (2.5.15), $\boldsymbol{\mu}=(\mu_1,\ldots,\mu_d)$, $\boldsymbol{\nu}=(\nu_1,\ldots,\nu_d)$, and $\boldsymbol{\mu},\boldsymbol{\nu}\in([0,N-1]^d\bigcap\mathbb{Z}^d)$. The inverse transform of multivariate DFRFT is

$$x_{\nu_1,\ldots,\nu_d}=\sum_{\mu_d=0}^{N-1}\cdots\sum_{\mu_1=0}^{N-1}X^{\alpha_1,\ldots,\alpha_d}(\mu_1,\ldots,\mu_d)\prod_{j=1}^{d}F^{-\alpha_j}(\mu_j,\nu_j).$$

2.6 Space–Frequency Distribution

In multivariate time series analysis, one important topic is to research how the dominant frequencies of the variations of time series analysis change with time/spatial scales. In this section, we will introduce various transforms which can give information about time series simultaneously in the time/spatial domain and the frequency domain, including windowed Fourier transform, Wigner–Ville distribution, page distribution, Levin distribution.

2.6.1 Multivariate Windowed Fourier Transform

The multivariate Fourier transform can provide only global frequency information. In order to overcome this disadvantage, multivariate windows are used to modify Fourier transform. One always assumes that these multivariate windows are even, real-valued. For example, Gaussian window function:

$$W(\mathbf{t})=\left(\frac{1}{2\sqrt{\pi\alpha}}\right)^d e^{-\frac{\|\mathbf{t}\|^2}{4\alpha}}\qquad(\alpha>0),$$

is the most used multivariate window. Other multivariate windows include:

(a) Rectangular window: $\prod_{k=1}^{N}\chi_{-\frac{1}{2},\frac{1}{2}}(t_k)$;

(b) Hamming window: $\prod_{k=1}^{N}(0.54+0.46\cos(2\pi t_k))\chi_{[-\frac{1}{2},\frac{1}{2}]}(t_k)$;

(c) Hanning window: $\prod_{k=1}^{N}\cos^2(\pi t_k)\chi_{-\frac{1}{2},\frac{1}{2}}(t_k)$;

(d) Blackman window: $\prod_{k=1}^{N}(0.42+0.5\cos(2\pi t_k)+0.08\cos(4\pi t_k))\chi_{[-\frac{1}{2},\frac{1}{2}]}(t_k)$,

where $\chi_{[-\frac{1}{2},\frac{1}{2}]}(t)$ is the characteristic function of $[-\frac{1}{2},\frac{1}{2}]$.

The *windowed Fourier transform* is defined as

$$S_f^W(\mathbf{t}, \boldsymbol{\omega}) = \int_{\mathbb{R}^d} e^{-2\pi i \boldsymbol{\tau}\cdot\boldsymbol{\omega}} f(\boldsymbol{\tau}) W(\boldsymbol{\tau} - \mathbf{t}) \mathrm{d}\boldsymbol{\tau},$$

where $f \in L(\mathbb{R}^d)$ and W is a window function. When the window function is Gaussian window, the corresponding windowed Fourier transform is also called a *Gabor transform* or *short-time Fourier transform*. If the window function W satisfies $\int_{\mathbb{R}^d} W(\mathbf{t})\mathrm{d}\mathbf{t} = 1$, the *inversion formula* is

$$f(t) = \int_{\mathbb{R}^d}\int_{\mathbb{R}^d} e^{2\pi i \mathbf{t}\cdot\boldsymbol{\omega}} S_f^W(\boldsymbol{\tau}, \boldsymbol{\omega}) W(\mathbf{t} - \boldsymbol{\tau}) \mathrm{d}\boldsymbol{\tau}\mathrm{d}\boldsymbol{\omega}.$$

Denote by $\rho_f^W(\mathbf{t}, \boldsymbol{\omega})$ the squared magnitude of the windowed Fourier transform, i.e.,

$$\rho_f^W(\mathbf{t}, \boldsymbol{\omega}) = |S_f^W(\mathbf{t}, \boldsymbol{\omega})|^2.$$

The $\rho_f^W(\mathbf{t}, \boldsymbol{\omega})$ is called the *spectrogram*. By the definition of the windowed Fourier transform,

$$\begin{aligned}\rho_f^W(\mathbf{t}, \boldsymbol{\omega}) &= \left|\int_{\mathbb{R}^d} e^{-2\pi i \boldsymbol{\tau}\cdot\boldsymbol{\omega}} f(\boldsymbol{\tau}) W(\boldsymbol{\tau} - \mathbf{t}) \mathrm{d}\boldsymbol{\tau}\right|^2 \\ &= \int_{\mathbb{R}^d} e^{-2\pi i \boldsymbol{\tau}\cdot\boldsymbol{\omega}} f(\boldsymbol{\tau}) W(\boldsymbol{\tau} - \mathbf{t}) \mathrm{d}\boldsymbol{\tau}\ \overline{\int_{\mathbb{R}^d} e^{-2\pi i \boldsymbol{\tau}\cdot\boldsymbol{\omega}} f(\boldsymbol{\tau}) W(\boldsymbol{\tau} - \mathbf{t}) \mathrm{d}\boldsymbol{\tau}}\end{aligned} \tag{2.6.1}$$

Let $\lambda_f(\mathbf{t}, \boldsymbol{\tau})$ be the inverse Fourier transform of $\rho_f^W(\mathbf{t}, \boldsymbol{\omega})$, i.e.,

$$\lambda_f(\mathbf{t}, \boldsymbol{\tau}) = \int_{\mathbb{R}^d} \rho_f^W(\mathbf{t}, \boldsymbol{\omega})\, e^{2\pi i \boldsymbol{\tau}\cdot\boldsymbol{\omega}} \mathrm{d}\boldsymbol{\omega}.$$

Taking the inverse Fourier transform on both sides of (2.6.1) and using properties of Fourier transform, we get

$$\begin{aligned}\lambda_f(\mathbf{t}, \boldsymbol{\tau}) &= (f(\boldsymbol{\tau})W(\boldsymbol{\tau} - \mathbf{t})) * (\overline{f}(-\boldsymbol{\tau})\overline{W}(-\boldsymbol{\tau} - \mathbf{t})) \\ &= \int_{\mathbb{R}^d} f(\boldsymbol{\alpha})W(\boldsymbol{\alpha} - \mathbf{t})\overline{f}(\boldsymbol{\alpha} - \boldsymbol{\tau})\overline{W}(\boldsymbol{\alpha} - \boldsymbol{\tau} - \mathbf{t})\mathrm{d}\boldsymbol{\alpha} \\ &= \int_{\mathbb{R}^d} f\left(\mathbf{u} + \tfrac{\boldsymbol{\tau}}{2}\right) W\left(\mathbf{u} - \mathbf{t} + \tfrac{\boldsymbol{\tau}}{2}\right) \overline{f}\left(\mathbf{u} - \tfrac{\boldsymbol{\tau}}{2}\right) \overline{W}\left(\mathbf{u} - \mathbf{t} - \tfrac{\boldsymbol{\tau}}{2}\right) \mathrm{d}\mathbf{u},\end{aligned}$$

and the last equality uses the substitution $\boldsymbol{\alpha} = \mathbf{u} + \frac{\boldsymbol{\tau}}{2}$. Let

$$\Gamma(\mathbf{t}, \boldsymbol{\tau}) = \overline{W}\left(\mathbf{t} + \frac{\boldsymbol{\tau}}{2}\right) W\left(\mathbf{t} - \frac{\boldsymbol{\tau}}{2}\right).$$

Since the window function $W(\mathbf{t})$ is even, real-valued, then $\Gamma(\mathbf{t}, \boldsymbol{\tau})$ is the instantaneous autocorrelation function of the window function, and $\Gamma(\mathbf{t} - \mathbf{u}, \boldsymbol{\tau}) = W\left(\mathbf{u} - \mathbf{t} + \frac{\boldsymbol{\tau}}{2}\right) \overline{W}\left(\mathbf{u} - \mathbf{t} - \frac{\boldsymbol{\tau}}{2}\right)$. This implies that

$$\lambda_f(\mathbf{t},\boldsymbol{\tau}) = \int_{\mathbb{R}^d} \Gamma(\mathbf{t}-\mathbf{u},\boldsymbol{\tau}) f\left(\mathbf{u}+\frac{\boldsymbol{\tau}}{2}\right)\overline{f}\left(\mathbf{u}-\frac{\boldsymbol{\tau}}{2}\right)\mathrm{d}\mathbf{u} = \Gamma(\mathbf{t},\boldsymbol{\tau}) * K_f(\mathbf{t},\boldsymbol{\tau}),$$

where $K_f(\mathbf{t},\boldsymbol{\tau}) = f\left(\mathbf{t}+\frac{\boldsymbol{\tau}}{2}\right)\overline{f}\left(\mathbf{t}-\frac{\boldsymbol{\tau}}{2}\right)$. Taking Fourier transforms on both sides, the spectrogram is

$$\rho_f^W(\mathbf{t},\boldsymbol{\omega}) = \int_{\mathbb{R}^d} (\Gamma(\mathbf{t},\boldsymbol{\tau}) * K_f(\mathbf{t},\boldsymbol{\tau}))\, e^{-2\pi i \boldsymbol{\tau}\cdot\boldsymbol{\omega}} \mathrm{d}\boldsymbol{\tau},$$

where $K_f(\mathbf{t},\boldsymbol{\tau})$ is stated as above and is called the *signal kernel*.

2.6.2 General Form

For a function $f(t)(t \in \mathbb{R})$, if the Cauchy's principal value:

$$\widetilde{f}(t) = \frac{1}{\pi}\lim_{\epsilon\to 0}\int_{|t-\tau|>\epsilon} \frac{f(\tau)}{t-\tau}\mathrm{d}\tau$$

exists, then $\widetilde{f}(t)$ is called the *Hilbert transform* of $f(t)$. Let f be a signal and $\widetilde{f}$ be its Hilbert transform. The $z(t) = f(t) + i\widetilde{f}(t)$ is called *analytic associate* of f. Fourier transform of z vanishes for the negative frequency, i.e.,

$$\widehat{z}(\omega) = \begin{cases} 2\widehat{f}(\omega), & \omega \ge 0, \\ 0, & \omega < 0. \end{cases}$$

Let $\mathbf{f}(t) = (f_1(t), \ldots, f_m(t))^T$ be a real vector and $\mathbf{z}(t) = (z_1(t), \ldots, z_m(t))^T$ be the analytic associate vector of $\mathbf{f}$. The *general form of space–frequency distribution* is defined as

$$P_{\mathbf{zz}}(t,\omega) = \int_{\mathbb{R}^2} G(t-u,\tau) K_{\mathbf{zz}}(u,\tau)\, e^{-2\pi i \omega\tau} \mathrm{d}u\mathrm{d}\tau,$$

where $G(t,\tau)$ is the *time-lag kernel* and the signal kernel matrix is

$$K_{\mathbf{zz}}(u,\tau) = (K_{z_iz_j}(u,\tau))_{m\times m} = \begin{pmatrix} K_{z_1z_1}(u,\tau) & K_{z_1z_2}(u,\tau) & \cdots & K_{z_1z_m}(u,\tau) \\ K_{z_2z_1}(u,\tau) & K_{z_2z_2}(u,\tau) & \cdots & K_{z_2z_m}(u,\tau) \\ \vdots & \vdots & \ddots & \vdots \\ K_{z_mz_1}(u,\tau) & K_{z_mz_2}(u,\tau) & \cdots & K_{z_mz_m}(u,\tau) \end{pmatrix}$$

$$= \begin{pmatrix} z_1\left(u+\frac{\tau}{2}\right)\overline{z}_1\left(u-\frac{\tau}{2}\right) & z_1\left(u+\frac{\tau}{2}\right)\overline{z}_2\left(u-\frac{\tau}{2}\right) & \cdots & z_1\left(u+\frac{\tau}{2}\right)\overline{z}_m\left(u-\frac{\tau}{2}\right) \\ z_2\left(u+\frac{\tau}{2}\right)\overline{z}_1\left(u-\frac{\tau}{2}\right) & z_2\left(u+\frac{\tau}{2}\right)\overline{z}_2\left(u-\frac{\tau}{2}\right) & \cdots & z_2\left(u+\frac{\tau}{2}\right)\overline{z}_m\left(u-\frac{\tau}{2}\right) \\ \vdots & \vdots & \ddots & \vdots \\ z_m\left(u+\frac{\tau}{2}\right)\overline{z}_1\left(u-\frac{\tau}{2}\right) & z_m\left(u+\frac{\tau}{2}\right)\overline{z}_2\left(u-\frac{\tau}{2}\right) & \cdots & z_m\left(u+\frac{\tau}{2}\right)\overline{z}_m\left(u-\frac{\tau}{2}\right) \end{pmatrix},$$

i.e., $K_{\mathbf{zz}}(u,\tau) = \mathbf{z}\left(u+\frac{\tau}{2}\right)\overline{\mathbf{z}}^T\left(u-\frac{\tau}{2}\right)$. So the class of space–frequency distribution is

$$P_{\mathbf{zz}}(t,\omega) = \int_{\mathbb{R}^2} G(t-u,\tau)\mathbf{z}\left(u+\frac{\tau}{2}\right)\overline{\mathbf{z}}^T\left(u-\frac{\tau}{2}\right)e^{-2\pi i\omega\tau}\mathrm{d}u\mathrm{d}\tau, \qquad (2.6.2)$$

2.6.3 Popular Distributions

Various popular space–frequency distributions can be derived directly from the general form (2.6.2).

(a) Let $G(t,\tau) = W\left(t+\frac{\tau}{2}\right)\overline{W}\left(t-\frac{\tau}{2}\right)$, where W is a univariate window function. By (2.6.2),

$$P_{\mathbf{zz}}(t,\omega) = \int_{\mathbb{R}^2} W\left(t-u+\frac{\tau}{2}\right)\overline{W}\left(t-u-\frac{\tau}{2}\right)\mathbf{z}\left(u+\frac{\tau}{2}\right)\overline{\mathbf{z}}^T\left(u-\frac{\tau}{2}\right)e^{-2\pi i\omega\tau}\mathrm{d}u\mathrm{d}\tau$$

which is called a *space spectrogram*. It can be rewritten as $P_{\mathbf{zz}}(t,\omega) = (P_{z_iz_j}(t,\omega))_{m\times m}$, where

$$P_{z_iz_j}(t,\omega) = \int_{\mathbb{R}^2} W\left(t-u+\frac{\tau}{2}\right)\overline{W}\left(t-u-\frac{\tau}{2}\right)z_i\left(u+\frac{\tau}{2}\right)\overline{z}_j\left(u-\frac{\tau}{2}\right)e^{-2\pi i\omega\tau}\mathrm{d}u\mathrm{d}\tau.$$

If $i=j$, then

$$P_{z_iz_j}(t,\omega) = \left|\int_{\mathbb{R}} e^{-2\pi i\omega\tau}z_i(\tau)W(\tau-t)\mathrm{d}\tau\right|^2 = |S_{z_i}^W(t,\omega)|^2,$$

where $S_{z_i}^W(t,\omega)$ is the windowed Fourier transform of $z_i(\tau)$.

(b) Let $G(t,\tau) = \delta(t)$, where δ is Dirac function. Note that $\delta(-t) = \delta(t)$. Since

$$\int_{\mathbb{R}} \delta(t-u)\mathbf{z}\left(u+\frac{\tau}{2}\right)\overline{\mathbf{z}}^T\left(u-\frac{\tau}{2}\right)\mathrm{d}u = \mathbf{z}\left(t+\frac{\tau}{2}\right)\overline{\mathbf{z}}^T\left(t-\frac{\tau}{2}\right),$$

we get

$$P_{\mathbf{zz}}(t,\omega) = \int_{\mathbb{R}} \left(\int_{\mathbb{R}} \delta(t-u)\mathbf{z}\left(u+\tfrac{\tau}{2}\right)\overline{\mathbf{z}}^T\left(u-\tfrac{\tau}{2}\right)\mathrm{d}u \right) e^{-2\pi i\omega\tau}\mathrm{d}\tau$$
$$= \int_{\mathbb{R}} \mathbf{z}\left(t+\tfrac{\tau}{2}\right)\overline{\mathbf{z}}^T\left(t-\tfrac{\tau}{2}\right) e^{-2\pi i\tau\omega}\mathrm{d}\tau$$

which is called a *space Wigner–Ville distribution.*
(c) Let $G(t,\tau)=\delta(t-\frac{\tau}{2})$. Note that $\delta(t-u-\frac{\tau}{2})=\delta(u-t+\frac{\tau}{2})$. Then,

$$P_{\mathbf{zz}}(t,\omega) = z(t)\overline{Z}^T(\omega)\, e^{-2\pi i\tau\omega}$$

is called the *space Rihaczek distribution*, where $Z(\omega)$ is the Fourier transform of $z(t)$.
(d) Let $G(t,\tau)=\delta(t-\frac{\tau}{2})W(t)$, where $W(t)$ is a window function. The function:

$$P_{\mathbf{zz}}(t,\omega) = \mathbf{z}(t)\int_{\mathbb{R}} \overline{\mathbf{z}}^T(t-\tau)W(\tau)\, e^{-2\pi i\tau\omega}\mathrm{d}\tau$$
$$= \mathbf{z}(t)\, e^{-2\pi i t\omega}\left(\int_{\mathbb{R}} \overline{\mathbf{z}}^T(\tau)W(t-\tau)\, e^{-2\pi i\tau\omega}\mathrm{d}\tau\right)$$

is called a *space W–Rihaczek distribution.*
(e) Let $G(t,\tau)=\frac{1}{2}\left(\delta\left(t+\frac{\tau}{2}\right)+\delta\left(t-\frac{\tau}{2}\right)\right)$. Then,

$$P_{\mathbf{zz}}(t,\omega) = \mathrm{Re}\left(\mathbf{z}(t)\overline{\mathbf{Z}}^T(\omega)\, e^{-2\pi i t\omega}\right)$$

which is called a *space Levin distribution.*
(f) Let $G(t,\tau)=\frac{W(\tau)}{2}\left(\delta\left(t+\frac{\tau}{2}\right)+\delta\left(t-\frac{\tau}{2}\right)\right)$, where $W(\tau)$ is a window function. Then,

$$P_{\mathbf{zz}}(t,\omega) = \mathrm{Re}\left(\int_{\mathbb{R}} \overline{\mathbf{z}}^T(\tau)W(t-\tau)\, e^{-2\pi i\tau\omega}\mathrm{d}\tau\right) e^{-2\pi i t\omega}$$

which is called a *space W–Levin distribution.*
(g) Let $G(t,\tau)=v(\tau)\delta\left(t-\frac{\tau}{2}\right)+v(-\tau)\delta\left(t+\frac{\tau}{2}\right)$, where $v(\tau)$ is the unit step function: $v(\tau)=1\ (\tau\geq 0)$ and $v(\tau)=0\,(\tau<0)$. Then,

$$P_{\mathbf{zz}}(t,\omega) = \frac{\partial}{\partial t}\left|\int_{-\infty}^{t} \mathbf{z}(\tau)\, e^{-2\pi i\omega\tau}\mathrm{d}\tau\right|^2,$$

which is called a *space Page distribution.*
(h) Let $G(t,\tau)=W(\tau)\,\mathrm{rect}\left(\frac{\alpha}{2\tau}\right)$. The $P_{\mathbf{zz}}(t,\omega)$ is called a *Zhao-Atlas-Marks distribution.*
(i) Let $G(t,\tau)=|\tau|^{\beta}\cosh^{-2\beta}t$. The $P^W_{\mathbf{zz}}(t,\omega)$ is called a *B-distribution.*
(j) Let $G(t,\tau)=\frac{\sqrt{\pi\sigma}}{|\tau|}\, e^{-\frac{\pi^2\sigma t^2}{\tau^2}}$. The $P_{\mathbf{zz}}(t,\omega)$ is called an *E-distribution.*

2.7 Multivariate Interpolation

Long-term multivariate time series always contain missing data or data with different sampling intervals. In this section, we introduce new advances in multivariate interpolation, including polynomial interpolation, positive definite function interpolation, radial function interpolation, and interpolation on sphere. These interpolation methods are widely applied in climate and environmental time series.

2.7.1 Multivariate Polynomial Interpolation

Let $\mathbf{x} = (x_1, \ldots, x_d)$ be a d−dimensional real-valued vector and $\boldsymbol{\alpha} = (\alpha_1, \ldots, \alpha_d)$ $(\alpha_i \in \mathbb{Z}_+ \ (i = 1, \ldots, d))$, denoted by $\mathbf{x} \in \mathbb{R}^d$ and $\boldsymbol{\alpha} \in \mathbb{Z}_+^d$. Let $|\boldsymbol{\alpha}| = \alpha_1 + \cdots + \alpha_d$. Define $\mathbf{x}^\alpha = x_1^{\alpha_1} x_2^{\alpha_2} \cdots x_d^{\alpha_d}$. The polynomial $P_k(\mathbf{x}) = \sum_{|\alpha| \le k} c_\alpha \mathbf{x}^\alpha$ is called a *polynomial of degree k*.

The multivariate interpolation of polynomials is different from the univariate interpolation of polynomials. In the one-dimensional case, a polynomial $\sum_{k=0}^n c_k x^k$ $(x \in \mathbb{R})$ of degree $\le n$ is a linear combinations of $1, x, \ldots, x^n$. It can interpolate arbitrary data $\lambda_0, \ldots, \lambda_n$ on any set of $n+1$ distinct nodes $t_0, \ldots, t_n \in \mathbb{R}$. Namely, the coefficients $c_0, \ldots, c_n$ can be chosen so that $\lambda_i = \sum_{k=0}^n c_k t_i^k$ $(i = 0, \ldots, n)$ since the Vandermonde determinant:

$$\begin{vmatrix} 1 & t_1 & \cdots & t_1^{n-1} \\ 1 & t_2 & \cdots & t_1^{n-1} \\ \vdots & \vdots & \ddots & \vdots \\ 1 & t_n & \cdots & t_n^{n-1} \end{vmatrix} = \prod_{1 \le j < i \le n} (t_i - t_j) > 0.$$

This results in the well-known Lagrange interpolation formula. However, in the high-dimensional case, there do not exist $n+1$ multivariate continuous functions $g_k(\mathbf{x})(k = 0, \ldots, n)$ (not only polynomials) such that their linear combination $\sum_{k=0}^n c_k g_k(\mathbf{x})$ $(\mathbf{x} \in \mathbb{R}^d)$ can interpolate arbitrary data $\lambda_0, \ldots, \lambda_n$ on any set of $n+1$ distinct nodes $x_0, \ldots, x_n$ in $\mathbb{R}^d$. In fact, if there exist such $\{g_k\}_{k=0,\ldots,n}$, the determinant $\det(g_k(\mathbf{t}_i)) \ne 0$ for any set of $n+1$ distinct nodes $\{\mathbf{t}_k\}_{k=0,\ldots,n}$ in $\mathbb{R}^d$. However, this is not possible. Select a closed path containing $\mathbf{t}_1$ and $\mathbf{t}_2$ in $\mathbb{R}^d$ but no other nodes. Continuously moving $\mathbf{t}_1$ and $\mathbf{t}_2$ in the same direction in the closed path can exchange positions of $\mathbf{t}_1$ and $\mathbf{t}_2$, so the determinant will change sign. Since the determinant is a continuous function, it will vanish at some stage. This is a contradiction.

An efficient method is on a Cartesian grid. For example, for $d = 3$, we interpolate $F(x, y, z)$ in the grid:

$$\{(x_i, y_j, z_k) : \quad 1 \le i \le n, \ 1 \le j \le m, \ 1 \le k \le l\}.$$

The interpolation formula is

$$L(F(x, y, z)) = \sum_{i=1}^{n} \sum_{j=1}^{m} \sum_{k=1}^{l} F(x_i, y_j, z_k) u_i(x) v_j(y) w_k(z),$$

where

$$u_i(x) = \prod_{\substack{\nu=1 \\ \nu \ne i}}^{n} \frac{x - x_i}{x_\nu - x_i}, \qquad v_j(y) = \prod_{\substack{\nu=1 \\ \nu \ne j}}^{m} \frac{y - y_j}{y_\nu - y_j}, \qquad w_k(z) = \prod_{\substack{k=1 \\ \nu \ne k}}^{l} \frac{z - z_k}{z_\nu - z_k}.$$

This interpolation formula is a direct generalization of the univariate case.

Consider a set N consists of $\frac{1}{2}(k+1)(k+2)$ nodes in $\mathbb{R}^d$ and there are k hyperplanes $H_0, \dots, H_k \subset \mathbb{R}^d$ such that $N \subset \bigcup_{i=0}^{k} H_i$ and just $i+1$ nodes lie in $H_i (0 \le i \le k)$, then arbitrary data on the set N can be interpolated by a $d-$variate polynomial of degree k. For example, if the set N consists of six nodes $\mathbf{t}_0, \mathbf{t}_1, \dots, \mathbf{t}_5$ in $\mathbb{R}^2$ and there are three straight lines $L_0, L_1, L_2 \subset \mathbb{R}^2$ such that $\mathbf{t}_0 \in L_0, \mathbf{t}_1, \mathbf{t}_2 \in L_1$, and the remaining $\mathbf{t}_3, \mathbf{t}_4, \mathbf{t}_5 \in L_2$, then arbitrary data on the set N can be interpolated by a bivariate polynomial of degree 2.

2.7.2 Schoenberg Interpolation

A complex-valued function f in $\mathbb{R}^d$ is said to be *strictly positive definite* if, for any finite set $\mathbf{x}_1, \dots, \mathbf{x}_n \in \mathbb{R}^d$, the $n \times n$ matrix $M_{jk} = f(\mathbf{x}_j - \mathbf{x}_k)$ is positive definite, i.e., for all $\mathbf{v} = (v_1, \dots, v_n)^T$,

$$\overline{\mathbf{v}}^T M \mathbf{v} = \sum_{j=1}^{n} \sum_{k=1}^{n} \overline{v}_j v_k M_{jk} > 0,$$

where $M = (M_{jk})_{n \times n}$. A linear combination of translation of a strictly positive definite function, $\sum_{k=1}^{n} a_k f(\mathbf{x} - \mathbf{x}_k)$, can interpolate arbitrary data on any distinct nodes $\mathbf{x}_1, \dots, \mathbf{x}_n \in \mathbb{R}^d$.

A function $f(t)$ is said to be *completely monotone* on $[0, \infty)$ if $f(t)$ is infinitely differentiable on $(0, \infty)$ and is continuous at $t = 0$ and $(-1)^k f^{(k)}(t) \ge 0\ (t > 0, k = 0, 1, \dots)$. For example, the following three functions:

$$f(t) = (\alpha + t)^{-\beta} \quad (\alpha > 0,\ \beta > 0),$$

$$f(t) = e^{-\alpha t} \quad (\alpha > 0),$$

$$f(t) = t^{-1}(1 - e^{t})$$

are all completely monotone.

Schoenberg Interpolation Theorem. *Schoenberg Interpolation Theorem. If f is completely monotone but not constant on $[0, \infty)$, then, for any distinct nodes $\{\mathbf{x}_k\}_{k=1,\ldots,n} \in \mathbb{R}^d$, the linear combination:*

$$\sum_{k=1}^{n} c_k f(\| \mathbf{x} - \mathbf{x}_k \|^2) = \sum_{k=1}^{n} c_k f\left((x_1 - x_{k1})^2 + \cdots + (x_d - x_{kd})^2\right),$$

where $\mathbf{x} = (x_1, \ldots, x_d)$ and $\mathbf{x}_k = (x_{k1}, \ldots, x_{kd})$, can interpolate arbitrary data on these nodes.

For any distinct $n + 1$ nodes $\{\mathbf{x}_k\}_{k=1,\ldots,n} \in \mathbb{R}^d$, Schoenberg interpolation theorem shows that the following linear combinations:

$$\sum_{k=1}^{n} c_k(\alpha + \| \mathbf{x} - \mathbf{x}_k \|^2)^{-\beta} \quad (\alpha > 0,\ \beta > 0),$$

$$\sum_{k=1}^{n} c_k\, e^{-\alpha\|\mathbf{x}-\mathbf{x}_k\|^2} \quad (\alpha > 0),$$

$$\sum_{k=1}^{n} c_k \| \mathbf{x} - \mathbf{x}_k \|^{-2} \left(1 - e^{-\|\mathbf{x}-\mathbf{x}_k\|^2}\right)$$

can interpolate arbitrary data $\lambda_1, \ldots, \lambda_n$ on nodes $\mathbf{x}_1, \ldots, \mathbf{x}_n$.

2.7.3 Micchelli Interpolation

There exist more functions such that the following form of interpolation holds

$$\sum_{k=0}^{\infty} c_k f(\| \mathbf{x} - \mathbf{x}_k \|^2).$$

Micchelli Interpolation Theorem. Let f be a positive-valued continuous function on $[0, \infty)$ and the derivative f' be completely monotone but not constant on $[0, \infty)$. Then, the combination $\sum_{k=1}^{n} c_k f(\| \mathbf{x} - \mathbf{x}_k \|^2)$ can interpolate arbitrary data $\lambda_1, \ldots, \lambda_n$ on any distinct nodes $\mathbf{x}_1, \ldots, \mathbf{x}_n$ in $\mathbb{R}^d$.

Proof Let $\mathbf{v} = (v_1, \ldots, v_n)^T$ be a nonzero real-valued vector, where $\sum_{k=1}^n v_k = 0$, and let $M = (M_{jk})_{n\times n}$, where $M_{jk} = f(\| \mathbf{x}_j - \mathbf{x}_k \|^2)$. Note that

$$\mathbf{v}^T M\mathbf{v} = \sum_{j=1}^n \sum_{k=1}^n v_j v_k f(\| \mathbf{x}_j - \mathbf{x}_k \|^2).$$

Since $f(t)$ is infinitely differentiable in $(0, \infty)$ and continuous at $t = 0$, $f(t)$ can be expressed by a Stieltjes integral:

$$f(t) = f(0) + \int_0^\infty \frac{1 - e^{-st}}{s} \mathrm{d}\alpha(s),$$

where $\alpha(s)$ is a non-decreasing bounded function and $\int_0^\infty \frac{\mathrm{d}\alpha(s)}{s} < \infty$. So

$$\mathbf{v}^T M\mathbf{v} = \sum_{j=1}^n \sum_{k=1}^n v_j v_k \left(f(0) + \int_0^\infty \frac{1-e^{-s\|\mathbf{x}_j - \mathbf{x}_k\|^2}}{s} \mathrm{d}\alpha(s) \right)$$

$$= - \sum_{j=1}^n \sum_{k=1}^n v_j v_k \int_0^\infty \frac{e^{-s\|\mathbf{x}_j - \mathbf{x}_k\|^2}}{s} \mathrm{d}\alpha(s).$$

Since the matrix $(e^{-s\|\mathbf{x}_j - \mathbf{x}_k\|^2})_{n\times n}$ is positive definite, $\mathbf{v}^T(-M)\mathbf{v} > 0$, and then M is non-singular. From this, the desired result is obtained.

The functions $\sqrt{t}$ and $\log(1+t)$ satisfy the conditions of Micchelli interpolation theorem. The function $\sqrt{1+t}$ used often in geophysics also satisfies the conditions of Micchelli interpolation theorem. Therefore, the following three linear combinations:

$$\sum_{k=1}^n c_k \| \mathbf{x} - \mathbf{x}_k \|,$$

$$\sum_{k=1}^n c_k \log(1+ \| \mathbf{x} - \mathbf{x}_k \|^2),$$

$$\sum_{k=1}^n c_k \sqrt{1+ \| \mathbf{x} - \mathbf{x}_k \|^2}$$

can interpolate arbitrary data $\lambda_1, \ldots, \lambda_n$ on any distinct nodes $\mathbf{x}_1, \ldots, \mathbf{x}_n$ in $\mathbb{R}^d$.

2.7.4 Interpolation on Spheres

Interpolation on spheres is often used in geography, climate, and environment, whose information is gathered over the Earth's surface. Denote by $\mathbb{S}^d$ the $d-$dimensional unit sphere in $\mathbb{R}^{d+1}$, i.e.,

$$\mathbb{S}^d = \{\mathbf{x} \in \mathbb{R}^{d+1} : \ \| \mathbf{x} \| = 1\}.$$

Ultraspherical polynomials are introduced to establish the interpolation formula on spheres. The ultraspherical polynomial $P_n^{(\lambda)}(x)$ is a polynomial of degree n which is determined by the equation:

$$(1 - 2rx + r^2)^{-\lambda} = \sum_{n=0}^{\infty} r^n P_n^{(\lambda)}(x) \qquad (\lambda > 0).$$

For any $\lambda > -\frac{1}{2}$, the ultraspherical polynomial family $\{P_n^{(\lambda)}\}_{n=0,1,\ldots}$ satisfies the conditions:

$$\int_{-1}^{1} P_j^{(\lambda)}(x) P_k^{(\lambda)}(x)(1 - x^2)^{\lambda-\frac{1}{2}} \mathrm{d}x = 0 \ (j \neq k),$$

$$P_n^{(\lambda)}(1) = C_{n+2\lambda-1}^n = \frac{(n+2\lambda-1)!}{n!(2\lambda-1)!}.$$

Interpolation Theorem on Spheres. Let $\mathbf{x}_1, \ldots, \mathbf{x}_n$ be distinct points on the unit sphere $\mathbb{S}^d$. For the ultraspherical polynomial series of $f(t)$ on $[-1,1]$,

$$f(t) = \sum_{j=0}^{\infty} \alpha_j P_j^{(\frac{d-1}{2})}(t),$$

if $\alpha_j \geq 0$, $\alpha_k > 0\,(0 \leq k < n)$, and $\sum_{j=0}^{\infty} \alpha_j P_j^{(\frac{d-1}{2})}(1) < \infty$, then the linear combination $\sum_{k=1}^{n} c_k f((\mathbf{x}, \mathbf{x}_k))$ can interpolate arbitrary values on nodes $\mathbf{x}_1, \ldots, \mathbf{x}_n$, where $(\mathbf{x}, \mathbf{x}_k)$ is the inner product of $\mathbf{x}$ and $\mathbf{x}_k$.

2.8 Sparse Approximation

Sparse approximation has attracted considerable interest recently because of its application in many aspects of data analysis. The crucial rule in sparse approximation is to seek a good approximation using as few elementary signals as possible, so the amount of space needed to store large multivariate spatial datasets would be reduced to a fraction of what was originally needed.

2.8.1 Approximation Kernels

Let $H_n(x)$ $(n \in \mathbb{Z}_+)$ be kernel functions on $\mathbb{R}^d$ satisfying

$$\int_{\mathbb{R}^d} H_n(\mathbf{x})\mathrm{d}\mathbf{x} = 1,$$

$$\sup_{n\in\mathbb{Z}_+} \int_{\mathbb{R}^d} |H_n(\mathbf{x})|\mathrm{d}\mathbf{x} < \infty,$$

$$\lim_{n\to\infty} \int_{\|\mathbf{x}\|\geq\delta} |H_n(\mathbf{x})|\mathrm{d}\mathbf{x} = 0 \quad \text{for any } \delta > 0,$$

and let f be any bonded continuous function on $\mathbb{R}^d$. Then,

$$\lim_{n\to\infty} (H_n * f)(\mathbf{x}) = f(\mathbf{x}) \ (\mathbf{x} \in \mathbb{R}^d),$$

where $H_n * f$ represents the convolution of H_n and f on $\mathbb{R}^d$.

Main kernel functions used often are as follows:

(a) Polynomial Kernel:

$$H_n(\mathbf{x}) = \begin{cases} c_n(1- \| \mathbf{x} \|^2)^n, & \| \mathbf{x} \| \leq 1, \\ 0, & \| \mathbf{x} \| > 1, \end{cases}$$

where $\| \mathbf{x} \|$ is the Euclidean norm of $\mathbf{x} \in \mathbb{R}^d$ and c_n is determined by $\int_{\mathbb{R}^d} H_n(\mathbf{x})$ $\mathrm{d}\mathbf{x} = 1$.

(b) Dilation Kernels: $H_n(\mathbf{x}) = n^d H(n\mathbf{x})$ $(\mathbf{x} \in \mathbb{R}^d,\ n \in \mathbb{Z}_+)$, where $\int_{\mathbb{R}^d} H(\mathbf{x})$ $\mathrm{d}\mathbf{x} = 1$.

(c) Weierstrass Kernel: $W(\mathbf{x}) = \pi^{-\frac{d}{2}} e^{-\|\mathbf{x}\|^2}$ $(\mathbf{x} \in \mathbb{R}^d)$.

(d) Cauchy Kernel: $C(\mathbf{x}) = \frac{\tau}{1+\|\mathbf{x}\|^2}$ $(\mathbf{x} \in \mathbb{R}^d)$, where $\tau > 0$ is determined by $\int_{\mathbb{R}^d} C(\mathbf{x})\mathrm{d}\mathbf{x} = 1$.

If a function $f(\mathbf{x})$ on $\mathbb{R}^d$ has the form:

$$f(\mathbf{x}) = \sum_{\substack{\mathbf{k}\in\mathbb{Z}^d \\ |k_1|+\cdots+|k_d|\leq n}} c_{\mathbf{k}}\, e^{2\pi i(\mathbf{k},\mathbf{x})},$$

then f is called an *multivariate trigonometric polynomial* of degree at most n. If

$$\int_{\mathbb{R}^d} |H(\mathbf{x})|\mathrm{d}\mathbf{x} = 1,$$

$$\widehat{H}(\mathbf{x}) = 0 \qquad (|x_1| + \cdots + |x_d| \geq 1),$$

and if f is an integer-periodic continuous function on $\mathbb{R}^d$, it can be proved that $H_\lambda * f$ is an multivariate trigonometric polynomial of degree at most $\lambda - 1$ and $\lim_{k\to\infty}(H_\lambda * f) = f$, where $H_\lambda(\mathbf{x}) = \lambda^d H(\lambda\mathbf{x})$.

Let $\boldsymbol{\alpha} = (\alpha_1, \ldots, \alpha_d)$. Define

$$(D^\alpha f)(\mathbf{x}) = \frac{\partial^{|\alpha|} f(\mathbf{x})}{\partial x_1^{\alpha_1} \cdots \partial x_d^{\alpha_d}} \qquad (|\boldsymbol{\alpha}| = \alpha_1 + \cdots + \alpha_d).$$

If $D^\alpha f$ is continuous and bounded on $\mathbb{R}^d$ for all $|\boldsymbol{\alpha}| \le k$, we say $f \in C^k(\mathbb{R}^d)$. If a kernel function H satisfies $\int_{\mathbb{R}^d} |H(\mathbf{x})\mathbf{x}^\alpha|\mathrm{d}\mathbf{x} < \infty$ and its dilation kernel $H_\lambda(\mathbf{x})$ satisfies $H_\lambda(\mathbf{x}) * \frac{\mathbf{x}^\alpha}{\alpha!} = \frac{\mathbf{x}^\alpha}{\alpha!}$ $(|\boldsymbol{\alpha}| < k)$, it can be proved that for $f \in C^k(\mathbb{R}^d)$,

$$|(H_\lambda * f)(\mathbf{x}) - f(\mathbf{x})| \le M|f|_k\lambda^{-k} \qquad (\mathbf{x} \in \mathbb{R}^d),$$

where $M = \sum_{|\alpha|=k} \frac{1}{\alpha!} \int_{\mathbb{R}^d} |H(\mathbf{y})||\mathbf{y}^\alpha|\mathrm{d}\mathbf{y}$ and $|f|_k = \max_{|\alpha|=k,\, \mathbf{x}\in\mathbb{R}^d} |D^\alpha f(\mathbf{x})|$. It is not difficult to construct this kind of kernel functions, for example,

$$H(\mathbf{x}) = \sum_{i=1}^{k} a_i h\left(\frac{\mathbf{x}}{t_i}\right) t_i^{-d} \qquad (\mathbf{x} \in \mathbb{R}^d),$$

where $t_1, \ldots, t_k \in \mathbb{R}$ are different nonzero points, $a_i = \prod_{j=1,\ldots,k;\, j\neq i} \frac{-t_j}{t_i - t_j}$ $(i = 1, \ldots, k)$, and h satisfies

$$\int_{\mathbb{R}^d} h(\mathbf{y})\mathrm{d}\mathbf{y} = 1, \qquad \int_{\mathbb{R}^d} |h(\mathbf{y})\mathbf{y}^\alpha|\mathrm{d}\mathbf{y} < \infty \quad (|\boldsymbol{\alpha}| < k).$$

2.8.2 Sparse Schemes

For a smooth function f on the cube $[0, 1]^d$, we extend it to an integer-periodic function on $\mathbb{R}^d$ and then expand it into Fourier series. Its Fourier coefficients decay slowly due to the discontinuity on the boundary, so we need a lot of Fourier coefficients to reconstruct it. In this subsection, we give a sparse scheme to reconstruct the smooth functions on the cube $[0, 1]^d$.

Let $f(x_1, \ldots, x_d)$ be defined on $[0, 1]^d$ and $\frac{\partial^{i_1+\cdots+i_d} f}{\partial x_1^{i_1} \cdots \partial x_d^{i_d}}$ $(i_1, \ldots, i_d = 0, 1, 2)$ be continuous on $[0, 1]^d$. The *reconstruction scheme* consists of five steps.

Step 1. Start from the $(d-1)-$variate boundary functions of $f(x_1, \ldots, x_d)$

$$f(0, x_2, \ldots, x_d), \quad f(x_1, 0, \ldots, x_d), \quad \ldots, \quad f(x_1, x_2, \ldots, 0),$$

$$f(1, x_2, \ldots, x_d), \quad f(x_1, 1, \ldots, x_d), \quad \ldots, \quad f(x_1, x_2, \ldots, 1).$$

Step 2. Construct a $d-$variate function $g(x_1, \dots, x_d)$, which is a combination of these $(d-1)-$variate boundary functions and the factors $x_1, \dots, x_d, (1-x_1), \dots, (1-x_d)$, to satisfy a condition:

$$g(x_1, \dots, x_d) = f(x_1, \dots, x_d) \quad \text{on} \quad \partial([0,1]^d).$$

Step 3. Let $h(x_1, \dots, x_d) = f(x_1, \dots, x_d) - g(x_1, \dots, x_d)$. Compute the Fourier coefficients $c_{n_1,\dots,n_d}(h)$ of h, where $|n_1 n_2 \cdots n_d| \le N$ and $|n_1|, |n_2|, \dots, |n_d| \le N$.
Step 4. Compute the hyperbolic cross truncations of Fourier series of $f(x_1, \dots, x_d)$ by the formula:

$$s_N^{(h,c)}(f; x_1, \dots, x_d) = \sum_{\substack{|n_1 \cdots n_d| \le N \\ |n_1|, |n_2|, \dots, |n_d| \le N}} c_{n_1,\dots,n_d}(h)\, e^{2\pi i (n_1 x_1 + \cdots + n_d x_d)}.$$

Step 5. The reconstruction formula of $f(x_1, \dots, x_d)$ is

$$f(x_1, \dots, x_d) \approx g(x_1, \dots, x_d) + s_N^{(h,c)}(x_1, \dots, x_d). \qquad (2.8.1)$$

The square error $e_N^{(d)}$ is

$$(e_N^{(d)})^2 = O\left(\frac{\log^{d-1} N}{N^3}\right),$$

and the total number of Fourier coefficients used is equivalent to $N \log^{d-1} N$.

In the reconstruction formula (2.8.1), the term $g(x_1, \dots, x_d)$ is a combination of these $(d-1)-$variate boundary functions:

$$f(0, x_2, \dots, x_d), \quad f(x_1, 0, \dots, x_d), \quad \dots, \quad f(x_1, \dots, x_{d-1}, 0),$$

$$f(1, x_2, \dots, x_d), \quad f(x_1, 1, \dots, x_d), \quad \dots, \quad f(x_1, \dots, x_{d-1}, 1)$$

and the factors $x_1, \dots, x_d$, $(1-x_1), \dots, (1-x_d)$. However, each of the above $(d-1)-$variate boundary functions can be processed again and again by Steps 1–5 until the involved boundary functions are univariate functions. Finally, $f(x_1, \dots, x_d)$ on $[0,1]^d$ can be reconstructed by few Fourier coefficients with small approximation error and its values at the vertexes of $[0,1]^d$. Such approximation scheme by using few coefficients is just *sparse approximation*.

Compared with this approximation scheme, if the function $f(x_1, \dots, x_d)$ is reconstructed by the partial sum of its Fourier series:

$$s_N(f; x_1, \dots, x_d) = \sum_{|n_1| \le N} \cdots \sum_{|n_d| \le N} c_{n_1,\dots,n_d}(f)\, e^{2\pi i (n_1 x_1 + \cdots + n_d x_d)},$$

then the corresponding square error $\widetilde{e}_N^{(d)}$ decays slowly,

$$(\widetilde{e}_N^{(d)})^2 = O(\widetilde{N}^{-\frac{1}{d}}),$$

where $\widetilde{N}$ is the total number of the Fourier coefficients and $\widetilde{N} \approx N^d$.

Now, we give the detail algorithm for the univariate, bivariate, and triple functions used often.

(i) Let $f(x)$ be a univariate function defined on $[0, 1]$ and $f''(x)$ be continuous on $[0, 1]$. We reconstruct $f(x)$ as follows. Clearly, two boundary values are $f(0)$ and $f(1)$. Construct a univariate function:

$$g(x) = f(x) - f(0)(1-x) - f(1)x \qquad (0 \le x \le 1).$$

Clearly, $g(0) = f(0)$ and $g(1) = f(1)$. Find the Fourier coefficients of $g(x)$ by the formula:

$$c_n(g) = \int_0^1 g(x)\, e^{-2\pi i n x} \mathrm{d}x.$$

The reconstruction formula of $f(x)$ is

$$f(x) \approx f(0)(1-x) + f(1)x + \sum_{n=-N}^{N} c_n(g)\, e^{2\pi i n x} \qquad (0 \le x \le 1)$$

and the square errors $e_N^{(1)}$ satisfy

$$(e_N^{(1)})^2 = O\left(\frac{1}{N^3}\right).$$

In the reconstruction formula, $f(x)$ can be reconstructed by $f(0)$, $f(1)$, and coefficients $\{c_n(g)\}(\,|n| \le N\,)$, and the total number of Fourier coefficients is equivalent to N.

(ii) Let $f(x, y)$ be a bivariate function defined on $[0, 1]^2$ and $\frac{\partial^{i+j} f}{\partial x^i \partial y^j}$ $(i, j = 0, 1, 2)$ be continuous on $[0, 1]^2$. We reconstruct $f(x, y)$ as follows. Clearly, four univariate boundary functions are

$$f(x, 0), \quad f(x, 1) \qquad (0 \le x \le 1),$$

$$f(0, y), \quad f(1, y) \qquad (0 \le y \le 1).$$

where each boundary function can be reconstructed by values at the end points and $2N + 1$ univariate Fourier coefficients. Based on these four boundary functions, we construct a bivariate function:

$$g(x, y) = f(x, 0)(1-y) + f(x, 1)y + f(0, y)(1-x) + f(1, y)x$$

$$-f(0, 0)(1-x)(1-y) - f(0, 1)(1-x)y - f(1, 0)x(1-y) - f(1, 1)xy.$$

Clearly, $g(x, y)$ is a combination of four boundary functions and four factors $x, y, (1-x), (1-y)$ and satisfies

$$g(x,0) = f(x,0), \qquad g(x,1) = f(x,1) \qquad (0 \le x \le 1),$$

$$g(0,y) = f(0,y), \qquad g(1,y) = f(1,y) \qquad (0 \le y \le 1).$$

Let

$$h(x, y) = f(x, y) - g(x, y).$$

Find Fourier coefficients of $h(x, y)$: $\{ c_{mn}(h) \} (|mn| \le N, \ |m|, |n| \le N)$ and then find the hyperbolic cross truncations of Fourier series of $h(x, y)$ by the formula:

$$s_N^{(h,c)}(x, y) = \sum_{\substack{|mn| \le N \\ |m|,|n| \le N}} c_{mn}(h)\, e^{2\pi i (mx+ny)}.$$

The reconstruction formula is

$$f(x, y) \approx g(x, y) + s_N^{(h,c)}(x, y) \qquad ((x, y) \in [0, 1]^2)$$

Since four univariate boundary functions, which are used to construct $g(x, y)$, can be reconstructed as those in (i), the total number of the Fourier coefficients used in the reconstruction of $f(x, y)$ is equivalent to $N \log N$, and the square errors $e_N^{(2)}$ satisfy

$$(e_N^{(2)})^2 = O\left(\frac{\log N}{N^3} \right).$$

(iii) Let $f(x, y, z)$ be a triple function on $[0, 1]^3$ and $\frac{\partial^{i+j+k} f}{\partial x^i \partial y^j \partial z^k}$ $(i, j, k = 0, 1, 2)$ be continuous on $[0, 1]^3$. We reconstruct f as follows. Clearly, six bivariate boundary functions are

$$f(0, y, z), \quad f(1, y, z) \qquad (0 \le y \le 1, \ 0 \le z \le 1),$$

$$f(x, 0, z), \quad f(x, 1, z) \qquad (0 \le x \le 1, \ 0 \le z \le 1),$$

$$f(x, y, 0), \quad f(x, y, 1) \qquad (0 \le x \le 1, \ 0 \le y \le 1).$$

Based on these six bivariate boundary functions, construct a triple function as follows:

$$g(x, y, z) = g_1(x, y, z) + g_2(x, y, z) + g_3(x, y, z),$$

where

$$\begin{aligned}
g_1(x, y, z) &= f(x, y, 0)(1 - z) + f(x, y, 1)z + f(x, 0, z)(1 - y) + f(x, 1, z)y \\
&\quad + f(0, y, z)(1 - x) + f(1, y, z)x, \\
g_2(x, y, z) &= -f(x, 0, 0)(1 - y)(1 - z) - f(x, 0, 1)(1 - y)z \\
&\quad - f(x, 1, 0)y(1 - z) - f(x, 1, 1)yz - f(0, y, 0)(1 - x)(1 - z) \\
&\quad - f(0, y, 1)(1 - x)z - f(1, y, 0)x(1 - z) - f(1, y, 1)xz \\
&\quad - f(0, 0, z)(1 - x)(1 - y) - f(0, 1, z)(1 - x)z \\
&\quad - f(1, 0, z)x(1 - y) - f(1, 1, z)xy, \\
g_3(x, y, z) &= f(0, 0, 0)(1 - x)(1 - y)(1 - z) + f(1, 0, 0)x(1 - y)(1 - z) \\
&\quad + f(0, 1, 0)(1 - x)y(1 - z) + f(0, 0, 1)(1 - x)(1 - y)z \\
&\quad + f(1, 1, 0)xy(1 - z) + f(1, 0, 1)x(1 - y)z \\
&\quad + f(0, 0, 1)(1 - x)yz + f(1, 1, 1)xyz.
\end{aligned}$$

Clearly, $g(x, y, z) = f(x, y, z)$ on $\partial([0, 1]^3)$. Let $h(x, y, z) = f(x, y, z) - g(x, y, z)$. Find Fourier coefficients $\{c_{n_1n_2n_3}(h)\}$ ($|n_1n_2n_3| \le N$, $|n_1|, |n_2|, |n_3| \le N$) and find the hyperbolic cross truncations of Fourier series of $h(x, y, z)$ by the formula:

$$s_N^{(h,c)}(x, y, z) = \sum_{\substack{|n_1n_2n_3| \le N \\ |n_1|, |n_2|, |n_3| \le N}} c_{n_1n_2n_3}(h)\, e^{2\pi i(n_1x+n_2y+n_3z)}.$$

The reconstruction formula is

$$f(x, y, z) \approx g(x, y, z) + s_N^{(h,c)}(x, y, z), \qquad (x, y, z) \in [0, 1]^2$$

Since six bivariate boundary functions, which are used to construct $g(x, y, z)$, can be reconstructed as those in (ii), the total number of the Fourier coefficients in the reconstruction of $f(x, y, z)$ is equivalent to $N \log^2 N$, and the square errors $e_N^{(3)}$ satisfy $(e_N^{(3)})^2 = O(\frac{\log^2 N}{N^3})$.

2.8.3 Greedy Algorithm

An arbitrary subset of $L^2(\mathbb{R}^d)$ is called a *dictionary*. One approximates to f by a linear combination of N functions in a dictionary D. For example, the translations of Gabor functions $g_{\alpha,\beta}(t) = e^{-i\alpha t}\, e^{-\beta t^2}$ $(\alpha, \beta \in \mathbb{R})$ can generate a dictionary $\{g_{\alpha,\beta}(t - \gamma) : \alpha, \beta, \gamma \in \mathbb{R}\}$ in $L^2(\mathbb{R})$.

(a) Pure Greedy Algorithm

Let D be a dictionary in $L^2(\mathbb{R}^d)$. For $f \in L^2(\mathbb{R}^d)$, let $h = h(f) \in D$ be such that $(f, h(f)) = \sup_{h\in D}(f, h)$, where $(f, g) = \int_{\mathbb{R}^d} f(\mathbf{t})\overline{g}(\mathbf{t})d\mathbf{t}$. Define

$$G_1(f) = (f, h(f))h(f), \qquad R_1(f) = f - G_1(f),$$

$$G_2(f) = G_1(f) + G_1(R_1(f)), \qquad R_2(f) = f - G_2(f).$$

For $m \geq 2$, inductively define

$$G_m(f) = G_{m-1}(f) + G_1(R_{m-1}(f)), \qquad R_m(f) = f - G_m(f).$$

For the dictionary D generated by an orthogonal basis, $G_m(f)$ is the best $m-$term approximation of f from D. However, in general, this algorithm is not ideal.
(b) Relaxed Greedy Algorithm
Let $G_1(f)$ and $R_1(f)$ be stated in (a). Define

$$G_0^{(r)}(f) = 0, \qquad R_0^{(r)}(f) = f,$$

$$G_1^{(r)}(f) = G_1(f),\ R_1^{(r)}(f) = R_1(f).$$

For a function g, let $h = h(g) \in D$ be such that $(g, h) = \sup_{\tau \in D}(g, \tau)$. Inductively define

$$G_m^{(r)}(f) = \left(1 - \tfrac{1}{m}\right) G_{m-1}^{(r)}(f) + \tfrac{1}{m} h(R_{m-1}^{(r)}(f)),$$

$$R_m^{(r)}(f) = f - G_m^{(r)}(f).$$

In general, $G_m^{(r)}(f)$ can approximate to f better than $G_m(f)$.

2.9 Spherical Harmonics

Spherical harmonics are a frequency–space basis for representing multivariate time series defined over the sphere. Based on the rotation and spherical harmonics, a square integrable function can be expanded into an orthogonal series whose term is a product of a radial function and a solid spherical harmonic function of degree k. The structure of each term is invariant under Fourier transform.

2.9.1 Spherical Harmonic Functions

A function u defined on a domain D on $\mathbb{R}^d$ is called a *harmonic function* if it satisfies the Laplace equation $\Delta u = \frac{\partial^2 u}{\partial t_1^2} + \cdots + \frac{\partial^2 u}{\partial t_d^2} = 0$ on D.

A polynomial of the form $p(\mathbf{t}) = \sum_{|\alpha|=k} C_\alpha \mathbf{t}^\alpha$ $(k \in \mathbb{Z}_+)$, where

$$\boldsymbol{\alpha} = (\alpha_1, \ldots, \alpha_d), \quad \text{each } \alpha_k = 0, 1, \ldots,$$

$$|\boldsymbol{\alpha}| = \alpha_1 + \alpha_2 + \cdots + \alpha_d,$$

$$\mathbf{t}^{\alpha} = t_1^{\alpha_1} t_2^{\alpha_2} \cdots t_d^{\alpha_d},$$

is called a *homogeneous polynomial* of degree k on $\mathbb{R}^d$.

A homogeneous harmonic polynomial of degree k is called a *solid spherical harmonics* of degree k, and the restriction of such function on the surface of the unit sphere is called a *surface spherical harmonics* of degree k. For example, the polynomial $p(x, y) = x^3 - 3xy^2$ is a solid spherical harmonics of degree 3 on $\mathbb{R}^2$. Let $x = r\cos\theta,\ y = r\sin\theta$. The restriction of $p(x, y)$ on the unit circle is $\cos(3\theta)$. So $\cos(3\theta)$ is a surface spherical harmonics of degree 3 on $\mathbb{R}^2$.

2.9.2 Invariant Subspace under Fourier Transform

(i) Radial Function

Let f be a function on $\mathbb{R}^d$. If

$$f(\rho\mathbf{t}) = f(\mathbf{t}), \tag{2.9.1}$$

where ρ is any orthogonal transform on $\mathbb{R}^d$, then $f(\mathbf{t})$ is called a *radial function*. If $f(\mathbf{t}) = \xi(\|\ \mathbf{t}\ \|)$, i.e., $f(\mathbf{t})$ on $\mathbb{R}^d$ only depends on the norm of $\mathbf{t}$, then $f(\mathbf{t})$ must be a radial function.

(a) Fourier transform and orthogonal transform are exchangeable

Let $f(\mathbf{x})$ be an integrable function and ρ be a $d \times d$ orthogonal matrix. Fourier transform of $f(\rho\mathbf{t})$ is

$$(f(\rho\mathbf{t}))^{\wedge}(\boldsymbol{\omega}) = \int_{\mathbb{R}^d} f(\rho\mathbf{t})\, e^{-2\pi i(\boldsymbol{\omega},\, \mathbf{t})} d\mathbf{t}.$$

Since ρ is an orthogonal matrix, $\rho^{-1} = \rho^T$ and $\det(\rho) = \pm 1$. Making the substitution $\mathbf{x} = \rho\mathbf{t}$, we get

$$(f(\rho\mathbf{t}))^{\wedge}(\boldsymbol{\omega}) = \int_{\mathbb{R}^d} f(\mathbf{x})\, e^{-2\pi i(\boldsymbol{\omega},\, \rho^T\mathbf{x})} d\mathbf{x}. \tag{2.9.2}$$

The inner product can be rewritten into the product of matrices $(\boldsymbol{\omega},\ \rho^T\mathbf{x}) = \boldsymbol{\omega}^T\rho^T\mathbf{x}$. The transpose of 1×1 matrix is equal to itself and $(\boldsymbol{\omega},\ \rho^T\mathbf{x})$ is a 1×1 matrix, so $(\boldsymbol{\omega},\ \rho^T\mathbf{x}) = \mathbf{x}^T\rho\boldsymbol{\omega} = (\mathbf{x},\ \rho\boldsymbol{\omega}) = (\rho\boldsymbol{\omega},\ \mathbf{x})$. From this and (2.9.2), we get

$$(f(\rho\mathbf{t}))^{\wedge}(\boldsymbol{\omega}) = \int_{\mathbb{R}^d} f(\mathbf{x})\, e^{-2\pi i(\rho\boldsymbol{\omega},\, \mathbf{x})} d\mathbf{x} = \widehat{f}(\rho\boldsymbol{\omega}). \tag{2.9.3}$$

In detail, $f(\mathbf{t}) \xrightarrow{\rho} f(\rho\mathbf{t}) \xrightarrow{F} (f(\rho\mathbf{t}))^\wedge(\omega)$ is equivalent to $f(\mathbf{t}) \xrightarrow{F} \widehat{f}(\omega) \xrightarrow{\rho} \widehat{f}(\rho\omega)$, where F is the Fourier transform.

(b) Fourier transforms of radial functions are still radial functions

If $f(\mathbf{t})$ is a radial function on $\mathbb{R}^d$, ρ is an orthogonal transform on $\mathbb{R}^d$, by (2.9.1) and (2.9.3), we get

$$\widehat{f}(\omega) = (f(\rho\mathbf{t}))^\wedge(\omega) = \widehat{f}(\rho\omega).$$

So $\widehat{f}(\omega)$ is also a radial function on $\mathbb{R}^d$.

(ii) 2−Dimensional Invariant Subspace

The polar coordinate representation of Fourier transform on $\mathbb{R}^2$:

$$\widehat{f}(\omega) = \int_0^\infty \int_0^{2\pi} f(r\,e^{i\theta})\,e^{-2\pi i Rr\cos(\varphi-\theta)}\,r\,\mathrm{d}r\,\mathrm{d}\theta \qquad (\omega = R\,e^{i\varphi}). \tag{2.9.4}$$

The subspace $D_k^{(2)} (k \in \mathbb{Z})$ is the set of functions f on $\mathbb{R}^2$ satisfying $f(\mathbf{z}) = g(r)\,e^{ik\theta}$, where $\mathbf{z} = r\,e^{i\theta}$ and $g(r)$ only depends on r. If $f \in D_k^{(2)}$, then $f(\mathbf{z}) = g(r)\,e^{ik\theta}(\mathbf{z} = re^{i\theta})$. Let $h(\mathbf{z}) = f(e^{i\varphi}\mathbf{z})$ for a fixed φ. Then,

$$h(\mathbf{z}) = f(e^{i\varphi}\mathbf{z}) = g(r)\,e^{ik(\theta+\varphi)} = e^{ik\varphi}f(\mathbf{z}).$$

This implies that

$$\widehat{h}(!) = e^{ik\varphi}\widehat{f}(!). \tag{2.9.5}$$

On the other hand, since Fourier transform and orthogonal transform are exchangeable and $e^{i\varphi}\mathbf{z}$ is a rotation, it follows from $h(\mathbf{z}) = f(e^{i\varphi}\mathbf{z})$ that $\widehat{h}(\omega) = \widehat{f}(e^{i\varphi}\omega)$. Comparing this with (2.9.5) gives $\widehat{f}(e^{i\varphi}\omega) = e^{ik\varphi}\widehat{f}(\omega)$. Let $\omega = R$. Then,

$$\widehat{f}(R\,e^{i\varphi}) = \widehat{f}(R)\,e^{ik\varphi}, \tag{2.9.6}$$

i.e., $\widehat{f} \in D_k^{(2)}$. Hence, $D_k^{(2)}$ is a invariant subspace under Fourier transform.

In the invariant subspace $D_k^{(2)}$, if $k \neq l$, let $\xi \in D_k^{(2)}$ and $\eta \in D_l^{(2)}$. Then, $\xi(t) = \mu(r)\,e^{ik\theta}$, $\eta(t) = \tau(r)\,e^{il\theta}$. Since $\int_0^{2\pi} e^{i(k-l)\theta}\mathrm{d}\theta = 0\,(k \neq l)$,

$$(\xi,\,\eta) = \int_{\mathbb{R}} \mu(r)\tau(r)\left(\int_0^{2\pi} e^{i(k-l)}\mathrm{d}\theta\right) r\mathrm{d}r = 0,$$

So $D_k^{(2)} \perp D_l^{(2)}$, i.e., $D_k^{(2)}$ and $D_l^{(2)}$ are orthogonal.

This means that the sequence of subspaces $\{D_k^{(2)}\}_{k\in\mathbb{Z}}$ is an orthogonal sequence. Any $f(\mathbf{z}) = f(r\,e^{i\theta})$ is a $2\pi-$periodic function of θ. Expand it into Fourier series:

$$f(\mathbf{z}) = f(r\,e^{i\theta}) = \sum_{k\in\mathbb{Z}} f_k(r)\,e^{ik\theta},$$

where Fourier coefficient $f_k(r) = \frac{1}{2\pi}\int_0^{2\pi} f(r\,e^{i\theta})\,e^{-ik\theta}\mathrm{d}\theta$. Clearly, each $f_k(r)\,e^{ik\theta}$ is a function in $D_k^{(2)}$. Hence, $L^2(\mathbb{R}^2)$ is an orthogonal sum of $D_k^{(2)}$ $(k \in \mathbb{Z})$,

$$L^2(\mathbb{R}^2) = \bigoplus_{k=0}^{\infty} D_k^{(2)}$$

and each subspace $D_k^{(2)}$ is invariant under the Fourier transform. So any function $f \in L^2(\mathbb{R}^2)$ can be decomposed into $f(\mathbf{z}) = \sum\limits_k F_k$, where $F_k \in D_k^{(2)}$.

If $f \in D_k^{(2)}$ and $f(\mathbf{z}) = g(r)\,e^{ik\theta}$, then, by (2.9.4),

$$\widehat{f}(\omega) = \int_0^{\infty}\int_0^{2\pi} g(r)\,e^{ik\theta}\,e^{-2\pi i Rr\cos(\varphi-\theta)}r\mathrm{d}r\mathrm{d}\theta \qquad (\omega = R\,e^{i\varphi}).$$

Let $\varphi = 0$ and then let $\theta = \frac{\pi}{2} - \theta^*$. Then,

$$\begin{aligned}\widehat{f}(R) &= \textstyle\int_0^{\infty} g(r)\left(\int_0^{2\pi} e^{-2\pi i Rr\cos\theta}\,e^{ik\theta}\,\mathrm{d}\theta\right)r\mathrm{d}r\\ &= 2\pi(-i)^k \textstyle\int_0^{\infty} g(r)\left(\frac{1}{2\pi}\int_0^{2\pi} e^{-2\pi i Rr\sin\theta^*}\,e^{-ik\theta^*}\mathrm{d}\theta^*\right)r\mathrm{d}r.\end{aligned}$$

Note that Bessel function $J_k(t)$ has the integral representation: $J_k(t) = \frac{1}{2\pi}\int_0^{2\pi} e^{it\sin\theta}$ $e^{-ik\theta}\mathrm{d}\theta$ $(k \in \mathbb{Z})$. So

$$\widehat{f}(R) = 2\pi(-i)^k \int_0^{\infty} g(r)J_k(2\pi Rr)r\mathrm{d}r.$$

From this and (2.9.6),

$$\widehat{f}(R\,e^{i\varphi}) = 2\pi(-i)^k\left(\int_0^{\infty} g(r)J_k(2\pi Rr)r\mathrm{d}r\right)e^{ik\varphi}.$$

(iii) $d-$Dimensional Invariant Subspace

The *subspace* $D_k^{(d)}$ is the set of all linear combinations of functions of the form $\tau(\|\,\mathbf{t}\,\|)p(\mathbf{t})$, where $\tau(\|\,\mathbf{t}\,\|)$ ranges over all radial functions and $p(\mathbf{t})$ over all solid spherical harmonic functions of degree k.

If $f(\mathbf{t}) = \tau(\|\,\mathbf{t}\,\|)$, where $\tau(\|\,\mathbf{t}\,\|)$ is a radial function, then its Fourier transform $\widehat{f}(\omega) = F_0(|\omega|)$, where

$$F_0(|\omega|) = F_0(R) = 2\pi R^{-\frac{d-2}{2}}\int_0^{\infty} \tau(s)J_{\frac{d-2}{2}}(2\pi Rs)s^{\frac{d}{2}}\mathrm{d}s \qquad (|\omega| = R)$$

is also a radial function, where $J_\nu(x)$ is the $\nu-$order Bessel function and

$$J_\nu(x) = \frac{1}{2\pi}\int_{-\pi}^{\pi} e^{i(\nu\theta - x\sin\theta)}\mathrm{d}\theta.$$

If $f(\mathbf{t}) = \tau(\|\ \mathbf{t}\ \|)\ p(\mathbf{t})$, where $p(\mathbf{t})$ is a (solid) spherical harmonic function of degree k, then its Fourier transform $\widehat{f}(\omega) = F_k(|\omega|)p(\omega)$, where

$$F_k(|\omega|) = F_k(R) = 2\pi(i)^{-k}R^{-\frac{d-2}{2}-k}\int_0^\infty \tau(s)J_{\frac{d-2}{2}+k}(2\pi Rs)s^{\frac{d}{2}+k}\mathrm{d}s \quad (|\omega| = R).$$

This implies that each $D_k^{(d)}$ is an invariant subspace in $L^2(\mathbb{R}^d)$ and the Fourier transform in $D_k^{(d)}$ corresponds to a Bessel transform.

Similarly, if $k \neq l$, then $D_k^{(d)} \perp D_l^{(d)}$, i.e., $D_k^{(d)}$ and $D_l^{(d)}$ are orthogonal. Hence, $L^2(\mathbb{R}^2)$ is an orthogonal sum of $D_k^{(d)}$ $(k = 0, 1, \ldots)$, i.e.,

$$L^2(\mathbb{R}^d) = \bigoplus_{k=0}^{\infty} D_k^{(d)}.$$

For $f \in L^2(\mathbb{R}^d)$, there exists a sequence $\{f_k\}_{k=0,1,\ldots}$ of functions such that

$$f(\mathbf{t}) = \sum_{k=0}^{\infty} f_k(\mathbf{t}),$$

where $f_k \in D_k^{(d)}$ and $f_k(\mathbf{t}) = \tau_k(\|\ \mathbf{t}\ \|)p_k(\mathbf{t})$, i.e., f can be expanded into an orthogonal series:

$$f(\mathbf{t}) = \sum_{k=0}^{\infty} \tau_k(\|\ \mathbf{t}\ \|)p_k(\mathbf{t}),$$

where $\tau_k(\|\ \mathbf{t}\ \|)$ is a radial function and $p_k(\mathbf{t})$ is a solid spherical harmonic function of degree k.

2.10 Harmonic Analysis on General Domains

More and more data are collected in a distributed and irregular manner. In this section, we will analyze the spatial frequency information of multivariate time series on general domains.

2.10.1 Symmetric Kernels

Let $k(\mathbf{x}, \mathbf{t})$ be a continuous function on a domain $\Omega \times \Omega$, where $\mathbf{x} = (x_1, \ldots, x_d) \in \Omega$ and $\mathbf{t} = (t_1, \ldots, t_d) \in \Omega$. A continuous functional F on Ω is called a functional expressed by the kernel $k(\mathbf{x}, \mathbf{t})$ if it maps a continuous function $h(\mathbf{t})$ on Ω into

$$\int_\Omega k(\mathbf{x}, \mathbf{t}) h(\mathbf{t}) d\mathbf{t} \; (\mathbf{x} \in \Omega).$$

The kernel $k(\mathbf{x}, \mathbf{t})$ is said to be a *symmetric kernel* if it satisfies the condition $k(\mathbf{x}, \mathbf{t}) = k(\mathbf{t}, \mathbf{x})$ $(\mathbf{x}, \ \mathbf{t} \in \Omega)$. When $\lambda = \lambda_k$, if the integral equation:

$$\varphi(\mathbf{x}) = \lambda \int_\Omega k(\mathbf{x}, \mathbf{t}) \varphi(\mathbf{t}) d\mathbf{t}$$

has nonzero solutions, then λ_k is said to be an *eigenvalue of the kernel function* $k(\mathbf{x}, \mathbf{t})$ and the nonzero solutions, denoted by φ_k, of the integral equation are said to be *eigenfunctions corresponding to the eigenvalue* λ_k, i.e.,

$$\varphi_k(\mathbf{x}) = \lambda_k \int_\Omega k(\mathbf{x}, \mathbf{t}) \varphi_k(\mathbf{t}) d\mathbf{t} \qquad (\mathbf{x} \in \Omega).$$

Any function expressed by the symmetric kernel $k(\mathbf{x}, \mathbf{t})$ has infinite eigenvalues. Denote all eigenvalues by $|\lambda_1| \le |\lambda_2| \le |\lambda_3| \le \cdots$ and $|\lambda_k| \to \infty$ $(k \to \infty)$. The corresponding eigenfunctions $\{\varphi_k\}$ are normal orthogonal, i.e., $\int_\Omega \varphi_k(\mathbf{t}) \varphi_l(\mathbf{t}) d\mathbf{t} = \delta_{k,l}$, where $\delta_{k,l}$ is the Kronecker delta.

Let $f(\mathbf{x})$ be a function on the general domain Ω. Then, the Fourier series of f:

$$f(\mathbf{x}) = \sum_{k \in \mathbb{Z}_+} c_k \varphi_k(\mathbf{x}),$$

where $c_k = \int_\Omega f(\mathbf{x}) \varphi_k(\mathbf{x}) d\mathbf{x}$.

2.10.2 Smooth Extensions and Approximation

Smooth extensions of the higher-dimensional function are similar to those of the two-dimensional functions. However, the representation of the higher-dimensional case is too complicated. Therefore, we only consider the two-dimensional case.

(i) Smooth Extension

Let $\Omega \subset [0,1]^2$ be a planer domain and its boundary $\partial\Omega$ be a piecewise very smooth curve, and let f be a function on Ω and its all partial derivatives are continuous on Ω. Then, for $r \in \mathbb{Z}_+$, there exists a smooth function $F(x,y)$ on $[0,1]^2$ such that

(a) $F(x,y) = f(x,y)\ ((x,y) \in \Omega)$;
(b) $\frac{\partial^{i+j}F}{\partial x^i \partial y^j}(x,y)\ (i+j \le r)$ is continuous on $[0,1]^2$;
(c) $\frac{\partial^{i+j}F}{\partial x^i \partial y^j}(x,y) = 0\ (i+j \le r)$ on the boundary $\partial[0,1]^2$;
(d) $F(x,y)$ can be expressed locally in the form:

$$\sum_{j=0}^{L} \xi_j(x) y^j \quad \text{or} \quad \sum_{j=0}^{L} \eta_j(y) x^j \quad \text{or} \quad \sum_{i,j=0}^{L} c_{ij} x^i y^j$$

on the complement $[0,1]^2 \setminus \Omega$, where $L \in \mathbb{Z}_+$ and each c_{ij} is constant.

The construction of the smooth extension $F(x,y)$ of $f(x,y)$ from Ω to $[0,1]^2$ is stated as follows. Without loss of generality, the complement $[0,1]^2 \setminus \Omega$ is divided into some trapezoids with a curved side and some rectangles. For convenience of representation, assume that there are four points $(x_\nu, y_\nu) \in \partial\Omega\ (\nu = 1,2,3,4)$ such that $[0,1]^2 \setminus \Omega$ is divided into the following four rectangles:

$$H_1 = [0,x_1] \times [0,y_1], \qquad H_1 = [x_2,1] \times [0,y_2],$$

$$H_3 = [x_3,1] \times [y_3,1], \qquad H_4 = [0,x_4] \times [y_4,1]$$

and four trapezoids with a curved side:

$$E_1 = \{(x,y) : \quad x_1 \le x \le x_2, \quad 0 \le y \le g(x)\},$$

$$E_2 = \{(x,y) : \quad h(y) \le x \le 1, \quad y_2 \le y \le y_1\},$$

$$E_3 = \{(x,y) : \quad x_4 \le x \le x_3, \quad g^*(x) \le y \le 1\},$$

$$E_4 = \{(x,y) : \quad 0 \le x \le h^*(y), \quad y_1 \le y \le y_4\}.$$

The curve side of each trapezoid E_ν is on $\partial\Omega$, and the bottom side is on $\partial([0,1]^2)$.

First, we extend f smoothly to each trapezoid E_ν.

By similarity, we only state how to extend f from Ω to $\Omega \bigcup E_1$. Using two sequences $\{a_{k,1}(x,y)\}_{k=0,1,...}$ and $\{b_{k,1}(x,y)\}_{k=0,1,...}$, where

$$a_{0,1}(x,y) = \tfrac{y}{g(x)}, \qquad a_{k,1}(x,y) = \tfrac{(y-g(x))^k}{k!}\left(\tfrac{y}{g(x)}\right)^{k+1} \quad (k \in \mathbb{Z}_+),$$

$$b_{0,1}(x,y) = \tfrac{y-g(x)}{-g(x)}, \qquad b_{k,1}(x,y) = \tfrac{y^k}{k!}\left(\tfrac{y-g(x)}{-g(x)}\right)^{k+1} \quad (k \in \mathbb{Z}_+),$$

by induction, we define a sequence of functions $\{S_1^{(k)}(x,y)\}_{k=0,1,...}$ on E_1 as follows:

$$S_1^{(0)}(x,y) = f(x,g(x))\,a_{0,1}(x,y) \qquad (x_1 \le x \le x_2,\ \ 0 \le y \le g(x)).$$

In general, for $k \in \mathbb{Z}_+$,

$$S_1^{(k)}(x,y) = S_1^{(k-1)}(x,y) + a_{k,1}(x,y)\left(\frac{\partial^k f}{\partial y^k}(x,g(x)) - \frac{\partial^k S_1^{(k-1)}}{\partial y^k}(x,g(x))\right)$$

$$-b_{k,1}(x,y)\frac{\partial^k S_1^{(k-1)}}{\partial y^k}(x,0) \qquad (x_1 \le x \le x_2,\ \ 0 \le y \le g(x)).$$

Similarly, we construct $S_2^{(k)}(x,y)$, $S_3^{(k)}(x,y)$, and $S_4^{(k)}(x,y)$ on E_1, E_2, and E_3, respectively.

Next, we smoothly extend f to each rectangle $H_\nu(\nu = 1,2,3,4)$. By similarity, we only state the smooth extension of f on H_1. Let

$$\alpha_{k,11}(y) = \frac{(y-y_1)^k}{k!}\left(\frac{y}{y_1}\right)^{k+1},$$

$$\beta_{k,11}(y) = \frac{y^k}{k!}\left(\frac{y-y_1}{-y_1}\right)^{k+1} \qquad (k = 0,1,\ldots),$$

and let

$$M_1^{(0)}(x,y) = S_4^{(l_r)}(x,y_1)\,\alpha_{0,11}(y) \qquad (l_r = r(r+1)(r+2)(r+3)),$$

$$M_1^{(k)}(x,y) = M_1^{(k-1)}(x,y) + \alpha_{k,11}(y)\left(\frac{\partial^k S_4^{(l_r)}}{\partial y^k}(x,y_1) - \frac{\partial^k M_1^{(k-1)}}{\partial y^k}(x,y_1)\right)$$

$$-\beta_{k,11}(y)\frac{\partial^k M_1^{(k-1)}}{\partial y^k}(x,0)(k = 1,\ldots,\tau_r,\ (x,y) \in H_1),$$

where $\tau_r = r(r+2)$, and let

$$\alpha_{k,14}(x) = \frac{(x-x_1)^k}{k!}\left(\frac{x}{x_1}\right)^{k+1},$$

$$\beta_{k,14}(x) = \frac{x^k}{k!}\left(\frac{x-x_1}{-x_1}\right)^{k+1} \qquad (k = 0,1,\ldots),$$

and let

$$N_1^{(0)}(x,y) = (S_1^{(l_r)}(x_1,y) - M_1^{(\tau_r)}(x_1,y))\,\alpha_{0,14}(x),$$

$$N_1^{(k)}(x,y) = N_1^{(k-1)}(x,y) + \alpha_{k,14}(x)\left(\frac{\partial^k (S_1^{(l_r)} - M_1^{(\tau_r)})}{\partial x^k}(x_1,y) - \frac{\partial^k N_1^{(k-1)}}{\partial x^k}(x_1,y)\right)$$

$$-\beta_{k,14}(x)\frac{\partial^k N_1^{(k-1)}}{\partial x^k}(0,y)(k = 1,\ldots,r,\ (x,y) \in H_1).$$

Similarly, we construct $M_\nu^{(T_r)}(x, y)$ and $N_\nu^{(r)}(x, y)$ on $H_\nu (\nu = 2, 3, 4)$.

Finally, let

$$F(x, y) = \begin{cases} f(x, y), & (x, y) \in \Omega, \\ S_\nu^{(l_r)}(x, y), & (x, y) \in E_\nu \ \ (\nu = 1, 2, 3, 4), \\ M_\nu^{(T_r)}(x, y) + N_\nu^{(r)}(x, y), & (x, y) \in H_\nu \ \ (\nu = 1, 2, 3, 4). \end{cases} \tag{2.10.1}$$

Then, the function $F(x, y)$ is the desired smooth extension of f from Ω to $[0, 1]^2$.
(ii) Fourier Approximation
Let $f(x, y)$ and its all partial derivatives be continuous on $\Omega \subset [0, 1]^2$, and let $F(x, y)$ [see (2.10.1)] be its smooth extension from Ω to $[0, 1]^2$. We extend $F(x, y)$ to an one-periodic function on $\mathbb{R}^2$, still denoted by $F(x, y)$, and then expand $F(x, y)$ into a Fourier series. Note that $F(x, y) = f(x, y)$ on Ω. Then,

$$f(x, y) = \sum_{m,n\in\mathbb{Z}} c_{mn}(F)\, e^{2\pi i(mx+ny)} \qquad ((x, y) \in \Omega), \tag{2.10.2}$$

and the Fourier coefficients $c_{mn}(F)$ decay fast, where $c_{mn}(F) = \int_0^1 \int_0^1 F(x, y)$ $e^{-2\pi i(mx+ny)}\mathrm{d}x\mathrm{d}y$.

Take the hyperbolic cross truncations of the Fourier series (2.10.2):

$$s_N^{(h,c)}(f; x, y) = \sum_{\substack{|n_1 n_2|\leq N \\ |n_1|, |n_2|\leq N}} c_{mn}(F)\, e^{2\pi i(mx+ny)}.$$

So $f(x, y) \sim s_N^{(h,c)}(f; x, y)$ with a small error, i.e., $f(x, y)$ is reconstructed by few Fourier coefficients $c_{mn}(F)$.

Let $F^{(o)}(x, y)$ be an odd extension of $F(x, y)$ from $[0, 1]^2$ to $[-1, 1]^2$:

$$F^{(o)}(x, y) = \begin{cases} F(x, y), & (x, y) \in [0, 1]^2, \\ -F(-x, y), & (x, y) \in [-1, 0] \times [0, 1], \\ F(-x, -y), & (x, y) \in [-1, 0]^2, \\ -F(x, -y), & (x, y) \in [0, 1] \times [-1, 0]. \end{cases}$$

We extend $F^{(o)}(x, y)$ to a 2-periodic function, still denoted by $F^{(o)}(x, y)$, and then expand it into a Fourier sine series. Note that $F^{(o)}(x, y) = f(x, y)$ on Ω. Then,

$$f(x, y) = \sum_{m\in\mathbb{Z}_+} \sum_{n\in\mathbb{Z}_+} c_{mn}(F^{(o)}) \sin(m\pi x) \sin(n\pi y) \qquad ((x, y) \in \Omega),$$

where the Fourier sine coefficients are

$$c_{mn}(F^{(o)}) = \int_{-1}^{1}\int_{-1}^{1} F^{(o)}(x, y)\sin(m\pi x)\sin(n\pi y)\mathrm{d}x\mathrm{d}y$$

and $c_{mn}(F^{(o)})$ decay fast.

Let $F^{(e)}(x, y)$ be an even extension of $F(x, y)$ from $[0, 1]^2$ to $[-1, 1]^2$:

$$F^{(e)}(x, y) = \begin{cases} F(x, y), & (x, y) \in [0, 1]^2, \\ F(-x, y), & (x, y) \in [-1, 0] \times [0, 1], \\ F(-x, -y), & (x, y) \in [-1, 0]^2, \\ F(x, -y), & (x, y) \in [0, 1] \times [-1, 0]. \end{cases}$$

We extend $F^{(e)}(x, y)$ to a 2-periodic function, still denoted by $F^{(e)}(x, y)$, and then expand it into a Fourier cosine series. Note that $F^{(e)}(x, y) = f(x, y)$ on Ω. Then,

$$f(x, y) = \sum_{m=0}^{\infty}\sum_{n=0}^{\infty} c_{mn}(F^{(e)})\cos(m\pi x)\cos(n\pi y) \qquad ((x, y) \in \Omega),$$

where the Fourier cosine coefficients $c_{mn}(F^{(o)})$ decay fast.

If we consider the hyperbolic cross truncations of the Fourier sine (or cosine) series, $f(x, y)$ on Ω can be reconstructed by few Fourier sine (or cosine) coefficients $c_{mn}(F^{(o)})$ or $c_{mn}(F^{(e)})$.

2.11 Harmonic Analysis on Graphs

Harmonic analysis tools originally developed for Euclidean spaces and regular lattices are now being transferred to the general settings of graphs in order to analyze geometric and topological structures, as well as data measured on them. A graph G has n vertices $v_0, v_1, \ldots, v_{n-1}$. If there is an edge between two vertices v_i and v_j, we say that v_i and v_j are *adjacent*, denoted by $v_i \sim v_j$. Denote by d_i the number of adjacent vertices of the vertex v_i. The d_i is called the *degree* of the vertex v_i.

2.11.1 The Laplacian of a Graph

The Laplacian matrix $\mathcal{L}$ of a graph G is defined as $\mathcal{L} = (\mathcal{L}(v_i, v_j))_{i,j=0,1,\ldots,n-1}$, where

$$\mathcal{L}(v_i, v_j) = \begin{cases} 1 & \text{if } i = j \text{ and } d_i \neq 0, \\ -\frac{1}{\sqrt{d_i d_j}} & \text{if } v_i \text{ and } v_j \text{ are adjacent}, \\ 0 & \text{otherwise}. \end{cases} \tag{2.11.1}$$

Let g be a function defined at vertices $v_0, v_1, \ldots, v_{n-1}$. Then, $\mathbf{g} = (g(v_0), \ldots, g(v_{n-1}))^T$ is a column vector. By (2.11.1), it follows that $\mathcal{L}\mathbf{g} = (\gamma_0, \ldots, \gamma_{n-1})^T$, where

$$\gamma_i = \frac{1}{\sqrt{d_i}} \sum_{j=0}^{n-1} \left(\frac{g(v_i)}{\sqrt{d_i}} - \frac{g(v_j)}{\sqrt{d_j}} \right) \quad (i = 0, 1, \ldots, n-1). \tag{2.11.2}$$

Let $\mathbf{g}_0 = (\sqrt{d_0}, \ldots, \sqrt{d_{n-1}})^T$. By (2.11.2), $\mathcal{L}\mathbf{g}_0 = \mathbf{0}$ or $\mathcal{L}\mathbf{g}_0 = \lambda_0 \mathbf{g}_0$, where $\lambda_0 = 0$. This implies that λ_0 is a eigenvalue of the Laplacian matrix $\mathcal{L}$ and $\mathbf{g}_0$ is the corresponding eigenvector. Since the Laplacian matrix $\mathcal{L}$ is symmetric, its all eigenvalues are nonnegative real numbers, denoted by $0 = \lambda_0 \leq \lambda_1 \leq \cdots \leq \lambda_{n-1}$. The set of eigenvalues is called the *spectrum* of $\mathcal{L}$ (or the spectrum of the associated graph G).

Let T be a diagonal matrix:

$$T = \begin{pmatrix} d_0 & 0 & \cdots & 0 \\ 0 & d_1 & \cdots & 0 \\ \vdots & \vdots & \ddots & \vdots \\ 0 & 0 & \cdots & d_{n-1} \end{pmatrix},$$

and let $\mathbf{I}$ be a $n-$dimensional column vector: $\mathbf{I} = (1, 1, \ldots, 1)^T$. Then, $T\mathbf{I} = (d_0, d_1, \ldots, d_{n-1})^T$.

Let $\mathbf{f} = (f(v_0), f(v_1), \ldots, f(v_{n-1}))^T$. The notation $\mathbf{f} \perp T\mathbf{I}$ means that the inner product of two $n-$dimensional vectors $\mathbf{f}$ and $T\mathbf{I}$ vanishes, i.e., $\sum_{i=0}^{n-1} f(v_i) d_i = 0$. The eigenvalue λ_1 is

$$\lambda_1 = \inf_{\mathbf{f} \perp T\mathbf{I}} \frac{\sum\limits_{v_i \sim v_j} (f(v_i) - f(v_j))^2}{\sum\limits_{i=0}^{n-1} f^2(v_i)\, d_i}. \qquad (2.11.3)$$

If the function $\mathbf{f}_1$ at vertices attains the infimum, then $\mathbf{f}_1$ is a harmonic eigenfunction of $\mathcal{L}$. The largest eigenvalue is

$$\lambda_{n-1} = \sup_{\mathbf{f}} \frac{\sum\limits_{v_i \sim v_j} (f(v_i) - f(v_j))^2}{\sum\limits_{i=0}^{n-1} f^2(v_i)\, d_i}.$$

Let S_{k-1} be the subspace generated by the harmonic eigenfunctions corresponding to λ_i for $i \le k-1$. Then,

$$\lambda_k = \inf_{\mathbf{f}} \sup_{\mathbf{g} \in S_{k-1}} \frac{\sum\limits_{v_i \sim v_j} (f(v_i) - f(v_j))^2}{\sum\limits_{i=0}^{n-1} (f(v_i) - g(v_i))^2} = \inf_{\mathbf{f} \perp T S_{k-1}} \frac{\sum\limits_{v_i \sim v_j} (f(v_i) - f(v_j))^2}{\sum\limits_{i=0}^{n-1} f^2(v_i)\, d_i} \quad (k = 2, \ldots, n-2).$$

2.11.2 Eigenvalues and Eigenfunctions

From $(f(v_i) - f(v_j))^2 \le 2f^2(v_i) + 2f^2(v_j)$, it follows that

$$\sum_{v_i \sim v_j} (f(v_i) - f(v_j))^2 \le \sum_{i=0}^{n-1} f^2(v_i)\, d_i + \sum_{j=0}^{n-1} f^2(v_j)\, d_j = 2\sum_{i=0}^{n-1} f^2(v_i)\, d_i,$$

where in the sum of the left-hand side, $f(v_i) - f(v_j)$ occurs only one time. So

$$\frac{\sum\limits_{v_i \sim v_j} (f(v_i) - f(v_j))^2}{\sum\limits_{i=0}^{n-1} f^2(v_i)\, d_i} \le 2.$$

From this and the above expressions of λ_k, we get an estimate $0 \le \lambda_k \le 2\ (k = 0, 1, \ldots, n-1)$.

Now, we estimate the lower bound of λ_1. Define the *distance* between two vertices

v_i and v_j by the number of edges in a shortest path joining v_i and v_j, denoted by $\operatorname{dist}(v_i, v_j)$. Define the *diameter* of a graph G as $D = \max_{i,j=0,1,\ldots,n-1} \operatorname{dist}(v_i, v_j)$. Let G be a connected graph with the diameter D.

Suppose that $\mathbf{f}$ is a harmonic eigenfunction. By (2.11.3), it follows that $\mathbf{f} \perp T\mathbf{I}$ and

$$\lambda_1 = \frac{\sum\limits_{v_i \sim v_j} (f(v_i) - f(v_j))^2}{\sum\limits_{i=0}^{n-1} f^2(v_i) d_i}. \tag{2.11.4}$$

Let v_{k_0} be a vertex with $|f(v_{k_0})| = \max_{i=0,\ldots,n_1} |f(v_i)|$. Since $\mathbf{f} \perp T\mathbf{I}$, the sum $\sum_{i=0}^{n-1} f(v_i) d_i = 0$. So there exists a v_s such that $f(v_{k_0}) f(v_s) < 0$. Let P be the shortest path in G joining v_{k_0} and v_s. By (2.11.4),

$$\lambda_1 \geq \frac{\sum\limits_{(v_i, v_j) \in P} (f(v_i) - f(v_j))^2}{f^2(v_{k_0}) \operatorname{vol} G}, \tag{2.11.5}$$

where $\operatorname{vol} G = \sum\limits_{i=0}^{n-1} d_i$ is called the *volume* of G. Let $l + 1$ vertices $v_{k_0}, v_{k_1}, \ldots, v_{k_l} \in P$, where $v_{k_l} = v_s$ and $l = \operatorname{dist}(v_{k_0}, v_s)$. According to the Schwarz inequality and $v_{k_l} = v_s$, $l \leq D$, and $f(v_{k_0}) f(v_s) < 0$, it follows that

$$\sum_{(v_i, v_j) \in P} (f(v_i) - f(v_j))^2 \geq \frac{1}{l} \sum_{j=1}^{l} (f(v_{k_j}) - f(v_{k_{j-1}}))^2$$

$$\geq \frac{1}{l} (f(v_{k_l}) - f(v_{k_0}))^2 = \frac{(f(v_s) - f(v_{k_0}))^2}{\operatorname{dist}(v_{k_0}, v_s)} \geq \frac{f^2(v_{k_0})}{D}.$$

From this and (2.11.5),

$$\lambda_1 \geq (D \operatorname{vol} G)^{-1} = \left(D \sum_{i=0}^{n-1} d_i \right)^{-1}.$$

2.11.3 Fourier Expansions

Let G be a graph with vertices $v_0, \ldots, v_{n-1}$ and the set $E(G)$ be edges. The Laplacian matrix $\mathcal{L}$ of G is an $n \times n$ positive semi-definite matrix which is stated in (2.11.1). The eigenvectors $\mathbf{g}_0, \ldots, \mathbf{g}_{n-1}$ of $\mathcal{L}$ form a normal orthogonal basis for the space H_n of functions at vertices $v_0, \ldots, v_{n-1}$, i.e., for $i, j = 0, \ldots, n-1$,

$$(\mathbf{g}_i, \mathbf{g}_j) = \sum_{k=0}^{n-1} g_i(v_k) g_j(v_k) = \begin{cases} 1, \ i = j, \\ \\ 0, \ i \neq j, \end{cases}$$

where $\mathbf{g}_i = (g_i(v_0), \ldots, g_i(v_{n-1}))^T$ and $\mathcal{L}\mathbf{g}_i = \lambda_i \mathbf{g}_i$ $(i = 0, 1, \ldots, n-1)$. So any $\mathbf{f} = (f(v_0), \ldots, f(v_{n-1}))^T \in H_n$ can be expanded into Fourier series with respect to eigenvectors $\{\mathbf{g}_i\}_{i=0,\ldots,n-1}$:

$$f(v) = \sum_{i=0}^{n-1} c_i g_i(v) \qquad (v = v_0, \ldots, v_{n-1}),$$

where Fourier coefficients $c_i = (\mathbf{f}, \mathbf{g}_i) = \sum_{k=0}^{n-1} f(v_k) g_i(v_k)$ $(i = 0, \ldots, n-1)$, and they satisfy the Parseval identity:

$$\sum_{k=0}^{n-1} f^2(v_k) = \sum_{k=0}^{n-1} c_k^2.$$

Consider a *graph G (possibly with loops) with weighted function* $w(v_i, v_j)$ $(i, j = 0, \ldots, n-1)$ satisfying $w(v_i, v_j) = w(v_j, v_i)$ and $w(v_i, v_j) \geq 0$, where $v_0, \ldots, v_{n-1}$ are vertices of the graph G. Unweighted graphs are the special case where all weights are 0 or 1. The *degree* of a vertex v_i is defined as $d_i = \sum_{j=0}^{n-1} w(v_i, v_j)$. The Laplacian matrix $\mathcal{L}_w$ of the graph G is

$$\mathcal{L}_w(v_i, v_j) = \begin{cases} 1 - \frac{w(v_i, v_j)}{d_i} & \text{if } i = j \text{ and } d_i \neq 0, \\ \\ -\frac{w(v_i, v_j)}{\sqrt{d_i d_j}} & \text{if } i \text{ and } j \text{ are adjacent}, \\ \\ 0 & \text{otherwise}. \end{cases}$$

All definitions and results for graphs can be generalized to the weighted graphs. So the eigenvectors $\{\mathbf{g}_i^w\}_{i=0,\ldots,n-1}$ of the Laplacian matrix $\mathcal{L}_w$ form a normal orthogonal basis for H_n. Any function $\mathbf{f} = (f(v_0), \ldots, f(v_{n-1}))^T \in H_n$ at vertices of the graph G can be expanded into Fourier series:

$$f(v) = \sum_{i=0}^{n-1} c_i^w g_i^w(v) \qquad (v = v_0, \ldots, v_{n-1}),$$

where $c_i^w = (\mathbf{f}, \mathbf{g}_i^w) = \sum_{k=0}^{n-1} f(v_k) g_i^w(v_k)$. The *contraction* of a graph means that two distinct vertices v_i and v_j are identified into a vertex v^*. The *weights* of edges associated with v^* are defined as

$$w(v_k, v^*) = w(v_k, v_i) + w(v_k, v_j),$$

$$w(v^*, v^*) = w(v_i, v_i) + w(v_j, v_j) + 2w(v_i, v_j).$$

If a graph G^* is the contraction of the graph G, then the eigenvalue $\lambda_1^G \leq \lambda_1^{G^*}$.

Further Reading

J. Aalto, P. Pirinen, J. Heikkinen et al., Spatial interpolation of monthly climate data for Finland: comparing the performance of kriging and generalized additive models. Theor. Appl. Climatol. **112**, 99–111 (2013)

M.W. Ashiq, C. Zhao, J. Ni et al., GIS-based high-resolution spatial interpolation of precipitation in mountain-plain areas of Upper Pakistan for regional climate change impact studies. Theor. Appl. Climatol. **99**, 239 (2010)

N.M. Atakishiyev, L.E. Vicent, K.B. Wolf, Continuous vs. discrete fractional Fourier transforms. J. Comput. Appl. Math. **107**, 73–95 (1999)

B.G.J. Brooks, Applying Wavelet and Fourier Transform Analysis to Large Geophysical Datasets, in *Computational Science*, vol. 5545, Lecture Notes in Computer Science, ed. by G. Allen, et al. (Springer, Heidelberg, 2009)

C. Candan, M.A. Kutay, H.M. Ozaktas, The discrete fractional Fourier transform. IEEE Trans. Sig. Process. **48**, 1329–1337 (2000)

X. Chen, P. Xing, Y. Luo et al., Surface temperature dataset for North America obtained by application of optimal interpolation algorithm merging tree-ring chronologies and climate model output. Theor. Appl. Climatol. **127**, 533–549 (2017)

W. Cheney, W. Light, *A Course in Approximation Theory* (Brooks/Cole Publishing, Pacific Grove, 2000)

R.A. DeVore, Nonlinear approximation. Acta Numerica **7**, 51–150 (1998)

S. Eghdamirad, F. Johnson, A. Sharma, Using second-order approximation to incorporate GCM uncertainty in climate change impact assessments. Clim. Change **142**, 37–52 (2017)

J. Fan, J. Meng, X. Chen et al., Network approaches to climate science. Sci. China Phys. Mech. Astron. **60**, 010531 (2017)

I. Fountalis, A. Bracco, C. Dovrolis, Spatio-temporal network analysis for studying climate patterns. Clim. Dyn. **42**, 879–899 (2014)

E.D. Giuseppe, G.J. Lasinio, S. Esposito et al., Functional clustering for Italian climate zones identification. Theor. Appl. Climatol. **114**, 39–54 (2013)

S.V. Henriksson, P. Ralsanen, J. Silen et al., Quasiperiodic climate variability with a period of 50–70 years: Fourier analysis of measurements and earth system model simulations. Clim. Dyn. **39**, 1999–2011 (2012)

A.L. Kay, S.M. Crooks, H.N. Davies et al., Probabilistic impacts of climate change on flood frequency using response surfaces I: England and Wales. Reg. Environ. Change **14**, 1215–1227 (2014)

R. Kandel, Understanding and measuring earth's energy budget: from fourier, Humboldt, and Tyndall to CERES and Beyond. Surv. Geophys. **33**, 337–350 (2012)

K. Kikuchi, An introduction to combined Fourier-wavelet transform and its application to convectively coupled equatorial waves. Clim. Dyn. **43**, 1339–1356 (2014)

V.A. Narayanan, K.M.M. Prabhu, The fractional Fourier transform: theory, implementation and error analysis. Microprocess. Microsyst. **27**, 511–521 (2003)

M. Ogurtsov, G. Kocharov, M. Lindholm et al., Evidence of solar variation in tree-ring-based climate reconstructions. Solar Phys. **205**, 403–417 (2002)

S. Samanta, D.K. Pal, D. Lohar et al., Interpolation of climate variables and temperature modeling. Theor. Appl. Climatol. **107**, 35–45 (2012)

E. Sejdic, I. Djurovic, L. Stankovic, Fractional Fourier transform as a signal processing tool: an overview of recent developments. Sig. Process. **91**, 1351–1369 (2011)

E.M. Stein, G. Weiss, *Introduction to Fourier analysis on Euclidean spaces* (Princeton University Press, Princeton, 1971)

Z. Wu, E.K. Schneider, B.P. Kirtman et al., The modulated annual cycle: an alternative reference frame for climate anomalies. Clim. Dyn. **31**, 823–841 (2008)

Z. Zhang, Approximation of bivariate functions via smooth extensions. Sci. World J. **2014**, 102062 (2014)

F. Zwiers, S. Shen, Errors in estimating spherical harmonic coefficients from partially sampled GCM output. Clim. Dyn. **13**, 703–716 (1997)

Chapter 3
Multivariate Wavelets

Wavelet analysis offers an alternative to harmonic analysis of multivariate time series and is particularly useful when the amplitudes and periods of dominant cycles are time dependent. Wavelets can accommodate multiscale nonstationary information and minimize correlation and time dependency in data. In this chapter, we introduce wavelet representation for multivariate time series with time-dependent dominant cycles. Main techniques and methods include multiresolution analysis and wavelets, discrete wavelet transform, wavelet packet, wavelet variance, significant tests, wavelet shrinkage, and shearlets, bandelets and curvelets.

3.1 Multiresolution Analysis

The *multiresolution analysis* (MRA) concept was initiated by Meyer and Mallat, which provides a natural framework for the construction and understanding of wavelet bases.

3.1.1 Structure of MRA

The set of square integrable functions $f(\mathbf{t})$ on $\mathbb{R}^d$ is denoted by $L^2(\mathbb{R}^d)$, namely

$$L^2(\mathbb{R}^d) = \{f(\mathbf{t});\ \int_{\mathbb{R}^d} |f(\mathbf{t})|^2 < \infty\}.$$

A sequence $\{V_m\}_{m\in\mathbb{Z}}$ of closed subspaces in $L^2(\mathbb{R}^d)$ is called an orthogonal MRA if

Z. Zhang, *Multivariate Time Series Analysis in Climate and Environmental Research*, https://doi.org/10.1007/978-3-319-67340-0_3

(i) $V_m \subset V_{m+1}$ $(m \in \mathbb{Z})$, $\overline{\bigcup_{m\in\mathbb{Z}} V_m} = L^2(\mathbb{R}^d)$, $\bigcap_{m\in\mathbb{Z}} V_m = \{\mathbf{0}\}$, where $\overline{A}$ is the closure of A;
(ii) $f(\mathbf{t}) \in V_m$ if and only if $f(2\mathbf{t}) \in V_{m+1}$ $(m \in \mathbb{Z})$;
(iii) There is a function φ so that $\{\varphi(\mathbf{t}-\mathbf{n})\}_{\mathbf{n}\in\mathbb{Z}^d}$ is a normal orthogonal basis of V_0, where φ is called a *scaling function*.

Any $f \in V_0$ can be expanded into Fourier series with respect to $\{\varphi(\mathbf{t}-\mathbf{n})\}_{\mathbf{n}\in\mathbb{Z}^d}$:

$$f(\mathbf{t}) = \sum_{\mathbf{n}\in\mathbb{Z}^d} c_{\mathbf{n}}\varphi(\mathbf{t}-\mathbf{n}),$$

where Fourier coefficients $c_{\mathbf{n}} = (f(\mathbf{t}), \varphi(\mathbf{t}-\mathbf{n}))$ $(\mathbf{n} \in \mathbb{Z}^d)$. It is clear that the sequence $\{2^{\frac{md}{2}}\varphi(2^m\mathbf{t}-\mathbf{n})\}_{\mathbf{n}\in\mathbb{Z}^d}$ is a normal orthogonal basis of V_m for any $m \in \mathbb{Z}$. For $f \in L^2(\mathbb{R}^d)$, there exists a sufficiently large M such that

$$f(\mathbf{t}) \approx f_M(\mathbf{t}) = \sum_{\mathbf{n}\in\mathbb{Z}^d} c_{M,\mathbf{n}} 2^{\frac{Md}{2}}\varphi(2^M\mathbf{t}-\mathbf{n}),$$

where

$$c_{M,\mathbf{n}} = 2^{\frac{Md}{2}}\int_{\mathbb{R}^d} f_M(\mathbf{t})\varphi(2^M\mathbf{t}-\mathbf{n})\mathrm{d}\mathbf{t} = 2^{\frac{Md}{2}}\int_{\mathbb{R}^d} f(\mathbf{t})\varphi(2^M\mathbf{t}-\mathbf{n})\mathrm{d}\mathbf{t}$$

and f_M is the projection of f on the space V_M.

Below, we give some popular examples of orthogonal MRA.

(a) Piecewise-constant MRA. Let the space V_m be a set of all $g \in L^2(\mathbb{R}^d)$ such that $g(\mathbf{t})$ is constant on each cube:

$$\prod_{i=1}^d \left[\frac{n_i}{2^m}, \frac{n_i+1}{2^m}\right] \qquad \left(\mathbf{n} = (n_1, \ldots, n_d) \in \mathbb{Z}^d\right).$$

Then, $\{V_m\}_{m\in\mathbb{Z}}$ is an orthogonal MRA with a scaling function:

$$\varphi(\mathbf{t}) = \begin{cases} 1, & \mathbf{t} \in [0,1]^d, \\ 0, & \mathbf{t} \notin [0,1]^d. \end{cases}$$

(b) Shannon MRA. Let V_m be a set of all $g \in L^2(\mathbb{R}^d)$ with $\widehat{g}(\boldsymbol{\omega}) = 0$ $(\boldsymbol{\omega} \notin [-2^m\pi, 2^m\pi]^d)$, where $\widehat{g}(\boldsymbol{\omega})$ is the Fourier transform of $g(\mathbf{t})$. Then, $\{V_m\}_{m\in\mathbb{Z}}$ is an orthogonal MRA with a scaling function $g(\mathbf{t}) = \prod_{i=1}^d \frac{\sin(\pi t_i)}{\pi t_i}$.
(c) Meyer MRA. Let $\theta(\omega)$ $(i = 1, \ldots, d)$ be an infinitely differentiable, compactly supported even function and

$$\begin{aligned} \theta(\omega) &= \begin{cases} 1, \ \omega \in [-\frac{2\pi}{3}, \frac{2\pi}{3}], \\ 0, \ \omega \notin [-\frac{4\pi}{3}, \frac{4\pi}{3}], \end{cases} \\ \theta^2(\omega) &+ \theta^2(2\pi - \omega) = 1 \quad (\omega \in [0, 2\pi]) \end{aligned} \tag{3.1.1}$$

and $g(t_i)$ satisfies

$$\widehat{g}(\omega_i) = \cos\left(\frac{\pi}{2}\theta\left(\frac{3}{2\pi}|\omega_i| - 1\right)\right),$$

and $\varphi(\mathbf{t}) = g(t_1)\cdots g(t_d)$. Define V_m to be the closed subspace spanned by $\varphi(2^m\mathbf{t} - \mathbf{n})$ ($\mathbf{n} \in \mathbb{Z}^d$). Then, $\{V_m\}$ constitutes an orthogonal MRA with the scaling function $\varphi(\mathbf{t})$.

3.1.2 Scaling Functions

Assume that $\{V_m\}_{\mathbf{m}\in\mathbb{Z}^d}$ is an orthogonal MRA with the scaling function φ. Since $\{\varphi(\mathbf{t} - \mathbf{n})\}_{\mathbf{n}\in\mathbb{Z}^d}$ is a normal orthogonal system,

$$\int_{\mathbb{R}^d} \varphi(\mathbf{t} - \mathbf{k})\overline{\varphi}(\mathbf{t} - \mathbf{n})d\mathbf{t} = \delta_{\mathbf{k},\mathbf{n}}, \tag{3.1.2}$$

where $\delta_{\mathbf{k},\mathbf{n}}$ is the Kronecker Delta. By generalized Parseval identity of Fourier transforms,

$$\begin{aligned} &\int_{\mathbb{R}^d} \varphi(\mathbf{t} - \mathbf{k})\overline{\varphi}(\mathbf{t} - \mathbf{n})\, d\mathbf{t} \\ &= \frac{1}{(2\pi)^d} \int_{\mathbb{R}^d} \widehat{\varphi}(\boldsymbol{\omega})\overline{\widehat{\varphi}}(\boldsymbol{\omega})\, e^{-i(\mathbf{n}-\mathbf{k})\cdot\boldsymbol{\omega}}\, d\boldsymbol{\omega} \\ &= \frac{1}{(2\pi)^d} \sum_{\nu\in\mathbb{Z}^d} \int_{[0,2\pi]^d} |\widehat{\varphi}(\boldsymbol{\omega} + 2\nu\pi)|^2\, e^{-i(\mathbf{n}-\mathbf{k})\cdot\boldsymbol{\omega}} d\boldsymbol{\omega}. \end{aligned}$$

From this and (3.1.2), we get

$$\frac{1}{(2\pi)^d} \int_{[0,2\pi]^d} \left(\sum_{\nu\in\mathbb{Z}^d} |\widehat{\varphi}(\boldsymbol{\omega} + 2\nu\pi)|^2\right) e^{i(\mathbf{n}\cdot\boldsymbol{\omega})} d\boldsymbol{\omega} = \delta_{\mathbf{0},\mathbf{n}}.$$

So the corresponding Fourier series is

$$\sum_{\nu\in\mathbb{Z}^d} |\widehat{\varphi}(\boldsymbol{\omega} + 2\nu\pi)|^2 = \sum_{\mathbf{n}\in\mathbb{Z}^d} \delta_{\mathbf{0},\mathbf{n}}\, e^{-i(\mathbf{n}\cdot\boldsymbol{\omega})} \qquad (\boldsymbol{\omega} \in \mathbb{R}^d).$$

i.e., an orthogonal scaling function φ satisfies $\sum_{\nu\in\mathbb{Z}^d} |\widehat{\varphi}(\boldsymbol{\omega} + 2\nu\pi)|^2 = 1$ $(\boldsymbol{\omega} \in \mathbb{R}^d)$.

Since orthogonal scaling function φ belongs to V_0 and $V_{-1} \subset V_0$, $\frac{1}{2^d}\varphi\left(\frac{\mathbf{t}}{2}\right) \in V_{-1} \subset V_0$, and so $\frac{1}{2^d}\varphi\left(\frac{\mathbf{t}}{2}\right)$ can be expanded into a series with respect to $\{\varphi(\mathbf{t}-\mathbf{n})\}_{\mathbf{n}\in\mathbb{Z}^d}$:

$$\frac{1}{2^d}\varphi\left(\frac{\mathbf{t}}{2}\right) = \sum_{\mathbf{n}\in\mathbb{Z}^d} a_{\mathbf{n}}\varphi(\mathbf{t}\text{-}\mathbf{n})$$

which is called the *bi-scale equation* and $\{a_{\mathbf{n}}\}_{\mathbf{n}\in\mathbb{Z}^d}$ is interpreted as a *discrete filter*. Taking Fourier transform on both sides, $\widehat{\varphi}(2\boldsymbol{\omega}) = H(\omega)\widehat{\varphi}(\boldsymbol{\omega})$, where $H(\boldsymbol{\omega}) = \sum_{\mathbf{n}\in\mathbb{Z}^d} a_{\mathbf{n}}\, e^{-i(\mathbf{n}\cdot\omega)}$ is called *transfer function* associated with the scaling function φ. Namely, for an orthogonal scaling function $\varphi(\mathbf{t})$, there is a $2\pi\mathbb{Z}^d-$periodic function $H(\boldsymbol{\omega})$ so that

$$\widehat{\varphi}(2\boldsymbol{\omega}) = H(\boldsymbol{\omega})\widehat{\varphi}(\boldsymbol{\omega}) \qquad (\boldsymbol{\omega} \in \mathbb{R}^d). \tag{3.1.3}$$

Each $\boldsymbol{\alpha} \in \mathbb{Z}^d$ may be written into $\boldsymbol{\alpha} = 2\boldsymbol{\beta} + \boldsymbol{\nu}\ (\boldsymbol{\beta} \in \mathbb{Z}^d,\ \boldsymbol{\nu} \in \{0, 1\}^d)$, where the notation $\{0, 1\}^d$ consists of all vertexes of the unit cube $[0, 1]^d$. By (3.1.3), it follows that

$$\begin{aligned} &\sum_{\boldsymbol{\nu}\in\{0,1\}^d} |H(\boldsymbol{\omega} + \boldsymbol{\nu}\pi)|^2 \\ &= \sum_{\substack{\boldsymbol{\beta}\in\mathbb{Z}^d \\ \boldsymbol{\nu}\in\{0,1\}^d}} \left|H\left(\tfrac{2\boldsymbol{\omega}+4\boldsymbol{\beta}\pi+2\pi\boldsymbol{\nu}}{2}\right)\right|^2 \left|\widehat{\varphi}\left(\tfrac{2\boldsymbol{\omega}+4\boldsymbol{\beta}\pi+2\boldsymbol{\nu}\pi}{2}\right)\right|^2 \\ &= \sum_{\boldsymbol{\alpha}\in\mathbb{Z}^d} |\widehat{\varphi}(2\boldsymbol{\omega} + 2\boldsymbol{\alpha}\pi)|^2 = 1. \end{aligned}$$

Conversely, for any function $\varphi \in L^2(\mathbb{R}^d)$, if

(a) $\sum_{\boldsymbol{\nu}\in\mathbb{Z}^d} |\widehat{\varphi}(\boldsymbol{\omega} + 2\boldsymbol{\nu}\pi)|^2 = 1$;
(b) There exists a $2\pi\mathbb{Z}^d-$periodic function $H(\boldsymbol{\omega})$ such that $\widehat{\varphi}(2\boldsymbol{\omega}) = H(\boldsymbol{\omega})\widehat{\varphi}(\boldsymbol{\omega})$;
(c) $\widehat{\varphi}$ is a bounded continuous function and $\widehat{\varphi}(\mathbf{0}) \neq 0$;

then $\varphi(t)$ is a scaling function for an orthogonal MRA.

If a function f satisfies $f(\mathbf{t}) \neq 0\,(\mathbf{t} \in G)$ and $f(\mathbf{t}) = 0\,(\mathbf{t} \in \mathbb{R}^d \setminus G)$, we call G the *support* of f, denoted by $\operatorname{supp} f = G$. Let E be a domain. Denote

$$\begin{aligned} E + \mathbf{a} &:= \{\omega + \mathbf{a},\ \ \omega \in E\}, \\ \lambda E &:= \{\lambda\omega,\ \ \omega \in E\}, \\ E + 2\pi\mathbb{Z}^d &= \bigcup_{\boldsymbol{\nu}\in\mathbb{Z}^d} (E + 2\pi\boldsymbol{\nu}). \end{aligned}$$

Zhang (2007) showed that if φ is a scaling function and its Fourier transform has a bounded support G, then G must satisfy the following:

$$G \subset 2G,\quad \bigcup_{m\in\mathbb{Z}} 2^m G = \mathbb{R}^d,\quad G + 2\pi\mathbb{Z}^d = \mathbb{R}^d,\quad (G \setminus \frac{G}{2}) \bigcap (\frac{G}{2} + 2\pi\boldsymbol{\nu}) = \emptyset\ \ (\boldsymbol{\nu} \in \mathbb{Z}^d).$$

Conversely, if a bounded domain $G \subset \mathbb{R}^d$ satisfies these four conditions, then there exists a scaling function φ so that $\operatorname{supp}\widehat{\varphi} = G$.

The algorithm constructing orthogonal MRA from the transfer function:

Step 1. Choose the $2\pi\mathbb{Z}^d-$periodic continuous function $H(\boldsymbol{\omega})$ satisfying $\sum_{\boldsymbol{\nu}\in\{0,1\}^d} |H(\boldsymbol{\omega} + 2\pi\boldsymbol{\nu})|^2 = 1$ and $|H(2^{-m}\boldsymbol{\omega})| \geq K > 0\,(m > 1, \boldsymbol{\omega} \in E)$, where E and $[-\pi, \pi]^d$ is $2\pi\mathbb{Z}^d-$congruent.

Step 2. Define $\varphi(\mathbf{t})$ by the infinite product $\widehat{\varphi}(\boldsymbol{\omega}) = C \prod_{m\in\mathbb{Z}_+} H(2^{-m}\boldsymbol{\omega})$, where C is a constant such that $\widehat{\varphi}(\mathbf{0}) = 1$.

Step 3. Define $V_0 = \operatorname{span}\{\varphi(\mathbf{t} - \mathbf{n})\}_{\mathbf{n}\in\mathbb{Z}^d}$, i.e., V_0 is the closure of linear combination of $\{\varphi(\mathbf{t} - \mathbf{n})\}_{\mathbf{n}\in\mathbb{Z}^d}$. Again, define $V_m = \{f(2^m\mathbf{t}) : \; f \in V_0\}\ (m \in \mathbb{Z})$. An orthogonal MRA follows.

3.2 Multivariate Orthogonal Wavelets

A multivariate orthogonal wavelet basis is a normal orthogonal basis obtained by integral translations and dyadic dilations of several multivariate square integrable functions. Its construction is closely related to orthogonal MRAs.

3.2.1 Separable Wavelets

Let $\{V_m\}_{m\in\mathbb{Z}}$ be an one-dimensional MRA with an orthogonal scaling function $\varphi(t)$. The bi-scale equation and transfer function are, respectively,

$$\begin{aligned} \tfrac{1}{2}\varphi\left(\tfrac{t}{2}\right) &= \sum_{n\in\mathbb{Z}} a_n \varphi(t-n), \\ H_n(t) &= \sum_{n\in\mathbb{Z}} a_n\, e^{-int}. \end{aligned}$$

Then, the function $\psi(t)$ satisfying one of the following two equations:

$$\begin{aligned} \widehat{\psi}(\omega) &= e^{-\frac{i}{2}\omega}\overline{H}\left(\tfrac{\omega}{2} + \pi\right)\widehat{\varphi}\left(\tfrac{\omega}{2}\right), \\ \psi(t) &= -2\sum_{n\in\mathbb{Z}} (-1)^n \overline{a}_{1-n}\varphi(2t-n)) \end{aligned} \qquad (3.2.1)$$

is an *orthogonal wavelet* for W_0. Let

$$\psi_{m,n}(t) = 2^{\frac{m}{2}}\psi(2^m t - n) \qquad (m, n \in \mathbb{Z}).$$

Then, $\{\psi_{m,n}\}_{m,n\in\mathbb{Z}}$ is an *orthogonal wavelet basis* for $L^2(\mathbb{R})$.

There are two methods constructing tensor product wavelets.

Method 1. Let $\psi(t)$ be an one-dimensional orthogonal wavelet. Then, $\{\psi_{m,n}(t)\}_{m,n\in\mathbb{Z}}$ forms a normal orthogonal basis for $L^2(\mathbb{R})$. The tensor product of two one-dimensional wavelet bases is

$$\psi_{m_1,n_1,m_2,n_2}(t_1,t_2)=\psi_{m_1,n_1}(t_1)\psi_{m_2,n_2}(t_2)=2^{\frac{m_1+m_2}{2}}\psi(2^{m_1}t_1-n_1)\psi(2^{m_2}t_2-n_2).$$

Then, $\{\psi_{m_1,n_1,m_2,n_2}(t_1,t_2)\}_{m_1,m_2,n_1,n_2\in\mathbb{Z};t_1,t_2\in\mathbb{R}}$ is a *two-dimensional wavelet orthogonal basis* for $L^2(\mathbb{R}^2)$.

In general, the tensor product of d one-dimensional wavelet bases is

$$\psi_{\mathbf{m},\mathbf{n}}(\mathbf{t})=\prod_{k=1}^{d}\psi_{m_k,n_k}(t_k),$$

where $\mathbf{m}=(m_1,\ldots,m_d)$, $\mathbf{n}=(n_1,\ldots,n_d)$, and $\mathbf{t}=(t_1,\ldots,t_d)$. The $\{\psi_{\mathbf{m},\mathbf{n}}(\mathbf{t})\}_{\mathbf{m},\mathbf{n}\in\mathbb{Z}^d,\,\mathbf{t}\in\mathbb{R}^d}$ is a $d-$dimensional wavelet orthogonal basis for $L^2(\mathbb{R}^d)$.

Method 2. Let $\{V_m\}_{m\in\mathbb{Z}}$ be an one-dimensional MRA with orthogonal scaling function $\varphi(t)$ and wavelet $\psi(t)$. This leads us to define three wavelets:

$$\begin{aligned}\psi^{(1)}(\mathbf{t})&=\varphi(t_1)\psi(t_2),\\ \psi^{(2)}(\mathbf{t})&=\psi(t_1)\varphi(t_2),\\ \psi^{(3)}(\mathbf{t})&=\psi(t_1)\psi(t_2),\end{aligned}$$

where $\mathbf{t}=(t_1,t_2)$. Let

$$\psi_{m,\mathbf{n}}(\mathbf{t})=2^m\psi^{(\tau)}(2^m\mathbf{t}-\mathbf{n})\qquad (\mathbf{t}=(t_1,t_2),\ \mathbf{n}=(n_1,n_2),\ \tau=1,2,3).$$

Then, $\{\psi_{m,\mathbf{n}}(\mathbf{t})\}_{m\in\mathbb{Z},\,\mathbf{n}\in\mathbb{Z}^2,\tau=1,2,3}$ forms a wavelet orthogonal basis for $L^2(\mathbb{R}^2)$.

The one-dimensional wavelets used often are as follows.

(a) *Haar wavelet* is

$$\psi_H(t)=\begin{cases}-1 & \text{if } 0\le t<\frac{1}{2},\\ 1 & \text{if } \frac{1}{2}\le t<1,\\ 0 & \text{otherwise},\end{cases}\tag{3.2.2}$$

(b) *Shannon wavelet* is

$$\psi_S(t)=\frac{\sin 2\pi(t-\frac{1}{2})}{\pi(t-\frac{1}{2})}-\frac{\sin\pi(t-\frac{1}{2})}{\pi(t-\frac{1}{2})}.\tag{3.2.3}$$

The Shannon wavelet is infinitely differentiable but has a slow decay, $\psi(t)=O\left(\frac{1}{t}\right)$ as $t\to\infty$.

(c) The class of *Meyer wavelets* $\psi_M(t)$ satisfies

$$\widehat{\psi}_M(\omega)=\begin{cases}e^{-i\frac{\omega}{2}}\sin\left(\frac{\pi}{2}\theta(\frac{3}{2\pi}|\omega|-1)\right), & \frac{2\pi}{3}\le|\omega|\le\frac{4\pi}{3},\\ e^{-i\frac{\omega}{2}}\cos\left(\frac{\pi}{2}\theta(\frac{3}{4\pi}|\omega|-1)\right), & \frac{4\pi}{3}\le|\omega|\le\frac{8\pi}{3},\end{cases}\tag{3.2.4}$$

where $\theta(\omega)$ is stated in (3.1.1). Meyer wavelet is infinitely differentiable and fast decay.

(d) *Daubechies wavelet* is the more often used wavelet. It is a compactly supported and smooth wavelet. The construction of Daubechies wavelets is as follows.

Step 1. Construct a polynomial:

$$P(y) = \sum_{k=0}^{L-1} C_{L+k-1}^{k} y^k$$

where $L \in \mathbb{Z}_+$ and $C_m^n = \frac{m!}{n!(m-n)!}$.

Step 2. Use factorization method to construct a polynomial $Q(z)$ with real coefficients such that

$$|Q(e^{i\omega})|^2 = P(\sin^2 \frac{\omega}{2}).$$

Step 3. Construct a polynomial $m(z)$ such that

$$m(z) = \left(\frac{1+z}{2}\right)^L Q(z) = \sum_{n=0}^{N} a_n z^n \qquad (L \in \mathbb{Z}_+).$$

Step 4. Use iteration method to solve the bi-scale equation:

$$\frac{1}{2}\varphi\left(\frac{t}{2}\right) = \sum_{n \in \mathbb{Z}} a_n \varphi(t-n),$$

and find out the values of $\varphi(t)$ at dyadic decimals $\frac{k}{2^m}$, where $\sum_{k=1}^{N-1} \varphi(k) = 1$.

Step 5. Construct Daubechies wavelet:

$$\psi(t) = -2 \sum_{n=1-N}^{1} (-1)^n a_{1-n} \varphi(2t-n).$$

Find the values of $\psi(t)$ at dyadic decimals $\frac{k}{2^m}$.

Let $\{\psi_{m,n}\}_{m,n \in \mathbb{Z}}$ be an orthogonal wavelet basis. Then, for any $f \in L^2(\mathbb{R})$,

$$f(t) = \sum_{m \in \mathbb{Z}} \sum_{n \in \mathbb{Z}} d_{mn} \psi_{m,n}(t),$$

where $d_{mn} = (f, \psi_{mn}) = \int_{\mathbb{R}} f(t)\overline{\psi}_{mn}(t)\mathrm{d}t \ (m, n \in \mathbb{Z})$.

The wavelets have three key properties: *vanishing moment*, *support*, and *regularity*.

(a) Vanishing Moment

The wavelet ψ has p vanishing moments if $\int_{\mathbb{R}} t^k \psi(t)\mathrm{d}t = 0\,(0 \le k < p)$. Such wavelet ψ is orthogonal to any polynomial of degree $p-1$. If the scaling function and wavelet are both fast decay, then the following results are equivalent:

- ψ has p vanishing moments;
- $\widehat{\psi}(\omega)$ has a $p-$order zero at $\omega = 0$;
- $H(\omega)$ has a $p-$order zero at $\omega = \pi$.

(b) Support

If the transfer function is a trigonometric polynomial $H(\omega) = \sum_{n=N_1}^{N_2} a_n\, e^{-in\omega}$, then the corresponding scaling function and wavelet satisfy

$$\begin{aligned} \operatorname{supp} \varphi &= [N_1,\ N_2], \\ \operatorname{supp} \psi &= \left[\tfrac{N_1-N_2+1}{2},\ \tfrac{N_2-N_1+1}{2}\right]. \end{aligned}$$

If ψ has p vanishing moments, then its support is at least of size $2p-1$. Daubechies wavelets have a minimal size support for a given number of vanishing moments.

(c) Regularity

The smoothness of the wavelet ψ has an influence on the error introduced by thresholding or quantizing the wavelet coefficients.

The relationship among the smoothness, the decay (or the support), and the vanishing moment is as follows: If a wavelet ψ is $\gamma-$time continuously, differentiable, and $\psi(t) = O(1+|t|)^{-\gamma-1-\eta}$, then its vanishing moment $p \ge \gamma$.

Let $\{V_m\}_{m\in\mathbb{Z}}$ be an one-dimensional Haar MRA. Its scaling function is $\varphi_H(t) = \chi_{[0,1]}(t)$, where $\chi_{[0,1]}(t)$ is the characteristic function of $[0, 1]$. The Haar wavelet ψ_H is stated in (3.2.2). The corresponding two-dimensional tensor product Haar wavelets are

$$\begin{aligned} \psi_1^H(\mathbf{t}) &= \varphi_H(t_1)\psi_H(t_2), \\ \psi_2^H(\mathbf{t}) &= \psi_H(t_1)\varphi_H(t_2), \\ \psi_3^H(\mathbf{t}) &= \psi_H(t_1)\psi_H(t_2) \qquad (\mathbf{t} = (t_1, t_2)). \end{aligned}$$

Let $\{V_m\}_{m\in\mathbb{Z}}$ be an one-dimensional Shannon MRA. Its scaling function is $\varphi_S(t) = \frac{\sin(\pi t)}{\pi t}$. The Shannon wavelet $\psi_S(t)$ is stated in (3.2.3). The corresponding two-dimensional tensor product Shannon wavelets are

$$\begin{aligned} \psi_1^S(\mathbf{t}) &= \varphi_S(t_1)\psi_S(t_2), \\ \psi_2^S(\mathbf{t}) &= \psi_S(t_1)\varphi_S(t_2), \\ \psi_3^S(\mathbf{t}) &= \psi_S(t_1)\psi_S(t_2) \qquad (\mathbf{t} = (t_1, t_2)). \end{aligned}$$

Let $\{V_m\}_{m\in\mathbb{Z}}$ be an one-dimensional Meyer MRA with scaling function $\varphi_M(t)$ satisfying $\widehat{\varphi}_M(\omega) = \cos\left(\frac{\pi}{2}\theta(\frac{3}{2\pi}(|\omega|-1))\right)$, where $\theta(\omega)$ is stated in (3.1.1) and Meyer wavelet $\psi_M(t)$ is stated in (3.2.4). The corresponding two-dimensional tensor product Meyer wavelets are

$$\psi_1^M(\mathbf{t}) = \varphi_M(t_1)\psi_M(t_2),$$
$$\psi_2^M(\mathbf{t}) = \psi_M(t_1)\varphi_M(t_2),$$
$$\psi_3^M(\mathbf{t}) = \psi_M(t_1)\psi_M(t_2) \qquad (\mathbf{t} = (t_1, t_2)).$$

3.2.2 Non-separable Wavelets

Let $\{V_m\}_{m\in\mathbb{Z}}$ be a $d-$dimensional orthogonal MRA with scaling function $\varphi(\mathbf{t})$ ($\mathbf{t} \in \mathbb{R}^d$). The $\{\varphi(\mathbf{t}-\mathbf{k})\}_{\mathbf{k}\in\mathbb{Z}^d}$ forms a normal orthogonal basis for V_0. Let $V_1 = V_0 \bigoplus W_0$. It can be proved that there exist $2^d - 1$ functions $\psi_{\boldsymbol{\mu}}(\mathbf{t})$ $(\boldsymbol{\mu} \in \{0,1\}^d \setminus \{\mathbf{0}\})$ such that $\{\psi_{\mu}(\mathbf{t}-\mathbf{n})\}_{\mu,\mathbf{n}}$ $(\boldsymbol{\mu} \in \{0,1\}^d \setminus \{\mathbf{0}\},\ \ \mathbf{n} \in \mathbb{Z}^d)$ forms a normal orthogonal basis for W_0. The system $\{\psi_\mu\}$ is called a $d-$dimensional orthogonal wavelet, and their dyadic dilations and integral translations form a normal orthogonal basis for $L^2(\mathbb{R}^d)$. Below, we give two algorithms to find out the orthogonal wavelets ψ_μ $(\boldsymbol{\mu} \in \{0,1\}^d \setminus \{\mathbf{0}\})$.

Algorithm 1 Let $H_{\mathbf{0}}(\boldsymbol{\omega})$ be the transfer function corresponding to the orthogonal scaling function $\varphi(\mathbf{t})$, and $H_{\mathbf{0}}(\boldsymbol{\nu}\pi) = \delta_{\mathbf{0},\nu}$ $(\boldsymbol{\nu} \in \{0,1\}^d)$. Notice that the transfer function $H_{\mathbf{0}}(\boldsymbol{\omega})$ satisfies

$$\sum_{\nu\in\{0,1\}^d} |H_{\mathbf{0}}(\boldsymbol{\omega} + \pi\boldsymbol{\nu})|^2 = 1,$$

i.e., the $(H_{\mathbf{0}}(\boldsymbol{\omega} + \pi\boldsymbol{\nu}))_{\nu\in\{0,1\}^d}$ is a unit vector. Choose $(2^d - 1)$ periodic continuous functions H_μ $(\boldsymbol{\mu} \in \{0,1\}^d \setminus \{\mathbf{0}\})$ with period $2\pi\mathbb{Z}^d$ so that the matrix:

$$(H_\mu(\boldsymbol{\omega} + \boldsymbol{\nu}\pi))_{\mu,\nu} = \begin{pmatrix} H_{\mu_0}(\boldsymbol{\omega}) & H_{\mu_0}(\boldsymbol{\omega} + \boldsymbol{\nu}_1\pi) & \cdots & H_{\mu_0}(\boldsymbol{\omega} + \boldsymbol{\nu}_{2^d-1}\pi) \\ H_{\mu_1}(\boldsymbol{\omega}) & H_{\mu_1}(\boldsymbol{\omega} + \boldsymbol{\nu}_1\pi) & \cdots & H_{\mu_1}(\boldsymbol{\omega} + \boldsymbol{\nu}_{2^d-1}\pi) \\ \vdots & \vdots & \vdots & \vdots \\ H_{\mu_{2^d-1}}(\boldsymbol{\omega}) & H_{\mu_{2^d-1}}(\boldsymbol{\omega} + \boldsymbol{\nu}_1\pi) & \cdots & H_{\mu_{2^d-1}}(\boldsymbol{\omega} + \boldsymbol{\nu}_{2^d-1}\pi) \end{pmatrix}$$

is a unitary matrix, i.e.,

$$(H_\mu(\boldsymbol{\omega} + \boldsymbol{\nu}\pi))_{\mu,\nu}(H_\mu(\boldsymbol{\omega} + \boldsymbol{\nu}\pi))^*_{\mu,\nu} = I,$$

where I is the $d-$order unit matrix and $*$ represents the conjugate transposed matrix. Define wavelet functions $\{\psi_\mu\}_{\mu\in\{0,1\}^d\setminus\{\mathbf{0}\}}$ satisfying

$$\widehat{\psi}_\mu(\boldsymbol{\omega}) = H_\mu\left(\frac{\boldsymbol{\omega}}{2}\right)\widehat{\varphi}\left(\frac{\boldsymbol{\omega}}{2}\right) \qquad (\boldsymbol{\mu} \in \{0,1\}^d \setminus \{\mathbf{0}\}).$$

Then, $\{\psi_{\mu,m,\mathbf{n}}\}_{\mu\in\{0,1\}^d\setminus\{\mathbf{0}\},\, m\in\mathbb{Z},\mathbf{n}\in\mathbb{Z}^d}$ is an orthogonal wavelet basis associated with the orthogonal scaling function $\varphi(\mathbf{t})$.

Algorithm 2 Let $\varphi(\mathbf{t})$ be a $d-$dimensional orthogonal scaling function with skew symmetric about $\frac{\mathbf{c}}{2}$, i.e., $\varphi\left(\frac{\mathbf{c}}{2} + \mathbf{t}\right) = \overline{\varphi}\left(\frac{\mathbf{c}}{2} - \mathbf{t}\right)$. By the bi-scale equation,

$$\varphi(\mathbf{t}) = 2^d \sum_{\mathbf{n}\in\mathbb{Z}^d} a_{\mathbf{n}}\varphi(2\mathbf{t}-\mathbf{n})\ (\mathbf{t}\in\mathbb{R}^d).$$

Step 1. Let $\varphi_{\boldsymbol{\nu}}(\mathbf{t}) = 2^{\frac{d}{2}}\varphi(2\mathbf{t}-\boldsymbol{\nu})$ $(\boldsymbol{\nu}\in\{0,1\}^d)$.

Step 2. Construct a unit row rector $(p_{\mathbf{0},\boldsymbol{\nu}}(\boldsymbol{\omega}))_{\boldsymbol{\nu}\in\{0,1\}^d}$, where $p_{\mathbf{0},\boldsymbol{\nu}}(\boldsymbol{\omega}) = \sum_{\mathbf{k}\in\mathbb{Z}^d} 2^{\frac{d}{2}} a_{2\mathbf{k}-\boldsymbol{\nu}}\, e^{-i(\mathbf{k}\cdot\boldsymbol{\omega})}$.

Step 3. Denote $p_{\boldsymbol{\nu}}(\boldsymbol{\omega}) = p_{\mathbf{0},\boldsymbol{\nu}}(\boldsymbol{\omega})$. Let

$$P_{\boldsymbol{\mu},\boldsymbol{\nu}}(\boldsymbol{\omega}) = \begin{cases} \frac{\overline{p}_{\boldsymbol{\mu}}(1+p_{\mathbf{0}})}{1+\overline{p}_{\mathbf{0}}}, & \boldsymbol{\mu}\in\{0,1\}^d\setminus\{\mathbf{0}\},\ \boldsymbol{\nu}=\mathbf{0} \\ -\delta_{\boldsymbol{\mu},\boldsymbol{\nu}} + \frac{\overline{p}_{\boldsymbol{\mu}}p_{\boldsymbol{\nu}}}{1+\overline{p}_{\mathbf{0}}}, & \boldsymbol{\mu},\boldsymbol{\nu}\in\{0,1\}^d\setminus\{\mathbf{0}\}. \end{cases} \qquad (3.2.5)$$

Since $\varphi(\mathbf{t})$ is skew symmetric, it can be proved that the matrix $\{P_{\boldsymbol{\mu},\boldsymbol{\nu}}(\boldsymbol{\omega})\}$ $(\boldsymbol{\mu},\boldsymbol{\nu}\in\{0,1\}^d)$ is a unitary matrix.

Step 4. Define $\{\psi_{\boldsymbol{\mu}}(\mathbf{t})\}_{\boldsymbol{\mu}\in\{0,1\}^d\setminus\{\mathbf{0}\}}$ satisfying

$$\widehat{\psi}_{\boldsymbol{\mu}}(\boldsymbol{\omega}) = \sum_{\boldsymbol{\nu}\in\{0,1\}^d\setminus\{\mathbf{0}\}} P_{\boldsymbol{\mu},\boldsymbol{\nu}}(\boldsymbol{\omega})\widehat{\varphi}_{\boldsymbol{\nu}}(\boldsymbol{\omega}).$$

Then, $\{\psi_{\boldsymbol{\mu},m,\mathbf{n}}(\mathbf{t})\}_{\boldsymbol{\mu}\in\{0,1\}^d\setminus\{\mathbf{0}\},\ m\in\mathbb{Z},\ \mathbf{n}\in\mathbb{Z}^d}$ is the desired orthogonal wavelet basis associated with the orthogonal scaling function $\varphi(\mathbf{t})$.

Polyharmonic Box-Spline Wavelet. Let $g(\mathbf{t})$ be a polyharmonic box-spline satisfying

$$\widehat{g}(\boldsymbol{\omega}) = 2^{2r}\left(\sum_{\nu=1}^{d}\sin^2\frac{\omega_\nu}{2}\right)^r\left(\sum_{\nu=1}^{d}\omega_\nu^2\right)^{-r} \qquad (\boldsymbol{\omega}\in\mathbb{R}^d,\ r>\frac{d}{2}),$$

where $\boldsymbol{\omega} = (\omega_1,\ldots,\omega_d)$. Let $\varphi(\mathbf{t})$ satisfy

$$\widehat{\varphi}(\boldsymbol{\omega}) = \widehat{g}(\boldsymbol{\omega})\left(\sum_{\boldsymbol{\alpha}\in\mathbb{Z}^d}|\widehat{g}(\boldsymbol{\omega}+2\pi\boldsymbol{\alpha})|^2\right)^{-\frac{1}{2}} \qquad (\boldsymbol{\omega}\in\mathbb{R}^d).$$

Then, $\varphi(\mathbf{t})$ is an orthogonal scaling function. Note that $\varphi(\mathbf{t})$ is a skew symmetric function about $\mathbf{0}$. Use (3.2.5) to obtain real-valued orthogonal wavelets $\psi_{\boldsymbol{\mu}}(\mathbf{t})$ $(\boldsymbol{\mu}\in\{0,1\}^d\setminus\{\mathbf{0}\})$. So $\psi_{\boldsymbol{\mu}}(\mathbf{t})$ $(\boldsymbol{\mu}\in\{0,1\}^d\setminus\{\mathbf{0}\})$ decay fast and $\{\psi_{\boldsymbol{\mu},m,\mathbf{n}}(\mathbf{t})\}_{\boldsymbol{\mu}}\in\{0,1\}^d\setminus\{\mathbf{0}\}$, $m\in\mathbb{Z}$, $\mathbf{n}\in\mathbb{Z}^d$ is an orthogonal wavelet basis for $L^2(\mathbb{R}^d)$.

Algorithm 3 In the case of two or three dimensions, there is an one-to-one map $\mathcal{X}$ satisfying $\mathcal{X}(\mathbf{0}) = 0$ and $(\mathcal{X}(\boldsymbol{\mu})+\mathcal{X}(\boldsymbol{\nu}))(\boldsymbol{\mu}+\boldsymbol{\nu})$ is odd for $\boldsymbol{\mu},\boldsymbol{\nu}\in\{0,1\}^d$. Let $\varphi(\mathbf{t})$ be a scaling function for an orthogonal MRA and skew symmetric about $\frac{\mathbf{c}}{2}\in\frac{1}{2}\mathbb{Z}^d$, and let $H_0(\boldsymbol{\omega})$ be the corresponding transfer function. Define

$$H_{\boldsymbol{\mu}}(\boldsymbol{\omega}) = \begin{cases} e^{i\mathcal{X}(\boldsymbol{\mu})\cdot\boldsymbol{\omega}} H_{\mathbf{0}}(\boldsymbol{\omega}+\pi\boldsymbol{\mu}) & \text{if } \boldsymbol{\mu}, \mathbf{c} \text{ is even}, \quad \boldsymbol{\mu} \neq \mathbf{0}, \\ e^{i\mathcal{X}(\boldsymbol{\mu})\cdot\boldsymbol{\omega}} \overline{H}_{\mathbf{0}}(\boldsymbol{\omega}+\pi\boldsymbol{\mu}) & \text{if } \boldsymbol{\mu}, \mathbf{c} \text{ is odd}, \quad \boldsymbol{\mu} \neq \mathbf{0}. \end{cases}$$

It can be proved that $(H_{\boldsymbol{\mu}}(\boldsymbol{\omega}+\pi\boldsymbol{\nu}))_{\boldsymbol{\mu},\boldsymbol{\nu}\in\{0,1\}^d\setminus\{\mathbf{0}\}}$ is a unitary matrix. Let $\psi_{\boldsymbol{\mu}}(\mathbf{t})$ ($\boldsymbol{\mu} \in \{0,1\}^d \setminus \{\mathbf{0}\}$) satisfy $\widehat{\psi}_{\boldsymbol{\mu}}(\boldsymbol{\omega}) = H_{\boldsymbol{\mu}}\left(\frac{\boldsymbol{\omega}}{2}\right)\widehat{\varphi}\left(\frac{\boldsymbol{\omega}}{2}\right)$. Then, $\{\psi_{\boldsymbol{\mu},m,\mathbf{n}}(\mathbf{t})\}_{\boldsymbol{\mu}\in\{0,1\}^d\setminus\{\mathbf{0}\}}$; $m \in \mathbb{Z}$; $\mathbf{n} \in \mathbb{Z}^d$ is an orthogonal wavelet basis associated with the scaling function $\varphi(\mathbf{t})$ for $L^2(\mathbb{R}^d)$.

Algorithm 4 Zhang (2010) Start from an one-dimensional smooth scaling function $h(t,\alpha)$ with $\operatorname{supp}\widehat{h}(\omega,\alpha) = [-\alpha,\alpha]$. Arbitrarily choose two real-valued even functions $\beta_\nu(\eta)$ ($\nu = 1, 2$) satisfying the following:

$$\begin{array}{ll} \pi \leq \beta_\nu(\eta) \leq \frac{\pi}{3}\pi & (\eta \in \mathbb{R}), \\ \beta_\nu(\eta) = b_\nu & (|\eta| \geq \frac{2\pi}{3}), \end{array}$$

where each b_ν is a constant. Let $h(\alpha)$ satisfy the condition:

$$\widehat{h}(\eta,\alpha) = \cos\left(\frac{\pi}{2}\zeta_\alpha\left(\frac{|\eta|}{2\pi-\alpha} - 1\right)\right) \qquad (\eta \in \mathbb{R}),$$

where

$$\begin{array}{ll} \zeta_\alpha(\eta) = \begin{cases} 0, & \eta \leq 0, \\ 1, & \eta > \kappa(\alpha), \end{cases} & \kappa(\alpha) = \frac{2\alpha-2\pi}{2\pi-\alpha}, \\ 0 < \zeta_\alpha(\eta) < 1 & (0 < \eta < \kappa(\alpha)), \\ \zeta_\alpha(\eta) + \zeta_\alpha(\kappa(\alpha)-\eta) = 1 \ (\eta \in \mathbb{R}). & \end{array}$$

Define a non-tensor product scaling function φ as

$$\widehat{\varphi}(\omega) = \widehat{h}(\omega_1, \beta_1(\omega_2))\widehat{h}(\omega_2, \beta_2(\omega_1)) \qquad (\boldsymbol{\omega} = (\omega_1, \omega_2)).$$

For each $\boldsymbol{\mu} = (\mu_1, \mu_2) \in \{0,1\}^2 \setminus (0,0)$, define a wavelet $\psi_{\boldsymbol{\mu}}$ as

$$\widehat{\psi}_{\boldsymbol{\mu}}(\boldsymbol{\omega}) = e^{i\frac{1}{2}(\boldsymbol{\mu}\cdot\boldsymbol{\omega})}\widehat{\varphi}(|\omega_1| - 2\pi\mu_1, \ |\omega_2| - 2\pi\mu_2)\widehat{\varphi}\left(\frac{|\omega_1|}{2}, \frac{|\omega_2|}{2}\right) \quad (\boldsymbol{\omega} = (\omega_1, \omega_2) \in \mathbb{R}^2).$$

In more detail,

$$\begin{aligned} \widehat{\psi}_{(0,1)}(\boldsymbol{\omega}) &= e^{i\frac{\omega_2}{2}}\,\widehat{\varphi}(\omega_1, 2\pi - |\omega_2|)\,\widehat{\varphi}\left(\tfrac{\omega}{2}\right), \\ \widehat{\psi}_{(1,0)}(\boldsymbol{\omega}) &= e^{i\frac{\omega_1}{2}}\,\widehat{\varphi}(2\pi - |\omega_1|, |\omega_2|)\,\widehat{\varphi}\left(\tfrac{\omega}{2}\right), \\ \widehat{\psi}_{(1,1)}(\boldsymbol{\omega}) &= e^{i\frac{\omega_1+\omega_2}{2}}\,\widehat{\varphi}(2\pi - |\omega_1|, 2\pi - |\omega_2|)\,\widehat{\varphi}\left(\tfrac{\omega}{2}\right). \end{aligned}$$

Then, $\psi_{\boldsymbol{\mu}}$ ($\boldsymbol{\mu} \in \{0,1\}^2 \setminus (0,0)$) is the non-tensor product wavelets with high regularity and various symmetry (i.e., axial symmetry, central symmetry, and cyclic symmetry).

3.2.3 $p-$Band Wavelets

Let p be an integer and $p \geq 2$. A sequence of closed subspaces $\{V_m\}_{m\in\mathbb{Z}}$ in $L^2(\mathbb{R})$ is called an *$p-$band multiresolution analysis* if

(i) $V_m \subset V_{m+1}$ $(m \in \mathbb{Z})$, $\overline{\bigcup_{m\in\mathbb{Z}} V_m} = L^2(\mathbb{R})$, $\bigcap_{m\in\mathbb{Z}} V_m = \{0\}$;

(ii) $f(t) \in V_m$ if and only if $f(pt) \in V_{m+1}$ $(m \in \mathbb{Z})$;

(iii) There is a *scaling function* $\varphi^{(p)}$ so that $\{\varphi^{(p)}(t-n)\}_{\mathbf{n}\in\mathbb{Z}}$ is a normal orthogonal basis of V_0.

The $p-$band bi-scale equation is

$$\frac{1}{p}\varphi^{(p)}\left(\frac{t}{p}\right) = \sum_{n\in\mathbb{Z}} a_n \varphi^{(p)}(t-n).$$

Denote $H_0^{(p)}(\omega) = \sum_{n\in\mathbb{Z}} a_n\, e^{-in\omega}$. So $\sum_{l=0}^{p-1} |H_0^{(p)}(\omega + \frac{2\pi l}{p})|^2 = 1$, i.e., the vector $(H_0^{(p)}(\omega + \frac{2\pi l}{p}))_{l=0,\ldots,p-1}$ is a unit vector.

Construct $2\pi-$periodic functions $H_\mu^{(p)}(\omega)\,(\mu = 1, \ldots, p-1)$ such that the matrix:

$$\left(H_\mu^{(p)}(\omega + \frac{2\pi l}{p})\right)_{\mu,l=0,\ldots,p-1} = \begin{pmatrix} H_0^{(p)}(\omega) & H_0^{(p)}(\omega + \frac{2\pi}{p}) & \cdots & H_0^{(p)}(\omega + \frac{2\pi(p-1)}{p}) \\ H_1^{(p)}(\omega) & H_1^{(p)}(\omega + \frac{2\pi}{p}) & \cdots & H_1^{(p)}(\omega + \frac{2\pi(p-1)}{p}) \\ \vdots & \vdots & \ddots & \vdots \\ H_{p-1}^{(p)}(\omega) & H_{p-1}^{(p)}(\omega + \frac{2\pi}{p}) & \cdots & H_{p-1}^{(p)}(\omega + \frac{2\pi(p-1)}{p}) \end{pmatrix}$$

is a unitary matrix, i.e.,

$$\left(H_\mu^{(p)}(\omega + \frac{2\pi l}{p})\right)_{\mu,l=0,\ldots,p-1} \left(H_\mu^{(p)}(\omega + \frac{2\pi l}{p})\right)^*_{\mu,l=0,\ldots,p-1} = I,$$

where $*$ is conjugate transpose. Let

$$\widehat{\psi}_\mu^{(p)}(\omega) = H_\mu^{(p)}\left(\frac{\omega}{p}\right)\widehat{\varphi}_\mu^{(p)}\left(\frac{\omega}{p}\right) \qquad (\mu = 0, \ldots, p-1).$$

Then, $\{\psi_\mu^{(p)}(t)\}_{\mu=0,\ldots,p-1}$ is an orthogonal wavelet with dilation p. Let

$$\psi_{\mu,m,n}^{(p)}(t) = p^{-\frac{m}{2}}\psi_\mu^{(p)}(p^m t - n) \qquad (\mu = 0, \ldots, p-1;\ m, n \in \mathbb{Z}).$$

Then, $\{\psi_{\mu,m,n}^{(p)}(t)\}_{\mu=0,\ldots,p-1;m,n\in\mathbb{Z}}$ is a normal orthogonal wavelet basis for $L^2(\mathbb{R})$ with dilation p.

Example Let

$$H_0^{(3)}(\omega) = -\frac{1}{81}(1 + e^{-i\omega} + e^{-2i\omega})(1 + 2e^{-i\omega} - 15e^{-2i\omega} + 2e^{-3i\omega} + e^{-4i\omega}).$$

A compactly supported solution $\varphi^{(3)}(t)$ is obtained by solving the following three-band bi-scale equation:

$$\begin{cases} \widehat{\varphi}^{(3)}(\omega) = H_0^{(3)}\left(\frac{\omega}{3}\right)\widehat{\varphi}^{(3)}\left(\frac{\omega}{3}\right), \\ \widehat{\varphi}^{(3)}(0) = 1, \end{cases}$$

Let

$$V_m = \left\{ \sum_{n\in\mathbb{Z}} c_n 3^{\frac{m}{2}} \varphi^{(3)}(3^m t - n) : \sum_{n\in\mathbb{Z}} |c_n|^2 < \infty \right\}.$$

Then, $\{V_m\}_{m\in\mathbb{Z}}$ forms an MRA with dilation 3. Define

$$\begin{aligned} H_1^{(3)}(\omega) = \frac{1}{81\sqrt{2}} &\left(-5 - 20e^{-i\omega} + 40e^{-2i\omega} - 8e^{-3i\omega} - 14e^{-4i\omega} - 8e^{-5i\omega}\right. \\ &\left. + 40e^{-6i\omega} - 20e^{-7i\omega} - 5e^{-8i\omega}\right), \\ H_2^{(3)}(\omega) = \frac{\sqrt{6}}{54} &\left(-1 - 4e^{-i\omega} + 8e^{-2i\omega} - 8e^{-6i\omega} + 4e^{-7i\omega} + e^{-8i\omega}\right). \end{aligned}$$

Then, the matrix

$$\begin{pmatrix} H_0^{(3)}(\omega) & H_0\left(\omega + \frac{2\pi}{3}\right) & H_0^{(3)}\left(\omega + \frac{4\pi}{3}\right) \\ H_1^{(3)}(\omega) & H_1\left(\omega + \frac{2\pi}{3}\right) & H_1^{(3)}\left(\omega + \frac{4\pi}{3}\right) \\ H_2^{(3)}(\omega) & H_2\left(\omega + \frac{2\pi}{3}\right) & H_2^{(3)}\left(\omega + \frac{4\pi}{3}\right) \end{pmatrix}$$

is a unitary matrix. Define

$$\begin{aligned} \widehat{\psi}_1^{(3)}(\omega) &= H_1^{(3)}\left(\frac{\omega}{3}\right)\widehat{\varphi}^{(3)}\left(\frac{\omega}{3}\right), \\ \widehat{\psi}_2^{(3)}(\omega) &= H_2^{(3)}\left(\frac{\omega}{3}\right)\widehat{\varphi}^{(3)}\left(\frac{\omega}{3}\right). \end{aligned}$$

Then, $\psi_1^{(3)}(t)$ and $\psi_2^{(3)}(t)$ are two orthogonal wavelets with dilation 3. Let

$$\psi_{\mu,m,n}^{(3)}(t) = 3^{-\frac{m}{2}}\psi_\mu^{(3)}(3^m t - n) \qquad (\mu = 1, 2;\ m, n \in \mathbb{Z}).$$

Then, $\{\psi_{\mu,m,n}^{(3)}(t)\}_{\mu=1,2;m,n\in\mathbb{Z}}$ forms a normal orthogonal basis for $L^2(\mathbb{R})$, where $\psi_1^{(3)}(t)$ is a symmetric, compactly supported wavelet and $\psi_2^{(3)}(t)$ is an antisymmetric, compactly supported wavelet.

3.3 Biorthogonal Wavelets

The constructions of smooth, compactly supported orthogonal wavelets are rather difficult. Moreover, the smooth compactly supported 2-band orthogonal wavelets are not symmetry and antisymmetry. Hence, one turns to consider biorthogonal wavelets. Constructing biorthogonal wavelets allows more degrees of freedom than orthogonal wavelets.

3.3.1 *Univariate Biorthogonal Wavelets*

Let $\psi, \widetilde{\psi} \in L^2(\mathbb{R})$ and

$$\begin{aligned}\psi_{m,n}(t) &= 2^{\frac{m}{2}}\psi(2^m t - n),\\ \widetilde{\psi}_{m,n}(t) &= 2^{\frac{m}{2}}\widetilde{\psi}(2^m t - n).\end{aligned}$$

If $\{\psi_{m,n}\}_{m,n\in\mathbb{Z}}$, $\{\widetilde{\psi}_{m,n}\}_{m,n\in\mathbb{Z}}$ are both Riesz bases for $L^2(\mathbb{R})$ and

$$\int_{\mathbb{R}} \psi_{m,n}(t)\overline{\widetilde{\psi}}_{m',n'}(t)\mathrm{d}t = \delta_{m,m'}\delta_{n,n'} \qquad (m, m', n, n' \in \mathbb{Z}),$$

then ψ and $\widetilde{\psi}$ are *a pair of dual wavelets* and $\{\psi_{m,n}\}$, $\{\widetilde{\psi}_{m,n}\}$ are *a pair of biorthogonal wavelet bases.*

The construction of biorthogonal wavelets can begin with transfer functions. Assume that both trigonometric polynomials $H(\omega)$ and $\widetilde{H}(\omega)$ satisfy

$$\begin{aligned}&\overline{H}(\omega)\widetilde{H}(\omega) + \overline{H}(\omega+\pi)\widetilde{H}(\omega+\pi) = 1,\\ &H(0) = \widetilde{H}(0) = 1,\end{aligned}$$

and $\inf_{\omega\in[-\frac{\pi}{2},\frac{\pi}{2}]} |H(\omega)| > 0$ and $\inf_{\omega\in[-\frac{\pi}{2},\frac{\pi}{2}]} |\widetilde{H}(\omega)| > 0$. Define the scaling functions as

$$\widehat{\varphi}(\omega) = \prod_{\nu=1}^{\infty} H(2^{-\nu}\omega), \qquad \widehat{\widetilde{\varphi}}(\omega) = \prod_{\nu=1}^{\infty} \widetilde{H}(2^{-\nu}\omega).$$

If there exist two strictly positive trigonometric polynomials $P(e^{i\omega})$ and $\widetilde{P}(e^{i\omega})$ such that

$$\begin{aligned}|H\left(\tfrac{\omega}{2}\right)|^2 P(e^{i\frac{\omega}{2}}) + |H\left(\tfrac{\omega}{2}+\pi\right)|^2 P(e^{i\frac{\omega}{2}+\pi}) &= 2P(e^{i\omega}),\\ |\widetilde{H}\left(\tfrac{\omega}{2}\right)|^2 \widetilde{P}(e^{i\frac{\omega}{2}}) + |\widetilde{H}\left(\tfrac{\omega}{2}+\pi\right)|^2 \widetilde{P}(e^{i\frac{\omega}{2}+\pi}) &= 2\widetilde{P}(e^{i\omega}),\end{aligned}$$

then the functions $\widehat{\varphi}(\omega)$, $\widehat{\widetilde{\varphi}}(\omega) \in L^2(\mathbb{R})$ and φ, $\widetilde{\varphi}$ satisfy the biorthogonal relation $\int_{\mathbb{R}} \varphi(t)\overline{\widetilde{\varphi}}(t-n)\mathrm{d}t = \delta_{0,n}$.

Let $G(\omega) = e^{-i\omega}\overline{\widetilde{H}}(\omega+\pi)$ and $\widetilde{G}(\omega) = e^{-i\omega}\overline{H}(\omega+\pi)$. Define

$$\widehat{\psi}(\omega) = G\left(\frac{\omega}{2}\right) H\left(\frac{\omega}{2}\right), \qquad \widehat{\widetilde{\psi}}(\omega) = \widetilde{G}\left(\frac{\omega}{2}\right) \widetilde{H}\left(\frac{\omega}{2}\right).$$

Then, ψ, $\widetilde{\psi}$ are a pair of dual wavelets, and $\{\psi_{m,n}(t)\}_{m,n\in\mathbb{Z}}$ and $\{\widetilde{\psi}_{m,n}(t)\}_{m,n\in\mathbb{Z}}$ are biorthogonal wavelet bases for $L^2(\mathbb{R})$.

The biorthogonal scaling functions and wavelets have the following properties.

If $H(\omega) = \sum_{n=N_1}^{N_2} a_n\, e^{-int}$ and $\widetilde{H}(\omega) = \sum_{n=\widetilde{N}_1}^{\widetilde{N}_2} \widetilde{a}_n\, e^{-int}$, then

$$\begin{aligned} &\operatorname{supp} \varphi = [N_1, N_2], && \operatorname{supp} \widetilde{\varphi} = [\widetilde{N}_1,\ \widetilde{N}_2],\\ &\operatorname{supp} \psi = \left[\tfrac{N_1-\widetilde{N}_2+1}{2},\ \tfrac{N_2-\widetilde{N}_1+1}{2}\right], && \operatorname{supp} \widetilde{\psi} = \left[\tfrac{\widetilde{N}_1-N_2+1}{2},\ \tfrac{\widetilde{N}_2-N_1+1}{2}\right]. \end{aligned}$$

The number of vanishing moments of ψ is equal to the number of zeros of $\widetilde{H}(\omega)$ at π. Similarly, the number of vanishing moments of $\widetilde{\psi}$ is also equal to the number of zeros of $H(\omega)$ at π.

If $H(\omega)$, $\widetilde{H}(\omega)$ have a zero of orders p, $\widetilde{p}$ at π, respectively, and

$$M = \sup_{\omega\in[-\pi,\pi]} |H(\omega)|, \qquad \widetilde{M} = \sup_{\omega\in[-\pi,\pi]} |\widetilde{H}(\omega)|$$

and there is $2\pi-$periodic functions $S(\omega)$, $\widetilde{S}(\omega)$ such that

$$H(\omega) = \left(\frac{1+e^{-i\omega}}{2}\right)^p S(\omega), \qquad \widetilde{H}(\omega) = \left(\frac{1+e^{-i\omega}}{2}\right)^{\widetilde{p}} \widetilde{S}(\omega),$$

and if $\alpha \in \mathbb{Z}_+$ and $\widetilde{\alpha} \in \mathbb{Z}_+$ are such that $\alpha < p - \log_2 M - 1$ and $\widetilde{\alpha} < \widetilde{p} - \log_2 \widetilde{M} - 1$, then $\widetilde{\varphi}$ and $\widetilde{\psi}$ are α time and $\widetilde{\alpha}$ time continuously differentiable, respectively.

It is impossible to construct smooth, compactly supported 2-band orthogonal wavelets which are either symmetric or antisymmetric. However, this is possible for biorthogonal wavelets. If $H(\omega)$, $\widetilde{H}(\omega)$ have an odd number of nonzero coefficients and are symmetric about $n = 0$, then $\varphi(t)$, $\widetilde{\varphi}(t)$ are symmetric about $t = 0$ and $\psi(t)$, $\widetilde{\psi}(t)$ are symmetric with respect to a shifted center. If $H(\omega)$, $\widetilde{H}(\omega)$ have an even number of nonzero coefficients and are symmetric about $n = \frac{1}{2}$, then $\varphi(t)$, $\widetilde{\varphi}(t)$ are symmetric about $t = \frac{1}{2}$ and $\psi(t)$, $\widetilde{\psi}(t)$ are antisymmetric with respect to a shifted center.

Now, we construct a biorthogonal wavelet with minimal support.

A *pair of trigonometric polynomials* $H(\omega)$, $\widetilde{H}(\omega)$ are defined as

$$\begin{aligned} H(\omega) &= e^{-i\epsilon\frac{\omega}{2}} \left(\cos\tfrac{\omega}{2}\right)^p L(\omega),\\ \widetilde{H}(\omega) &= e^{-i\epsilon\frac{\omega}{2}} \left(\cos\tfrac{\omega}{2}\right)^{\widetilde{p}} \widetilde{L}(\omega), \end{aligned} \tag{3.3.1}$$

where $\epsilon = 0$ if p and $\widetilde{p}$ are even and $\epsilon = 1$ if p and $\widetilde{p}$ are odd, and

$$L(\cos\omega)\widetilde{L}(\cos\omega) = \sum_{k=0}^{q-1} C_{q-1+k}^{k} \sin^{2k}\frac{\omega}{2} \qquad \left(q = \frac{p+\widetilde{p}}{2}\right).$$

Let $H(\omega) = \sum_{n=-p}^{p} a_n\, e^{-in\omega}$ and $\widetilde{H}(\omega) = \sum_{n=-\widetilde{p}}^{\widetilde{p}} \widetilde{a}_n\, e^{-in\omega}$. Numerically solving the bi-scale equations:

$$\begin{aligned}
\tfrac{1}{2}\varphi(t) &= \sum_{n=-p}^{p} a_n \varphi(2t-n),\\
\tfrac{1}{2}\widetilde{\varphi}(t) &= \sum_{n=-\widetilde{p}}^{\widetilde{p}} \widetilde{a}_n \widetilde{\varphi}(2t-n),
\end{aligned}$$

we get a pair of dual scaling functions $\varphi(t), \widetilde{\varphi}(t)$. Finally, a pair of biorthogonal wavelets $\psi(t)$, $\widetilde{\psi}(t)$ with the minimal support are constructed by

$$\begin{aligned}
\psi(t) &= -2\sum_{n\in\mathbb{Z}} (-1)^n \overline{a}_{1-n} \varphi(2t-n),\\
\widetilde{\psi}(t) &= -2\sum_{n\in\mathbb{Z}} (-1)^n \overline{\widetilde{a}}_{1-n} \widetilde{\varphi}(2t-n).
\end{aligned}$$

In (3.3.1), taking

$$\begin{aligned}
L(\omega) &= 1,\\
\widetilde{L}(\omega) &= \textstyle\sum_{k=0}^{q-1} C_{q-1+k}^{k} C_{q-1+k}^{k} \sin^{2k}\frac{\omega}{2},
\end{aligned}$$

the scaling function $\varphi(t)$ satisfies

$$\widehat{\varphi}(\omega) = e^{-i\epsilon\frac{\omega}{2}} \left(\frac{\sin\frac{\omega}{2}}{\frac{\omega}{2}}\right)^p.$$

This is just a box-spline of degree $p-1$.

3.3.2 Multivariate Biorthogonal Wavelets

Let $\{\psi_{\mu}\}_{\mu\in\{0,1\}^d\setminus\{\mathbf{0}\}}$ and $\{\widetilde{\psi}_{\mu}\}_{\mu\in\{0,1\}^d\setminus\{\mathbf{0}\}} \in L^2(\mathbb{R}^d)$. If

$$\begin{aligned}
&\{\widetilde{\psi}_{\mu,m,\mathbf{n}}\}_{\mu\in\{0,1\}^d\setminus\{\mathbf{0}\}, m\in\mathbb{Z}, \mathbf{n}\in\mathbb{Z}^d},\\
&\{\psi_{\mu,m,\mathbf{n}}\}_{\mu\in\{0,1\}^d\setminus\{\mathbf{0}\}, m\in\mathbb{Z}, \mathbf{n}\in\mathbb{Z}^d}
\end{aligned}$$

are both Riesz bases for $L^2(\mathbb{R}^d)$, and

$$\int_{\mathbb{R}^d} \psi_{\mu,m,\mathbf{n}}(t)\widetilde{\psi}_{\mu',m',\mathbf{n}'}(t)\mathrm{d}t = \delta_{\mu,\mu'}\delta_{m,m'}\delta_{\mathbf{n},\mathbf{n}'},$$

we say $\{\psi_\mu\}_\mu \in \{0,1\}^d \setminus \{\mathbf{0}\}$ and $\{\widetilde{\psi}_\mu\}_{\mu\in\{0,1\}^d\setminus\{\mathbf{0}\}}$ are *multivariate biorthogonal wavelets* and

$$\{\psi_{\mu,m,\mathbf{n}}\}_{\mu\in\{0,1\}^d\setminus\{\mathbf{0}\},m\in\mathbb{Z},\mathbf{n}\in\mathbb{Z}^d},$$
$$\{\widetilde{\psi}_{\mu,m,\mathbf{n}}\}_{\mu\in\{0,1\}^d\setminus\{\mathbf{0}\},m\in\mathbb{Z},\mathbf{n}\in\mathbb{Z}^d}$$

are *multivariate biorthogonal wavelet bases.*

The multivariate biorthogonal wavelets are constructed as follows: Let $\{V_m\}_{m\in\mathbb{Z}}$ and $\{\widetilde{V}_m\}_{m\in\mathbb{Z}}$ be a pair of dual MRAs for $L^2(\mathbb{R}^d)$. Then, scaling functions $\varphi(\mathbf{t})$ and $\widetilde{\varphi}(\mathbf{t})$ are such that $\{\varphi(\mathbf{t}-\mathbf{k})\}_{\mathbf{k}\in\mathbb{Z}^d}$ and $\{\widetilde{\varphi}(\mathbf{t}-\mathbf{l})\}_{\mathbf{l}\in\mathbb{Z}^d}$ are Riesz bases for V_0 and $\widetilde{V}_0$, respectively, and

$$\int_{\mathbb{R}^d} \overline{\varphi}(\mathbf{t}-\mathbf{k})\widetilde{\varphi}(\mathbf{t}-\mathbf{l})d\mathbf{t} = \delta_{\mathbf{k},\mathbf{l}} \qquad (\mathbf{k},\mathbf{l}\in\mathbb{Z}^d).$$

So, there exist $2\pi\mathbb{Z}^d-$periodic functions $H_0(\boldsymbol{\mu})$ and $\widetilde{H}_0(\boldsymbol{\omega})$ such that

$$\widehat{\varphi}(\boldsymbol{\omega}) = H_0\left(\tfrac{\boldsymbol{\omega}}{2}\right)\widehat{\varphi}\left(\tfrac{\boldsymbol{\omega}}{2}\right),$$
$$\widehat{\widetilde{\varphi}}(\boldsymbol{\omega}) = \widetilde{H}_0\left(\tfrac{\boldsymbol{\omega}}{2}\right)\widehat{\widetilde{\varphi}}\left(\tfrac{\boldsymbol{\omega}}{2}\right),$$

and

$$\sum_{\boldsymbol{\nu}\in\{0,1\}^d} \overline{H}_0(\boldsymbol{\omega}+\pi\boldsymbol{\nu})\widetilde{H}_0(\boldsymbol{\omega}+\pi\boldsymbol{\nu}) = 1 \qquad (\boldsymbol{\omega}\in\mathbb{R}^d).$$

We extend vectors $(H_0(\boldsymbol{\omega}+\pi\boldsymbol{\nu}))_{\boldsymbol{\nu}\in\{0,1\}^d}$ and $(\widetilde{H}_0(\boldsymbol{\omega}+\pi\boldsymbol{\nu}))_{\boldsymbol{\nu}\in\{0,1\}^d}$ as the first row into the following two $2^d \times 2^d$ non-singular matrices:

$$H(\boldsymbol{\omega}) = \begin{pmatrix} H_0(\boldsymbol{\omega}) & H_0(\boldsymbol{\omega}+\boldsymbol{\nu}_1\pi) & \cdots & H_0(\boldsymbol{\omega}+\boldsymbol{\nu}_{2^d-1}\pi) \\ H_{\mu_1}(\boldsymbol{\omega}) & H_{\mu_1}(\boldsymbol{\omega}+\boldsymbol{\nu}_1\pi) & \cdots & H_{\mu_1}(\boldsymbol{\omega}+\boldsymbol{\nu}_{2^d-1}\pi) \\ \vdots & \vdots & \vdots & \vdots \\ H_{\mu_{2^d-1}}(\boldsymbol{\omega}) & H_{\mu_{2^d-1}}(\boldsymbol{\omega}+\boldsymbol{\nu}_1\pi) & \cdots & H_{\mu_{2^d-1}}(\boldsymbol{\omega}+\boldsymbol{\nu}_{2^d-1}\pi) \end{pmatrix}$$

$$\widetilde{H}(\boldsymbol{\omega}) = \begin{pmatrix} \widetilde{H}_0(\boldsymbol{\omega}) & \widetilde{H}_0(\boldsymbol{\omega}+\boldsymbol{\nu}_1\pi) & \cdots & \widetilde{H}_0(\boldsymbol{\omega}+\boldsymbol{\nu}_{2^d-1}\pi) \\ \widetilde{H}_{\mu_1}(\boldsymbol{\omega}) & \widetilde{H}_{\mu_1}(\boldsymbol{\omega}+\boldsymbol{\nu}_1\pi) & \cdots & \widetilde{H}_{\mu_1}(\boldsymbol{\omega}+\boldsymbol{\nu}_{2^d-1}\pi) \\ \vdots & \vdots & \vdots & \vdots \\ \widetilde{H}_{\mu_{2^d-1}}(\boldsymbol{\omega}) & \widetilde{H}_{\mu_{2^d-1}}(\boldsymbol{\omega}+\boldsymbol{\nu}_1\pi) & \cdots & \widetilde{H}_{\mu_{2^d-1}}(\boldsymbol{\omega}+\boldsymbol{\nu}_{2^d-1}\pi) \end{pmatrix}$$

which satisfy

$$H(\boldsymbol{\omega})\widetilde{H}(\boldsymbol{\omega})^* = I,$$

where I is the unit matrix. For $i = 1,\ldots,2^d-1$, let $\psi_{\mu_i}(\mathbf{t})$ and $\widetilde{\psi}_{\mu_i}(\mathbf{t})$ satisfy

$$\widehat{\psi}_{\mu_i}(\boldsymbol{\omega}) = H_{\mu_i}\left(\tfrac{\boldsymbol{\omega}}{2}\right)\widehat{\varphi}\left(\tfrac{\boldsymbol{\omega}}{2}\right),$$
$$\widehat{\widetilde{\psi}}_{\mu_i}(\boldsymbol{\omega}) = \widetilde{H}_{\mu_i}\left(\tfrac{\boldsymbol{\omega}}{2}\right)\widehat{\widetilde{\varphi}}\left(\tfrac{\boldsymbol{\omega}}{2}\right),$$

which are a pair of $d-$variate biorthogonal wavelets, and $\{\psi_{\mu_i,m,\mathbf{n}}\}_{i=1,...,2^d-1}$ and $\{\widetilde{\psi}_{\mu_i,m,\mathbf{n}}\}_{i=1,...,2^d-1}$, where $m \in \mathbb{Z}$ and $\mathbf{n} \in \mathbb{Z}^d$, are a pair of $d-$variate biorthogonal wavelet bases for $L^2(\mathbb{R}^d)$.

3.3.3 *p−Band Biorthogonal Wavelets*

Let $\psi_\mu, \widetilde{\psi}_\mu \in L^2(\mathbb{R})$. If $\{\psi^{(p)}_{\mu,m,n}(t)\}_{\mu=1,...,p-1;m,n\in\mathbb{Z}}$ and $\{\widetilde{\psi}^{(p)}_{\mu,m,n}(t)\}_{\mu=1,...,p-1;m,n\in\mathbb{Z}}$ form a pair of biorthogonal bases with dilation p for $L^2(\mathbb{R})$, then $\psi_\mu, \widetilde{\psi}_\mu$ ($\mu = 1, ..., p-1$) are called $p-$band biorthogonal wavelets.

The compactly supported biorthogonal wavelets are constructed as follows: Let $\{V_m\}_{m\in\mathbb{Z}}$ and $\{\widetilde{V}_m\}_{m\in\mathbb{Z}}$ be two $p-$band MRAs. If their scaling functions $\varphi_p(t)$ and $\widetilde{\varphi}_p(t)$ satisfy

$$\int_{\mathbb{R}} \varphi_p(t-k)\overline{\widetilde{\varphi}}_p(t-l)\mathrm{d}t = \delta_{k,l} \qquad (k,\ l \in \mathbb{Z}), \tag{3.3.2}$$

then $\{V_m\}_{m\in\mathbb{Z}}$ and $\{\widetilde{V}_m\}_{m\in\mathbb{Z}}$ are called *a pair of biorthogonal p− band MRAs* and $\varphi(t)$ and $\widetilde{\varphi}(t)$ are called *p−band biorthogonal scaling functions*. The bi-scale equations are

$$\begin{aligned} \tfrac{1}{p}\varphi_p\left(\tfrac{t}{p}\right) &= \sum_{n\in\mathbb{Z}} a_n \varphi_p(t-n), \\ \tfrac{1}{p}\widetilde{\varphi}_p\left(\tfrac{t}{p}\right) &= \sum_{n\in\mathbb{Z}} \widetilde{a}_n \widetilde{\varphi}_p(t-n). \end{aligned}$$

The transfer functions are

$$H_0^{(p)}(\omega) = \sum_{n\in\mathbb{Z}} a_n\, e^{-in\omega}, \qquad \widetilde{H}_0^{(p)}(\omega) = \sum_{n\in\mathbb{Z}} \widetilde{a}_n\, e^{-in\omega}.$$

From (3.3.2), it follows that

$$\sum_{\nu=0}^{p-1} \overline{H}_0^{(p)}\left(\omega + \frac{2\pi\nu}{p}\right) \widetilde{H}_0^{(p)}\left(\omega + \frac{2\pi\nu}{p}\right) = 1 \qquad (\omega \in \mathbb{R}).$$

Consider the matrix extensions for two vectors $(H_0(\omega + \frac{2\pi\nu}{p}))_{\nu=0,...,p-1}$ and $(\widetilde{H}_0(\omega + \frac{2\pi\nu}{p}))_{\nu=0,...,p-1}$. Precisely say, construct two $p \times p$ matrices of bounded $2\pi-$periodic functions:

$$H_p(\omega) = \begin{pmatrix} H_0^{(p)}(\omega) & H_0^{(p)}(\omega + \frac{2\pi}{p}) & \cdots & H_0^{(p)}(\omega + \frac{2\pi(p-1)}{p}) \\ H_1^{(p)}(\omega) & H_1^{(p)}(\omega + \frac{2\pi}{p}) & \cdots & H_1^{(p)}(\omega + \frac{2\pi(p-1)}{p}) \\ \vdots & \vdots & \ddots & \vdots \\ H_{p-1}^{(p)}(\omega) & H_{p-1}^{(p)}(\omega + \frac{2\pi}{p}) & \cdots & H_{p-1}^{(p)}(\omega + \frac{2\pi(p-1)}{p}) \end{pmatrix}$$

$$\widetilde{H}_p(\omega) = \begin{pmatrix} \widetilde{H}_0^{(p)}(\omega) & \widetilde{H}_0^{(p)}(\omega + \frac{2\pi}{p}) & \cdots & \widetilde{H}_0^{(p)}(\omega + \frac{2\pi(p-1)}{p}) \\ \widetilde{H}_1^{(p)}(\omega) & \widetilde{H}_1^{(p)}(\omega + \frac{2\pi}{p}) & \cdots & \widetilde{H}_1^{(p)}(\omega + \frac{2\pi(p-1)}{p}) \\ \vdots & \vdots & \ddots & \vdots \\ \widetilde{H}_{p-1}^{(p)}(\omega) & H_{p-1}^{(p)}(\omega + \frac{2\pi}{p}) & \cdots & \widetilde{H}_{p-1}^{(p)}(\omega + \frac{2\pi(p-1)}{p}) \end{pmatrix}$$

such that $H_p(\omega)\widetilde{H}_p(\omega)^* = I$, where I is the unit matrix. For $\mu = 1, \ldots, p-1$, define

$$\begin{aligned} \widehat{\psi}_\mu^{(p)}(\omega) &= H_\mu^{(p)}\left(\frac{\omega}{p}\right)\widehat{\varphi}_p\left(\frac{\omega}{p}\right), \\ \widehat{\widetilde{\psi}}_\mu^{(p)}(\omega) &= \widetilde{H}_\mu^{(p)}\left(\frac{\omega}{p}\right)\widehat{\widetilde{\varphi}}_p\left(\frac{\omega}{p}\right). \end{aligned}$$

Then, $\psi_\mu^{(p)}(t)$ and $\widetilde{\psi}_\mu^{(p)}(t)$ $(\mu = 1, \ldots, p-1)$ are a pair of biorthogonal wavelets. Let

$$\begin{aligned} \psi_{\mu,m,n}^{(p)} &= p^{\frac{m}{2}}\psi_\mu^{(p)}(p^m t - n), \\ \widetilde{\psi}_{\mu,m,n}^{(p)} &= p^{\frac{m}{2}}\widetilde{\psi}_\mu^{(p)}(p^m t - n). \end{aligned}$$

Then, $\{\psi_{\mu,m,n}^{(p)}(t)\}_{\mu=1,\ldots,p-1;m,n\in\mathbb{Z}}$ and $\{\widetilde{\psi}_{\mu,m,n}^{(p)}(t)\}_{\mu=1,\ldots,p-1;m,n\in\mathbb{Z}}$ form a pair of biorthogonal $p-$band wavelet bases.

3.3.4 Semi-orthogonal Wavelets

A biorthogonal wavelet $\psi \in L^2(\mathbb{R})$ is called a *semi-orthogonal wavelet* if

$$(\psi_{m,n},\ \psi_{m',n'}) = \int_{\mathbb{R}} \psi_{m,n}(t)\overline{\psi}_{m',n'}(t)\mathrm{d}t = 0 \qquad (m \neq m';\ m, m', n, n' \in \mathbb{Z}),$$

i.e., for $m \neq m'$, $\psi_{m,n}$, and $\psi_{m',n'}$ are orthogonal, where $\psi_{j,k} = 2^{\frac{j}{2}}\psi(2^j t - k)$.

For a generalized MRA, the scaling function $\varphi(t)$ is such that $\{\varphi(t-n)\}_{n\in\mathbb{Z}}$ is a Riesz basis for V_0. If the scaling function is a $B-$spline of order k, i.e., $\varphi(t) = N_k(t)$ $(k \in \mathbb{Z}_+)$, where

$$\begin{aligned} &N_1(t) = \chi_{[0,1]}(t), \\ &N_k(t) = \int_0^1 N_{k-1}(t-\tau)\mathrm{d}\tau \quad (k = 2, \ldots), \end{aligned}$$

then a compactly supported semi-orthogonal wavelet is given by

$$\psi_k(t) = 2\sum_{n=0}^{3k-2} q_k N_k(2t-n).$$

where $q_n = (-1)^n 2^{-k} \sum_{\nu=0}^{k} C_k^{\nu} N_{2k}(n+1-\nu)\ (n = 0, 1, \ldots, 3k-2)$ and $C_k^{\nu} = \frac{k!}{\nu!(k-\nu)!}$. The $\psi_k(t)$ is also called a *B−spline wavelet.*

The dual wavelet of $\psi_k(t)$ is

$$\widetilde{\psi}_k(t) = (-1)^{k+1} 2^{-k+1} \sum_{\nu\in\mathbb{Z}} d_{\nu}^{(k)} \psi_{I,k}(t-k-\nu+1),$$

where $\psi_{I,k}$ and $\{d_{\nu}^{(k)}\}_{\nu\in\mathbb{Z}}$ are determined by the following:

$$\begin{aligned} \psi_{I,k} &= 2\sum_{n\in\mathbb{Z}} \alpha_n^{(k)} N_k(2t-n), \\ \alpha_n^{(k)} &= \tfrac{1}{2}\sum_{l=0}^{k}(-1)^l C_k^l d_{k+n-1-l}^{(k)}, \\ \sum_{\nu\in\mathbb{Z}} d_{\nu}^{(k)} z^{\nu} &= \left(\sum_{\nu=-k+1}^{k-1} N_{2k}(k+\nu) z^{\nu}\right)^{-1}. \end{aligned}$$

For any k, $\{\psi_k(t), \widetilde{\psi}_k(t)\}$ is a pair of biorthogonal wavelets.

3.4 Wavelets on Domains

In this section, we analyze the target function on a general domain of $\mathbb{R}^2$ by wavelet methods. The target function is extended continuously to a square and vanished on the boundary of the square and then is expanded into a wavelet series on $\mathbb{R}^2$. Since the extension tool is polynomials, by the moment theorem, the sequence of wavelet coefficients obtained is sparse. Therefore, we can analyze the internal information of the function on a general domain very well by using a few wavelet coefficients.

3.4.1 Continuous Extension

Let Ω be a simply connected bounded domain in $\mathbb{R}^2$. Without loss of generality, we assume that $\Omega \subset [0, 1]^2$. Given a smooth target function $f(x, y)$ on Ω, we will construct the continuous extension $F(x, y)$ satisfying the conditions:

$$\begin{aligned} F(x, y) &= f(x, y), && (x, y) \in \Omega, \\ F(x, y) &= 0, && (x, y) \in \mathbb{R}^2 \setminus [0, 1]^2. \end{aligned}$$

On $[0,1]^2 \setminus \Omega$, $F(x,y)$ can be expressed locally in the forms:

$$\xi_0(x)+\xi_1(x)y \quad \text{or} \quad \eta_0(y)+\eta_1(y)x \quad \text{or} \quad c_{00}+c_{10}x+c_{01}y+c_{11}xy,$$

where $\xi_0, \xi_1, \eta_0, \eta_1 \in \text{lip } 1$ and c_{ij}s are constants $(i, j = 0, 1)$. The construction of $F(x,y)$ is stated as follows.

Step 1. Given a partition of the complement $[0,1]^2 \setminus \Omega$, assume that we can choose four points $(x_i, y_i) \in \partial\Omega$ $(i = 1, 2, 3, 4)$ such that $[0,1]^2 \setminus \Omega$ is divided into four rectangles:

$$\begin{aligned} H_1 &= [0, x_1] \times [0, y_1], \qquad & H_2 &= [x_2, 1] \times [0, y_2], \\ H_3 &= [x_3, 1] \times [y_3, 1], \qquad & H_4 &= [0, x_4] \times [y_4, 1] \end{aligned}$$

and four trapezoids with a curved side:

$$\begin{aligned} E_1 &= \{(x,y):\ x_1 \le x \le x_2,\ \ 0 \le y \le g(x)\}, \\ E_2 &= \{(x,y):\ h(y) \le x \le 1,\ \ y_2 \le y \le y_3\}, \\ E_3 &= \{(x,y):\ x_4 \le x \le x_3,\ \ g^*(x) \le y \le 1\}, \\ E_4 &= \{(x,y):\ 0 \le x \le h^*(y),\ \ y_1 \le y \le y_4\}, \end{aligned}$$

where the boundary functions $g(x), h(y), g^*(x), h^*(y)$ are continuously differentiable functions. So, the partition of the complement is

$$[0,1]^2 \setminus \Omega = \left(\bigcup_{i=1}^{4} E_i\right) \bigcup \left(\bigcup_{i=1}^{4} H_i\right).$$

Step 2. Define $Q_i(x,y)$ on each rectangles H_i $(i = 1, 2, 3, 4)$ as follows:

$$\begin{aligned} Q_1(x,y) &= \tfrac{f(x_1,y_1)}{x_1 y_1} xy, & (x,y) &\in H_1, \\ Q_2(x,y) &= \tfrac{f(x_2,y_2)}{(1-x_2)y_2}(1-x)y, & (x,y) &\in H_2, \\ Q_3(x,y) &= \tfrac{f(x_3,y_3)}{(1-x_3)(1-y_3)}(1-x)(1-y), & (x,y) &\in H_3, \\ Q_4(x,y) &= \tfrac{f(x_4,y_4)}{x_4(1-y_4)} x(1-y), & (x,y) &\in H_4. \end{aligned}$$

Step 3. Define $P_i(x,y)$ on each trapezoid E_i $(i = 1, 2, 3, 4)$ as follows:

$$\begin{aligned} P_1(x,y) &= f(x, g(x))\tfrac{y}{g(x)}, & (x,y) &\in E_1, \\ P_2(x,y) &= f(h(y), y)\tfrac{1-x}{1-h(y)}, & (x,y) &\in E_2, \\ P_3(x,y) &= f(x, g^*(x))\tfrac{1-y}{1-g^*(x)}, & (x,y) &\in E_3, \\ P_4(x,y) &= f(h^*(y), y)\tfrac{x}{h^*(y)}, & (x,y) &\in E_4. \end{aligned}$$

Step 4. Compute the smooth extension to $\mathbb{R}^2$. Define

$$F(x, y) = \begin{cases} f(x, y), & (x, y) \in \Omega, \\ P_i(x, y), & (x, y) \in E_i \ (i = 1, 2, 3, 4), \\ Q_i(x, y), & (x, y) \in H_i \ (i = 1, 2, 3, 4), \\ 0, & (x, y) \in \mathbb{R}^2 \setminus [0, 1]^2. \end{cases}$$

Then, $F(x, y) \in \text{lip}\, 1$ on $\mathbb{R}^2$ and $\text{supp}\, F(x, y) = [0, 1]^2$.

3.4.2 Wavelet Expansion

We will expand the smooth function $F(x, y)$ on Ω into wavelet series.

Let φ and ψ be a smooth compactly supported scaling function and the corresponding wavelet of $L^2(\mathbb{R})$. Then, functions

$$\begin{aligned} \psi^{(1)}(x, y) &= \varphi(x)\psi(y), \\ \psi^{(2)}(x, y) &= \psi(x)\varphi(y), \\ \psi^{(3)}(x, y) &= \psi(x)\psi(y) \end{aligned}$$

are the two-dimensional separable wavelets. For $m \in \mathbb{Z}$ and $\mathbf{n} = (n_1, n_2) \in \mathbb{Z}^2$, let

$$\begin{aligned} \psi^{(1)}_{m,\mathbf{n}}(x, y) &= \varphi_{m,n_1}(x)\psi_{m,n_2}(y), \\ \psi^{(2)}_{m,\mathbf{n}}(x, y) &= \psi_{m,n_1}(x)\varphi_{m,n_2}(y), \\ \psi^{(3)}_{m,\mathbf{n}}(x, y) &= \psi_{m,n_1}(x)\psi_{m,n_2}(y). \end{aligned}$$

Then, $\psi^{(1)}_{m,\mathbf{n}}(x, y)$, $\psi^{(2)}_{m,\mathbf{n}}(x, y)$, $\psi^{(3)}_{m,\mathbf{n}}(x, y)$ form a wavelet basis for $L^2(\mathbb{R}^2)$.

Expand $F(x, y)$ into wavelet series with respect to $\psi^{(\nu)}_{m,n}(x, y)$ $(\nu = 1, 2, 3)$:

$$F(x, y) = \sum_{\nu=1}^{3} \sum_{m\in\mathbb{Z}} \sum_{\mathbf{n}\in\mathbb{Z}^2} c^{(\nu)}_{m,\mathbf{n}} \psi^{(\nu)}_{m,\mathbf{n}}(x, y) \qquad ((x, y) \in \mathbb{R}^2),$$

where wavelet coefficients $c^{(\nu)}_{m,\mathbf{n}} = \int_{\mathbb{R}^2} F(x, y)\overline{\psi^{(\nu)}_{m,\mathbf{n}}(x, y)}\mathrm{d}x\mathrm{d}y$ $(m \in \mathbb{Z}, \mathbf{n} \in \mathbb{Z}^2, \nu = 1, 2, 3)$.

If $(x_0, y_0) \notin (\Omega \bigcup E_1 \bigcup E_2 \bigcup E_3 \bigcup E_4)$ and $(\frac{n_1}{2^m}, \frac{n_2}{2^m})$ is in a neighborhood of (x_0, y_0), where $m \in \mathbb{Z}$, $\mathbf{n} = (n_1, n_2) \in \mathbb{Z}^2$, then

$$c^{(\nu)}_{m,\mathbf{n}} = O(2^{-ml})\left(1 + (A_{m,\mathbf{n}}(x_0, y_0))^{\frac{1}{2}}\right) \qquad (\nu = 1, 2, 3),$$

where $A_{m,\mathbf{n}}(x_0, y_0) = (2^m x_0 - n_1)^2 + (2^m y_0 - n_2)^2$ and l is arbitrarily large. Due to vanish moments of wavelets, it can be proved easily that a lot of wavelet coefficients are zero. In detail, let

$$\begin{aligned}
M &= \max\{g(x)\ x_1 \le x \le x_2\}, \quad \tau = \min\{g(x),\ x_1 \le x \le x_2\},\\
M^* &= \max\{g^*(x)\ x_4 \le x \le x_3\}, \quad \tau^* = \min\{g^*(x),\ x_4 \le x \le x_3\},\\
N &= \max\{h(y)\ y_2 \le y \le y_3\}, \quad \lambda = \min\{h(y),\ y_2 \le y \le y_3\},\\
N^* &= \max\{h^*(y)\ y_1 \le y \le y_4\}, \quad \lambda^* = \min\{h^*(y),\ y_1 \le y \le y_4\},
\end{aligned}$$

and let $d_{m,\mathbf{n}} = \operatorname{supp}\varphi_{m,\mathbf{n}}$ and $D_{m,\mathbf{n}} = \operatorname{supp}\psi_{m,\mathbf{n}}$.

(a) If $d_{m,n_1} \bigcap [\lambda^*, N] = \emptyset$ or $(D_{m,n_2} \bigcap [\tau, M^*]) \bigcup \{0\} \bigcup \{1\} = \emptyset$, then $c^{(1)}_{m,\mathbf{n}} = 0$.

(b) If $d_{m,n_2} \bigcap [\tau, M^*] = \emptyset$ or $(D_{m,n_1} \bigcap [\lambda^*, N]) \bigcup \{0\} \bigcup \{1\} = \emptyset$, then $c^{(2)}_{m,\mathbf{n}} = 0$.

(c) If $(D_{m,n_1} \bigcap [\lambda^*, N]) \bigcup \{0\} \bigcup \{1\} = \emptyset$ or $(D_{m,n_2} \bigcap [\tau, M^*]) \bigcup \{0\} \bigcup \{1\} = \emptyset$, then $c^{(3)}_{m,\mathbf{n}} = 0$.

Therefore, we can analyze the smooth function on general domain by using few wavelet coefficients.

3.5 Discrete Wavelet Transforms

Discrete wavelet transform is the fast algorithm to compute wavelet coefficients. It was introduced by Mallat (Mallat (1989)), so discrete wavelet transform is also called Mallat's algorithm.

3.5.1 *Discrete Orthogonal Wavelet Transforms*

Let $\{V_m\}_{m\in\mathbb{Z}}$ be an orthogonal MRA for $L^2(\mathbb{R}^d)$. Denote by f_{m+1} the orthogonal projection of a function $f \in L^2(\mathbb{R}^d)$ on V_{m+1}. Since $\{\varphi_{m,\mathbf{n}}\}_{\mathbf{n}\in\mathbb{Z}^d}$ and $\{\psi_{\mu,m,\mathbf{n}}\}_{\mu\in\{0,1\}^d\setminus\{\mathbf{0}\};\mathbf{n}\in\mathbb{Z}^d}$ form a normal orthogonal basis for V_{m+1}, the wavelet expansion of f_{m+1} is

$$f_{m+1}(\mathbf{t}) = \sum_{\mathbf{n}\in\mathbb{Z}^d} c^{(m)}_{\mathbf{n}} \varphi_{m,\mathbf{n}}(\mathbf{t}) + \sum_{\mu\in\{0,1\}^d\setminus\{\mathbf{0}\}} \sum_{\mathbf{n}\in\mathbb{Z}^d} d^{(m)}_{\mu,\mathbf{n}} \psi_{\mu,m,\mathbf{n}}(\mathbf{t}) \qquad (m \in \mathbb{Z}),$$

where

$$\begin{aligned}
c^{(m)}_{\mathbf{n}} &= \int_{\mathbb{R}^d} f(\mathbf{t})\overline{\varphi}_{m,\mathbf{n}}(\mathbf{t})d\mathbf{t} \quad (\mathbf{n} \in \mathbb{Z}^d),\\
d^{(m)}_{\mu,\mathbf{n}} &= \int_{\mathbb{R}^d} f(\mathbf{t})\overline{\psi}_{\mu,m,\mathbf{n}}(\mathbf{t})d\mathbf{t} \quad (\mu \in \{0,1\}^d \setminus \{\mathbf{0}\},\ \mathbf{n} \in \mathbb{Z}^d).
\end{aligned}$$

It is difficult to compute these coefficients $c^{(m)}_{\mathbf{n}}$ and $d^{(m)}_{\mu,\mathbf{n}}$ fast by using these formulas.

Assume that $\operatorname{supp}\varphi = [0, N]^d$. For a very large m, it is clear that

$$\begin{aligned}
c^{(m)}_{\mathbf{n}} &= \int_{\mathbf{t}\in\frac{\mathbf{n}}{2^m}+[0,\frac{N}{2^m}]^d} f(\mathbf{t})\overline{\varphi}_{m,\mathbf{n}}(\mathbf{t})d\mathbf{t} \approx f\left(\frac{\mathbf{n}}{2^m}\right) \int_{\mathbf{t}\in\frac{\mathbf{n}}{2^m}+[0,\frac{N}{2^m}]^d} \overline{\varphi}_{m,\mathbf{n}}(\mathbf{t})d\mathbf{t}\\
&= f\left(\frac{\mathbf{n}}{2^m}\right) \int_{\mathbb{R}^d} \overline{\varphi}_{m,\mathbf{n}}(\mathbf{t})d\mathbf{t}.
\end{aligned}$$

It means that for a very large m, $c_{\mathbf{n}}^{(m)}$ can be computed approximately by using the samples of f. The corresponding error is very small.

With the help of bi-scale equation $\varphi(\mathbf{t}) = 2^d \sum_{\mathbf{n}\in\mathbb{Z}^d} a_{\mathbf{n}}\varphi(2\mathbf{t}-\mathbf{n})$ and the following wavelet formula:

$$\psi_\mu(\mathbf{t}) = 2^d \sum_{\mathbf{n}\in\mathbb{Z}^d} b_{\mu,\mathbf{n}}\varphi(2\mathbf{t}-\mathbf{n}) \qquad (\mu \in \{0,1\}^d \setminus \{\mathbf{0}\}),$$

the wavelet coefficients can be computed as follows:

The *decomposition formulas* to compute $\{c_{\mathbf{n}}^{(m)}\}$ and $\{d_{\mu,\mathbf{n}}^{(m)}\}_{\mu\in\{0,1\}^d\setminus\{\mathbf{0}\},\mathbf{n}\in\mathbb{Z}^d}$ using $\{c_{\mathbf{n}}^{(m+1)}\}_{\mathbf{n}\in\mathbb{Z}^d}$ are

$$\begin{aligned} c_{\mathbf{n}}^{(m)} &= 2^{\frac{d}{2}} \sum_{\mathbf{k}\in\mathbb{Z}^d} \overline{a}_{\mathbf{k}-2\mathbf{n}}\, c_{\mathbf{k}}^{(m+1)}, \\ d_{\mu,\mathbf{n}}^{(m)} &= 2^{\frac{d}{2}} \sum_{\mathbf{k}\in\mathbb{Z}^d} \overline{b}_{\mu,\mathbf{k}-2\mathbf{n}} c_{\mathbf{k}}^{(m+1)} \qquad (\mu \in \{0,1\}^d \setminus \{\mathbf{0}\}), \end{aligned}$$

The *reconstruction formula* to compute $\{c_{\mathbf{n}}^{(m+1)}\}_{\mathbf{n}\in\mathbb{Z}^d}$ using $\{c_{\mathbf{n}}^{(m)}\}_{\mathbf{n}\in\mathbb{Z}^d}$ and $\{d_{\mu,\mathbf{n}}^{(m)}\}_{\mathbf{n}\in\mathbb{Z}^d}$ is

$$c_{\mathbf{n}}^{(m+1)} = 2^{\frac{d}{2}} \left(\sum_{\mathbf{k}\in\mathbb{Z}^d} a_{\mathbf{n}-2\mathbf{k}}\, c_{\mathbf{k}}^{(m)} + \sum_{\mu\in\{0,1\}^d\setminus\{\mathbf{0}\}} \sum_{\mathbf{k}\in\mathbb{Z}^d} b_{\mu,\mathbf{n}-2\mathbf{k}}\, d_{\mu,\mathbf{k}}^{(m)} \right).$$

In the one-dimensional case, let $\{V_m\}_{m\in\mathbb{Z}}$ be an orthogonal MRA for $L^2(\mathbb{R})$. For $f \in L^2(\mathbb{R})$,

$$\begin{aligned} f_{m+1}(t) &= (P_{V_{m+1}} f)(t) = \sum_{n\in\mathbb{Z}} c_n^{(m+1)} \varphi_{m+1,n}(t), \\ f_{m+1}(t) &= \sum_{n\in\mathbb{Z}} c_n^{(m)} \varphi_{m,n}(t) + \sum_{n\in\mathbb{Z}} d_n^{(m)} \psi_{m,n}(t). \end{aligned}$$

The bi-scale equation and wavelet formula are

$$\begin{aligned} \varphi(t) &= 2 \sum_{n\in\mathbb{Z}} a_n \varphi(2t-n), \\ \psi(t) &= 2 \sum_{n\in\mathbb{Z}} b_n \varphi(2t-n), \end{aligned}$$

where $b_n = (-1)^{n+1}\overline{a}_{1-n}$.

The decomposition formulas are

$$\begin{aligned} c_n^{(m)} &= 2^{\frac{1}{2}} \sum_{n\in\mathbb{Z}} \overline{a}_{k-2n} c_k^{(m+1)}, \\ d_n^{(m)} &= 2^{\frac{1}{2}} \sum_{k\in\mathbb{Z}} \overline{b}_{k-2n} c_k^{(m+1)}. \end{aligned}$$

The reconstruction formulas are

$$c_n^{(m+1)} = 2^{\frac{1}{2}}\left(\sum_{k\in\mathbb{Z}} a_{n-2k}c_k^{(m)} + \sum_{k\in\mathbb{Z}} b_{n-2k}d_k^{(m)}\right).$$

Let

$$\begin{aligned} A &= (2^{\frac{1}{2}}a_{n-2k})_{n,k\in\mathbb{Z}}, \quad B = (2^{\frac{1}{2}}b_{n-2k})_{n,k\in\mathbb{Z}}, \\ C^{(m+1)} &= (c_n^{(m+1)})^T_{n\in\mathbb{Z}}, \quad D^{(m+1)} = (d_n^{(m+1)})^T_{n\in\mathbb{Z}}. \end{aligned}$$

The above decomposition and reconstruction formulas can be rewritten in the matrix forms:

$$\begin{aligned} C^{(m)} &= A^*C^{(m+1)}, \\ D^{(m)} &= B^*C^{(m+1)}, \\ C^{(m+1)} &= AC^{(m)} + BD^{(m)}, \end{aligned}$$

where A^* and B^* are the conjugate transposes of A and B, respectively.

3.5.2 Discrete Biorthogonal Wavelet Transforms

Let $\{V_m\}_{m\in\mathbb{Z}}$ be a $p-$band MRA with scaling function $\varphi(t)$ and $\{\widetilde{V}_m\}_{m\in\mathbb{Z}}$ be its dual $p-$band MRA with scaling function $\widetilde{\varphi}(t)$. Let $\{\psi_\mu\}_{\mu=1,\dots,p-1}$ and $\{\widetilde{\psi}_\mu\}_{\mu=1,\dots,p-1}$ be the corresponding biorthogonal $p-$band wavelets. From

$$\begin{aligned} \varphi(t) &= p\sum_{n\in\mathbb{Z}} p_n\varphi(pt-n), \quad \psi_\mu(t) = p\sum_{n\in\mathbb{Z}} q_n^{(\mu)}\varphi(pt-n), \\ \widetilde{\varphi}(t) &= p\sum_{n\in\mathbb{Z}} \widetilde{p}_n\widetilde{\varphi}(pt-n), \quad \widetilde{\psi}_\mu(t) = p\sum_{n\in\mathbb{Z}} \widetilde{q}_n^{(\mu)}\widetilde{\varphi}(pt-n), \end{aligned} \tag{3.5.1}$$

where $\mu = 1, \dots, p-1$. Let $f \in V_{m+1}$. Then,

$$\begin{aligned} f(t) &= \sum_{n\in\mathbb{Z}} c_n^{(m+1)}\varphi_{m+1,n}(t), \\ f(t) &= \sum_{n\in\mathbb{Z}} c_n^{(m)}\varphi_{m,n}(t) + \sum_{\mu=1}^{p-1}\sum_{n\in\mathbb{Z}} d_{\mu,n}^{(m)}\psi_{\mu,m,n}(t). \end{aligned}$$

The *decomposition formulas* are

$$\begin{aligned} c_n^{(m)} &= p^{\frac{1}{2}}\sum_{k\in\mathbb{Z}} \overline{\widetilde{p}}_{k-pn}c_k^{(m+1)}, \\ d_{\mu,n}^{(m)} &= p^{\frac{1}{2}}\sum_{k\in\mathbb{Z}} \overline{\widetilde{q}_{k-pn}^{(m)}}c_k^{(m+1)}. \end{aligned}$$

The *reconstruction formula* is

$$c_n^{(m+1)} = p^{\frac{1}{2}} \left(\sum_{k\in\mathbb{Z}} p_{n-pk} c_k^{(m)} + \sum_{\mu=1}^{p-1} \sum_{k\in\mathbb{Z}} q_{n-pk}^{(\mu)} d_{\mu,k}^{(m)} \right).$$

Now, we turn to the tensor product case.

Let $\{\varphi(t), \widetilde{\varphi}(t)\}$ be a pair of dual scaling functions and $\{\psi(t), \widetilde{\psi}(t)\}$ be a pair of dual wavelets. Then

(a) $\varphi(x)\varphi(y)$ and $\widetilde{\varphi}(x)\widetilde{\varphi}(y)$ are a pair of dual scaling functions;
(b) $\{\varphi(x)\psi(y),\ \psi(x)\varphi(y),\ \psi(x)\psi(y)\}$ and $\{\widetilde{\varphi}(x)\widetilde{\psi}(y),\ \widetilde{\psi}(x)\widetilde{\varphi}(y),\ \widetilde{\psi}(x)\widetilde{\psi}(y)\}$ are a pair of dual wavelets.

Let $f(x, y) \in V_{m+1}$. Then,

$$\begin{aligned} f(x,y) &= \sum_{n_1,n_2\in\mathbb{Z}} c_{n_1,n_2}^{(m+1)} \varphi_{m+1,n_1}(x)\, \varphi_{m+1,n_2}(y), \\ f(x,y) &= \sum_{n_1,n_2\in\mathbb{Z}} c_{n_1,n_2}^{(m)} \varphi_{m,n_1}(x)\, \varphi_{m,n_2}(y) \\ &\quad + \sum_{\mu=1}^{3} \sum_{n_1,n_2\in\mathbb{Z}} d_{\mu,n_1,n_2}^{(m)} \psi_{n_1,n_2}^{(\mu)}(x,y) \qquad (\mathbf{n} = (n_1, n_2)), \end{aligned}$$

where $\psi^{(1)}(x,y)=\widetilde{\varphi}(x)\widetilde{\psi}(y)$, $\psi^{(2)}(x,y)=\widetilde{\psi}(x)\widetilde{\varphi}(y)$, and $\psi^{(3)}(x,y) = \widetilde{\psi}(x)\widetilde{\psi}(y)$.

Let $\{p_{\mathbf{n}}\}_{\mathbf{n}\in\mathbb{Z}^2}$, $\{q_{\mathbf{n}}\}_{\mathbf{n}\in\mathbb{Z}^2}$, $\{\widetilde{p}_{\mathbf{n}}\}_{\mathbf{n}\in\mathbb{Z}^2}$, and $\{\widetilde{q}_{\mathbf{n}}\}_{\mathbf{n}\in\mathbb{Z}^2}$ satisfy the following:

$$\begin{aligned} \varphi(t) &= 2\sum_{n\in\mathbb{Z}} p_n \varphi(2t-n), & \widetilde{\varphi}(t) &= 2\sum_{n\in\mathbb{Z}} \widetilde{p}_n \widetilde{\varphi}(2t-n), \\ \psi(t) &= 2\sum_{n\in\mathbb{Z}} q_n \varphi(2t-n), & \widetilde{\psi}(t) &= 2\sum_{n\in\mathbb{Z}} \widetilde{q}_n \varphi(2t-n). \end{aligned}$$

Then, the decomposition formulas are

$$\begin{aligned} c_{n_1,n_2}^{(m)} &= 2\sum_{k_1,k_2\in\mathbb{Z}} \overline{\widetilde{p}}_{k_1-2n_1} \overline{\widetilde{p}}_{k_2-2n_2} c_{k_1,k_2}^{(m+1)}, \\ d_{1,n_1,n_2}^{(m)} &= 2\sum_{k_1,k_2\in\mathbb{Z}} \overline{\widetilde{p}}_{k_1-2n_1} \overline{\widetilde{q}}_{k_2-2n_2} c_{k_1,k_2}^{(m+1)}, \\ d_{2,n_1,n_2}^{(m)} &= 2\sum_{k_1,k_2\in\mathbb{Z}} \overline{\widetilde{q}}_{k_1-2n_1} \overline{\widetilde{p}}_{k_2-2n_2} c_{k_1,k_2}^{(m+1)}, \\ d_{3,n_1,n_2}^{(m)} &= 2\sum_{k_1,k_2\in\mathbb{Z}} \overline{\widetilde{q}}_{k_1-2n_1} \overline{\widetilde{q}}_{k_2-2n_2} c_{k_1,k_2}^{(m+1)} \end{aligned}$$

and the reconstruction formula is

$$\begin{aligned} c_{n_1,n_2}^{(m+1)} &= 2\sum_{k_1,k_2\in\mathbb{Z}} p_{n_1-2k_1} p_{n_2-2k_2} c_{k_1,k_2}^{(m)} + 2\sum_{k_1,k_2\in\mathbb{Z}} p_{n_1-2k_1} q_{n_2-2k_2} d_{1,k_1,k_2}^{(m)} \\ &\quad + 2\sum_{k_1,k_2\in\mathbb{Z}} q_{n_1-2k_1} p_{n_2-2k_2} d_{2,k_1,k_2}^{(m)} + 2\sum_{k_1,k_2\in\mathbb{Z}} q_{n_1-2k_1} q_{n_2-2k_2} d_{3,k_1,k_2}^{(m)}. \end{aligned}$$

3.5.3 Discrete Biorthogonal Periodic Wavelets Transforms

Let ψ, $\widetilde{\psi}$ be a pair of biorthogonal wavelets generated by scaling functions φ, $\widetilde{\varphi}$. Take the tensor products:

$$\varphi_0(x,y)=\varphi(x)\varphi(y),\quad \psi_1(x,y)=\varphi(x)\psi(y),\quad \psi_2(x,y)=\psi(x)\varphi(y),\quad \psi_3(x,y)=\psi(x)\psi(y),$$

$$\widetilde{\varphi}_0(x,y)=\widetilde{\varphi}(x)\widetilde{\varphi}(y),\quad \widetilde{\psi}_1(x,y)=\widetilde{\varphi}(x)\widetilde{\psi}(y),\quad \widetilde{\psi}_2(x,y)=\widetilde{\psi}(x)\widetilde{\varphi}(y),\quad \widetilde{\psi}_3(x,y)=\widetilde{\psi}(x)\widetilde{\psi}(y).$$

The families of functions:

$$\begin{aligned}\Psi^{per} &= \{\varphi_0^{per}\}\bigcup\{\psi_{\mu,m,\mathbf{n}}^{per}\}_{\mu=1,2,3,\ m\in\mathbb{Z}_+;\ \mathbf{n}=(n_1,n_2):\ n_1,n_2=0,\ldots,2^m-1},\\ \widetilde{\Psi}^{per} &= \{\widetilde{\varphi}_0^{per}\}\bigcup\{\psi_{\mu,m,\mathbf{n}}^{per}\}_{\mu=1,2,3,\ m\in\mathbb{Z}_+;\ \mathbf{n}=(n_1,n_2):\ n_1,n_2=0,\ldots,2^m-1}\end{aligned} \tag{3.5.2}$$

are a pair of biorthogonal bases for $L^2([-\frac{1}{2},\frac{1}{2}]^2)$, where

$$g^{per}(x,y)=\sum_{l_1,l_2\in\mathbb{Z}} g(x+l_1,y+l_2).$$

Ψ^{per} and $\widetilde{\Psi}^{per}$ are called *a pair of biorthogonal periodic wavelet bases*.

Let $f\in L^2([-\frac{1}{2},\frac{1}{2}]^2)$. Then, $f(x,y)$ can be expanded into a biorthogonal series:

$$f(x,y)=c_0+\sum_{\mu=1}^{3}\sum_{m=0}^{\infty}\sum_{n_1,n_2=0}^{2^m-1} d_{\mu\mathbf{n}}^{(m)}\,\widetilde{\psi}_{\mu,m,\mathbf{n}}^{per}(x,y). \tag{3.5.3}$$

Let the space $\widetilde{V}_J$ be the span of $\{\widetilde{\varphi}_{0,J,\mathbf{n}}^{per}\}_{\mathbf{n}=(n_1,n_2),\ n_1,n_2=0,\ldots,2^J-1}$. For $f\in\widetilde{V}_J$,

$$\begin{aligned} f(x,y) &= \sum_{n_1,n_2=0}^{2^{J-1}-1} c_{\mathbf{n}}^{(J-1)}\,\widetilde{\varphi}_{0,J-1,\mathbf{n}}^{per}(x,y)+\sum_{\mu=1}^{3}\sum_{n_1,n_2=0}^{2^{J-1}-1} d_{\mu,\mathbf{n}}^{(J-1)}\,\widetilde{\psi}_{\mu,J-1,\mathbf{n}}^{per}(x,y),\\ f(x,y) &= \sum_{n_1,n_2=0}^{2^J-1} c_{\mathbf{n}}^{(J)}\,\widetilde{\varphi}_{0,J,\mathbf{n}}^{per}(x,y)\qquad (\mathbf{n}=(n_1,n_2)).\end{aligned} \tag{3.5.4}$$

Denote the coefficients of the filter banks by

$$\begin{aligned} a_k &= 2^{\frac{1}{2}}\int_{\mathbb{R}}\varphi(t)\overline{\widetilde{\varphi}}(2t-k)\mathrm{d}t,\\ b_k &= 2^{\frac{1}{2}}\int_{\mathbb{R}}\psi(t)\overline{\widetilde{\varphi}}(2t-k)\mathrm{d}t.\end{aligned}$$

Assume that $\varphi,\widetilde{\varphi},\psi,\widetilde{\psi}$ are all compactly supported. Without loss of generality, assume $a_n=b_{n+1}=0\,(|n|\ge N)$ for some $N\in\mathbb{Z}_+$. Define

$$\begin{aligned} a_n^* &= a_n, & b_{n+1}^* &= b_{n+1}\ (|n|\le 2^{J-1}),\\ a_{n+2^J}^* &= a_n^*, & b_{n+2^J}^* &= b_n^*\ \ (n\in\mathbb{Z}).\end{aligned}$$

For $2^{J-1} \geq N$, the fast algorithm is

$$c_{k_1,k_2}^{(J-1)} = \sum_{n_1=0}^{2^J-1} \sum_{n_2=0}^{2^J-1} \overline{a_{n_1-2k_1}^*} \, \overline{a_{n_2-2k_2}^*} \, c_{n_1,n_2}^{(J)},$$

$$d_{1,k_1,k_2}^{(J-1)} = \sum_{n_1=0}^{2^J-1} \sum_{n_2=0}^{2^J-1} \overline{a_{n_1-2k_1}^*} \, \overline{b_{n_2-2k_2}^*} \, c_{n_1,n_2}^{(J)},$$

$$d_{2,k_1,k_2}^{(J-1)} = \sum_{n_1=0}^{2^J-1} \sum_{n_2=0}^{2^J-1} \overline{b_{n_1-2k_1}^*} \, \overline{a_{n_2-2k_2}^*} \, c_{n_1,n_2}^{(J)},$$

$$d_{3,k_1,k_2}^{(J-1)} = \sum_{n_1=0}^{2^J-1} \sum_{n_2=0}^{2^J-1} \overline{b_{n_1-2k_1}^*} \, \overline{b_{n_2-2k_2}^*} \, c_{n_1,n_2}^{(J)}.$$

3.5.4 Discrete Harmonic Wavelet Transforms

Harmonic wavelet transform was introduced by Zhang and Saito in 2010. We first start from the continuous case. Assume that $F(x, y)$ is a function defined on $[0, \frac{1}{2}]^2$. Let

$$F(x, y) = u(x, y) + v(x, y) \qquad ((x, y) \in [0, \frac{1}{2}]^2),$$

where $u(x, y)$ is a harmonic function satisfying the Laplace's equation $\Delta u = \frac{\partial^2 u}{\partial x^2} + \frac{\partial^2 u}{\partial y^2} = 0$ $((x, y) \in [0, \frac{1}{2}]^2)$ and $u(x, y) = F(x, y)$ on the boundary of $[0, \frac{1}{2}]^2$. So, $v(x, y) = 0$ on the boundary of $[0, \frac{1}{2}]^2$. We do the odd extension v_{odd} of v to $[-\frac{1}{2}, \frac{1}{2}]$, i.e.,

$$\begin{array}{ll} v_{odd}(x, y) = v(x, y), & (x, y) \in \left[0, \frac{1}{2}\right]^2, \\ v_{odd}(-x, -y) = -v_{odd}(-x, y) = -v_{odd}(x, -y) = v_{odd}(x, y), & (x, y) \in \left[-\frac{1}{2}, \frac{1}{2}\right]^2, \end{array}$$

and then we do one-periodic extension of v_{odd}, denoted by f_{odd}, and expand f_{odd} into a biorthogonal periodic wavelet series with respect to Ψ^{per}, $\widetilde{\Psi}^{per}$ stated in (3.5.2). This process is called *harmonic wavelet transform*.

Assume that $F(x, y)$ is second-order continuously differentiable on $[0, \frac{1}{2}]^2$. By (3.5.3), the biorthogonal periodic wavelet series of $f_{odd}(x, y)$ is

$$f_{odd}(x, y) = c_{\mathbf{0}} + \sum_{\mu=1}^{3} \sum_{m=0}^{\infty} \sum_{n_1,n_2=0}^{2^m-1} d_{\mu,\mathbf{n}}^{(m)} \, \widetilde{\psi}_{\mu,m,\mathbf{n}}^{per}(x, y)$$

and the coefficients $d_{\mu,\mathbf{n}}^{(m)}(f_{odd}) = O(2^{-3m})$. The partial sum:

$$s_{2^M}(f_{odd}; x, y) = c_0 + \sum_{\mu=1}^{3} \sum_{m=0}^{M-1} \sum_{n_1,n_2=0}^{2^m-1} d_{\mu,\mathbf{n}}^{(m)} \widetilde{\psi}_{\mu,m,\mathbf{n}}^{per}(x, y),$$

satisfies

$$\|f_{odd} - s_{2^M}(f_{odd})\|_{L^2([-\frac{1}{2},\frac{1}{2}]^2)} = \left(\int\int_{[-\frac{1}{2},\frac{1}{2}]^2} |f_{odd}(x, y) - s_{2^M}(f_{odd}; x, y)|^2 \mathrm{d}x\mathrm{d}y \right)^{\frac{1}{2}}$$
$$= O(2^{-2M}).$$

It is clear that the periodic wavelet coefficients of f_{odd} decay fast and f_{odd} can be reconstructed by few periodic wavelet coefficients.

Compared with harmonic wavelet transform, if we directly expand $F(x, y)$ on $[0, \frac{1}{2}]^2$ into a biorthogonal periodic wavelet series, the decay rate of wavelet coefficients is $d_{\mu,\mathbf{n}}^{(m)}(F) = O(2^{-m})$, and this process is called *periodic wavelet transform*; if we do an even extension F_{even} of f from $[0, \frac{1}{2}]^2$ to $[\frac{1}{2}, \frac{1}{2}]^2$, and then expand F_{even} into a biorthogonal periodic wavelet series, the decay rate for wavelet coefficients is $d_{\mu,\mathbf{n}}^{(m)}(F_{even}) = O(2^{-2m})$, and this process is called *folded wavelet transform*. Therefore, harmonic wavelet transform has obvious advantage over the other two.

Now, we turn to the discrete harmonic wavelet transform.

Let $\varphi, \widetilde{\varphi}, \psi, \widetilde{\psi}$ be stated as above. Assume further that $\varphi, \widetilde{\varphi}, \psi, \widetilde{\psi}$ are compactly supported real-valued functions, $\varphi, \widetilde{\varphi}$ are both symmetric at $t = 0$, and $\psi, \widetilde{\psi}$ are symmetric at $t = \frac{1}{2}$ and $t = -\frac{1}{2}$, respectively. For some large J, take $(2^{J-1} + 1)^2$ sample points of $f(x, y)$ on $[0, \frac{1}{2}]^2$:

$$x_{n_1 n_2} = f\left(\frac{n_1}{2^J}, \frac{n_2}{2^J}\right) \qquad (n_1, n_2 = 0, \ldots, 2^{J-1}).$$

By the harmonic wavelet transform,

$$f(x, y) = u(x, y) + v(x, y) \qquad ((x, y) \in [0, \frac{1}{2}]^2),$$

where $u(x, y)$ satisfies the following:

$$\begin{cases} \Delta u(x, y) = 0, & (x, y) \in (0, \frac{1}{2})^2, \\ u(x, y) = f(x, y), & (x, y) \in \partial([0, \frac{1}{2}]^2). \end{cases}$$

The $u(x, y)$ is computed numerically by Averbuch et al. (1998) or similar methods. The residual component $v(x, y)$ is computed as follows.

Let $y_{n_1 n_2} = v(\frac{n_1}{2^J}, \frac{n_2}{2^J})$. Then,

$$y_{0 n_2} = y_{n_1 0} = y_{2^{J-1}, n_2} = y_{n_1, 2^{J-1}} = 0 \qquad (n_1, n_2 = 0, \ldots, 2^{J-1}).$$

Let v_{odd} be an odd extension of v to $[\frac{1}{2}, \frac{1}{2}]^2$. Denote $z_{n_1 n_2} = v_{odd}(\frac{n_1}{2^J}, \frac{n_2}{2^J})$. Then,

$$z_{n_1 n_2} = -z_{-n_1, n_2} = -z_{-2^{J-1}, n_2} = z_{n_1, -2^{J-1}} = 0.$$

Let v^* be a one-periodic extension of v_{odd} to $\mathbb{R}^2$. Denote $z^*_{n_1 n_2} = v^*(\frac{n_1}{2^J}, \frac{n_2}{2^J})$. Then,

$$\begin{aligned} z^*_{n_1 n_2} &= z_{n_1 n_2} \qquad (n_1, n_2 = 0, \pm 1, \ldots, \pm 2^{J-1}), \\ z^*_{n_1+2^J, n_2} &= z_{n_1 n_2} \ (n_1, n_2 \in \mathbb{Z}), \\ z^*_{n_1, n_2+2^J} &= z^*_{n_1 n_2} \ (n_1, n_2 \in \mathbb{Z}). \end{aligned}$$

Take J sufficiently large such that

$$v^*(x, y) \approx \sum_{n_1, n_2 = 0}^{2^J - 1} c^{(J)}_{n_1, n_2} \widetilde{\varphi}^{per}_{0, J, n_1, n_2}(x, y).$$

Since φ is real-valued and then $\hat{\varphi}(0) = 1$, the above coefficients satisfy

$$c^{(J)}_{n_1, n_2} \approx v^* \left(\frac{n_1}{2^J}, \frac{n_2}{2^J} \right) = z^*_{n_1 n_2}.$$

By (3.5.3),

$$v^*(x, y) \approx \sum_{n_1, n_2 = 0}^{2^{J-1} - 1} c^{(J-1)}_{n_1, n_2} \varphi^{per}_{0, J-1, n_1, n_2}(x, y) + \sum_{\mu=1}^{3} \sum_{n_1, n_2 = 0}^{2^{J-1} - 1} d^{(J-1)}_{\mu, n_1, n_2} \widetilde{\psi}^{per}_{\mu, J-1, n_1, n_2}(x, y), \tag{3.5.5}$$

where $\{c^{(J-1)}_{n_1, n_2}\}$ and $\{d^{(J-1)}_{\mu, n_1, n_2}\}$ can be fast computed by the decomposition formula (3.5.4). Due to the symmetry of scaling function and wavelets used, the coefficients $c^{J-1}_{n_1, n_2}$ and $d^{(J-1)}_{\mu, n_1, n_2}$ have the following symmetry property.

(i) $c^{(J-1)}_{2^{J-1} - k_1, k_2} = -c^{(J-1)}_{k_1, k_2}$, $c^{(J-1)}_{k_1, 2^{J-1} - k_2} = -c^{(J-1)}_{k_1, k_2}$, and $c^{(J-1)}_{2^{J-2}, k_2} = c^{(J-1)}_{0, k_2} = c^{(J-1)}_{k_1, 2^{J-2}} = c^{(J-1)}_{k_1, 0} = 0$. Then, the symmetry of the matrix $(c^{(J-1)}_{k_1, k_2})_{k_1, k_2 = 0, \ldots, 2^{J-1} - 1}$ can be described as

$$\begin{pmatrix}
0 & 0 & \cdots & 0 & 0 & 0 & \cdots & 0 \\
0 & \alpha_{1,1} & \cdots & \alpha_{1,2^{J-2}-1} & 0 & -\alpha_{1,2^{J-2}-1} & \cdots & -\alpha_{1,1} \\
\vdots & \vdots & \vdots & \vdots & \vdots & \vdots & \vdots & \vdots \\
0 & \alpha_{2^{J-2}-1,1} & \cdots & \alpha_{2^{J-2}-1,2^{J-2}-1} & 0 & -\alpha_{2^{J-2}-1,2^{J-2}-1} & \cdots & -\alpha_{2^{J-2}-1,1} \\
\vdots & \vdots & \vdots & \vdots & \vdots & \vdots & \vdots & \vdots \\
0 & -\alpha_{2^{J-2}-1,1} & \cdots & -\alpha_{2^{J-2}-1,2^{J-2}-1} & 0 & \alpha_{2^{J-2}-1,2^{J-2}-1} & \cdots & \alpha_{2^{J-2}-1,1} \\
\vdots & \vdots & \vdots & \vdots & \vdots & \vdots & \vdots & \vdots \\
0 & -\alpha_{1,1} & \cdots & -\alpha_{1,2^{J-2}-1} & 0 & \alpha_{1,2^{J-2}-1} & \cdots & \alpha_{1,1}
\end{pmatrix}$$

where $\alpha_{k_1,k_2} = c_{k_1,k_2}^{(J-1)}$ $(k_1, k_2 = 1, \dots, 2^{J-2} - 1)$.

(ii) $d_{1,2^{J-1}-k_1,k_2}^{(J-1)} = -d_{k_1,k_2}^{(J-1)}$, $d_{1,k_1,2^{J-1}-k_2-1}^{(J-1)} = -d_{1,k_1,k_2}^{(J-1)}$, and $d_{1,2^{J-2},k_2}^{(J-1)} = d_{1,0,k_2}^{(J-1)} = 0$. Then, the symmetry of the matrix $(d_{2,k_1,k_2}^{(J-1)})_{k_1,k_2=0,\dots,2^{J-1}-1}$ can be described as

$$\begin{pmatrix}
0 & 0 & \cdots & 0 & 0 & \cdots & 0 \\
\beta_{1,0} & \beta_{1,1} & \cdots & \beta_{1,2^{J-2}-1} & -\beta_{1,2^{J-2}-1} & \cdots & -\beta_{1,0} \\
\vdots & \vdots & \vdots & \vdots & \vdots & \vdots & \vdots \\
\beta_{2^{J-2}-1,0} & \beta_{2^{J-2}-1,1} & \cdots & \beta_{2^{J-2}-1,2^{J-2}-1} & -\beta_{2^{J-2}-1,2^{J-2}-1} & \cdots & -\beta_{2^{J-2}-1,0} \\
\vdots & \vdots & \vdots & \vdots & \vdots & \vdots & \vdots \\
-\beta_{2^{J-2}-1,0} & -\beta_{2^{J-2}-1,1} & \cdots & -\beta_{2^{J-2}-1,2^{J-2}-1} & -\beta_{2^{J-2}-1,2^{J-2}-1} & \cdots & \beta_{2^{J-2}-1,0} \\
\vdots & \vdots & \vdots & \vdots & \vdots & \vdots & \vdots \\
-\beta_{1,0} & -\beta_{1,1} & \cdots & -\beta_{1,2^{J-2}-1} & \beta_{1,2^{J-2}-1} & \cdots & \beta_{1,0}
\end{pmatrix}$$

where $\beta_{k_1,k_2} = d_{1,k_1,k_2}^{(J-1)}$ $(k_1 = 1, \dots, 2^{J-2} - 1;\ k_2 = 0, 1, \dots, 2^{J-2} - 1)$.

(iii) $d_{2,2^{J-1}-k_1-1,k_2}^{(J-1)} = -d_{2,k_1,k_2}^{(J-1)}$, $d_{2,k_1,2^{J-1}-k_2}^{(J-1)} = -d_{2,k_1,k_2}^{(J-1)}$, and $d_{2,k_1,2^{J-2}}^{(J-1)} = d_{2,k_1,0}^{(J-1)} = 0$. Then, the symmetry of the matrix $(d_{2,k_1,k_2}^{(J-1)})_{k_1,k_2=0,\dots,2^{J-1}-1}$ can be described as

$$\begin{pmatrix}
0 & \gamma_{0,1} & \cdots & \gamma_{0,2^{J-2}-1} & 0 & -\gamma_{2^{J-2}-1,0} & \cdots & -\gamma_{0,1} \\
0 & \gamma_{1,1} & \cdots & \gamma_{2^{J-2}-1,1} & 0 & -\gamma_{2^{J-2}-1,1} & \cdots & -\gamma_{1,1} \\
\vdots & \vdots & \vdots & \vdots & \vdots & \vdots & \vdots & \vdots \\
0 & \gamma_{2^{J-2}-1,1} & \cdots & \gamma_{2^{J-2}-1,2^{J-2}-1} & 0 & -\gamma_{2^{J-2}-1,2^{J-2}-1} & \cdots & -\gamma_{2^{J-2}-1,1} \\
0 & -\gamma_{2^{J-2}-1,1} & \cdots & -\gamma_{2^{J-2}-1,2^{J-2}-1} & 0 & \gamma_{2^{J-2}-1,2^{J-2}-1} & \cdots & \gamma_{2^{J-2}-1,1} \\
\vdots & \vdots & \vdots & \vdots & \vdots & \vdots & \vdots & \vdots \\
0 & -\gamma_{0,1} & \cdots & \gamma_{0,2^{J-2}-1} & 0 & \gamma_{2^{J-2}-1,0} & \cdots & \gamma_{0,1}
\end{pmatrix}$$

where $\gamma_{k_1,k_2} = d_{2,k_1,k_2}^{(J-1)}$ $(k_1 = 0, 1, \dots, 2^{J-2} - 1;\ k_2 = 1, \dots, 2^{J-2} - 1)$.

(iv) $d_{3,2^{J-1}-k_1-1,k_2}^{(J-1)} = -d_{3,k_1,k_2}^{(J-1)}$ and $d_{3,k_1,2^{J-1}-k_2-1}^{(J-1)} = -d_{3,k_1,k_2}^{(J-1)}$. Then, the symmetry of the matrix $(d_{3,k_1,k_2}^{(J-1)})_{k_1,k_2=0,\dots,2^{J-1}-1}$ can be described as

$$\begin{pmatrix} \delta_{0,0} & \delta_{0,1} & \cdots & \delta_{0,2^{J-2}-1} & -\delta_{0,2^{J-2}-1} & \cdots & -\delta_{0,0} \\ \vdots & \vdots & \vdots & \vdots & \vdots & \vdots & \vdots \\ \delta_{2^{J-2}-1,0} & \delta_{2^{J-2}-1,1} & \cdots & \delta_{2^{J-2}-1,2^{J-2}-1} & -\delta_{2^{J-2}-1,2^{J-2}-1} & \cdots & -\delta_{2^{J-2}-1,0} \\ -\delta_{2^{J-2}-1,0} & -\delta_{2^{J-2}-1,1} & \cdots & -\delta_{2^{J-2}-1,2^{J-2}-1} & \delta_{2^{J-2}-1,2^{J-2}-1} & \cdots & \delta_{2^{J-2}-1,0} \\ \vdots & \vdots & \vdots & \vdots & \vdots & \vdots & \vdots \\ -\delta_{0,0} & -\delta_{0,1} & \cdots & -\delta_{2^{J-2}-1,2^{J-2}-1} & \delta_{2^{J-2}-1,2^{J-2}-1} & \cdots & \delta_{0,0} \end{pmatrix}$$

where $\delta_{k_1,k_2} = d^{(J-1)}_{3,k_1,k_2}$ $(k_1, k_2 = 0, 1, \ldots, 2^{J-2} - 1)$.

From (i)–(iv), we see that for $k_1, k_2 = 0, 1, \ldots, 2^{J-1} - 1$, coefficients $\{c^{(J-1)}_{k_1,k_2}\}_{k_1,k_2}$, $\{d^{(J-1)}_{1,k_1,k_2}\}_{k_1,k_2}$, $\{d^{(J-1)}_{2,k_1,k_2}\}_{k_1,k_2}$, and $\{d^{(J-1)}_{3,k_1,k_2}\}_{k_1,k_2}$ are determined by $(2^{J-2} - 1)^2$, $2^{J-2}(2^{J-2} - 1)$, $2^{J-2}(2^{J-2} - 1)$, and 2^{2J-4} values, respectively. Hence, to determine $v^*(x, y)$, the number of required values is

$$(2^{J-2} - 1)^2 + 2^{J-2}(2^{J-2} - 1) + 2^{J-2}(2^{J-2} - 1) + 2^{2J-4} = (2^{J-1} - 1)^2.$$

In the decomposition (3.5.5), to obtain the harmonic function $u(x, y)$, we need $4(2^{J-1} - 1) + 4$ boundary sample points of $f(x, y)$ on the boundary of $[0, \frac{1}{2}]^2$. Since $(2^{J-1} - 1)^2 + 4(2^{J-1} - 1) + 4 = (2^{J-1} + 1)^2$ is exactly equal to the number of sampling points of the target function $f(x, y)$ on $[0, \frac{1}{2}]^2$, the discrete harmonic wavelet transform is not a redundant transform.

3.6 Wavelet Packets

The wavelet packets introduced by Coifman, Meyer, and Wickerhauser in 1992 play an important role in the applications of wavelet analysis. The main tool in obtaining wavelet packets is the so-called splitting trick of wavelet bases.

3.6.1 Continuous Wavelet Packets

For an MRA with scaling function φ and transfer function $H_0(\omega)$, if the $2\pi\mathbb{Z}^d$–periodic functions $H_{\boldsymbol{\mu}}(\omega)$ $(\boldsymbol{\mu} \in \{0, 1\}^d \setminus \{\mathbf{0}\})$ are such that the matrix $(H_{\boldsymbol{\mu}}(\omega + \boldsymbol{\nu}\pi))_{\boldsymbol{\mu},\boldsymbol{\nu}}$ is a unitary matrix, then the corresponding wavelets $\psi_{\boldsymbol{\mu}}$ $(\boldsymbol{\mu} \in \{0, 1\}^d \setminus \{\mathbf{0}\})$ satisfying $\widehat{\psi}_{\boldsymbol{\mu}}(\omega) = H_{\boldsymbol{\mu}}\left(\frac{\omega}{2}\right) \widehat{\varphi}\left(\frac{\omega}{2}\right)$ are orthogonal wavelets.

The algorithm constructing wavelet packet bases:

Let $H_{\boldsymbol{\mu}_l}(\omega) = \sum_{\mathbf{k}\in\mathbb{Z}^d} u_{\boldsymbol{\mu}_l,\mathbf{k}}\, e^{-i(\mathbf{k}\cdot\omega)}$ $(l \in \mathbb{Z}_+)$.

Step 1. Let $\varphi = \psi_0$. Define $\psi_{\boldsymbol{\mu}_1}(\mathbf{t})$ $(\boldsymbol{\mu}_1 \in \{0, 1\}^d)$ as

$$\psi_{\mu_1}(\mathbf{t}) = 2^d \sum_{\mathbf{k}\in\mathbb{Z}^d} u_{\mu_1,\mathbf{k}}\, \psi_0(2\mathbf{t}-\mathbf{k}),$$
$$\widehat{\psi}_{\mu_1}(\boldsymbol{\omega}) = H_{\mu_1}\left(\tfrac{\omega}{2}\right) \widehat{\psi}_0\left(\tfrac{\omega}{2}\right) \qquad (\boldsymbol{\mu}_1 \in \{0,1\}^d),$$

Step 2. Replace ψ_0 by ψ_{μ_1} in Step 1. Define $\psi_{\mu_1,\mu_2}(\mathbf{t})$ as

$$\psi_{\mu_1,\mu_2}(\mathbf{t}) = 2^d \sum_{\mathbf{k}\in\mathbb{Z}^d} u_{\mu_2,\mathbf{k}}\, \psi_{\mu_1}(2\mathbf{t}-\mathbf{k}),$$
$$\widehat{\psi}_{\mu_1,\mu_2}(\boldsymbol{\omega}) = H_{\mu_1}\left(\tfrac{\omega}{2}\right) H_{\mu_2}\left(\tfrac{\omega}{4}\right) \widehat{\psi}_0\left(\tfrac{\omega}{4}\right) \qquad (\boldsymbol{\mu}_1, \boldsymbol{\mu}_2 \in \{0,1\}^d).$$

Again replace ψ_0 by ψ_{μ_1,μ_2} in Step 1. Define $\psi_{\mu_1,\mu_2,\mu_3}(\mathbf{t})$ as

$$\psi_{\mu_1,\mu_2,\mu_3}(\mathbf{t}) = 2^d \sum_{\mathbf{k}\in\mathbb{Z}^d} u_{\mu_3,\mathbf{k}}\, \psi_{\mu_1,\mu_2}(2\mathbf{t}-\mathbf{k}),$$
$$\widehat{\psi}_{\mu_1,\mu_2,\mu_3}(\boldsymbol{\omega}) = H_{\mu_1}\left(\tfrac{\omega}{2}\right) H_{\mu_2}\left(\tfrac{\omega}{4}\right) H_{\mu_3}\left(\tfrac{\omega}{8}\right) \widehat{\psi}_0\left(\tfrac{\omega}{8}\right) \qquad (\boldsymbol{\mu}_1, \boldsymbol{\mu}_2, \boldsymbol{\mu}_3 \in \{0,1\}^d),$$

In general,

$$\psi_{\mu_1,\ldots,\mu_j}(\mathbf{t}) = 2^d \sum_{\mathbf{k}\in\mathbb{Z}^d} u_{\mu_j,\mathbf{k}}\, \psi_{\mu_1,\ldots,\mu_{j-1}}(2\mathbf{t}-\mathbf{k}), \tag{3.6.1}$$

$$\begin{aligned}\widehat{\psi}_{\mu_1,\ldots,\mu_j}(\boldsymbol{\omega}) &= H_{\mu_1}\left(\tfrac{\omega}{2}\right) H_{\mu_2}\left(\tfrac{\omega}{4}\right) \cdots H_{\mu_j}\left(\tfrac{\omega}{2^j}\right) \widehat{\psi}_0\left(\tfrac{\omega}{2^j}\right)\\ &= \widehat{\psi}_0\left(\tfrac{\omega}{2^j}\right) \prod_{\nu=1}^{j} H_{\mu_j}\left(\tfrac{\omega}{2^\nu}\right) \qquad (\boldsymbol{\mu}_1, \ldots, \boldsymbol{\mu}_j \in \{0,1\}^d),\end{aligned}$$

Step 3. Let the set $E = \{0, 1, \ldots, 2^d - 1\}$ and σ is a one-to-one map from $\{0,1\}^d$ to E. Then, each multi-index $(\boldsymbol{\mu}_1, \ldots, \boldsymbol{\mu}_j)$ corresponds a number $l = \sigma(\boldsymbol{\mu}_1) + 2^d\sigma(\boldsymbol{\mu}_2) + \cdots + 2^{(j-1)d}\sigma(\boldsymbol{\mu}_j)$ in $\mathbb{Z}_+$. Denote $W_l(\mathbf{t}) = \psi_{\mu_1,\ldots,\mu_j}(\mathbf{t})$.

Step 4. Denote the dyadic interval:

$$I_{l,m} = \{\tau \in \mathbb{Z}_+ : \; 2^{md} l \le \tau < 2^{md}(l+1)\},$$

where the pair of indices $(l, m) \in \mathbb{Z}_+ \times \mathbb{Z}_+$. If $\{I_{l,m}\}_{(l,m)\in S}$ is a partition for $\mathbb{Z}_+$, i.e.,

$$\bigcup_{(l,m)\in S} I_{l,m} = \mathbb{Z}_+, \qquad I_{l,m} \bigcap I_{l',m'} = \emptyset \quad ((l,m) \neq (l',m')),$$

then $\{2^{\frac{md}{2}} W_l(2^m\mathbf{t} - \mathbf{n})\}_{(l,m)\in S, \mathbf{n}\in\mathbb{Z}^d}$ is a normal orthogonal basis for $L^2(\mathbb{R}^d)$.

3.6.2 Discrete Wavelet Packets

Let $U_{\mu_1,\ldots,\mu_j} = \text{span}\{\psi_{\mu_1,\ldots,\mu_j}(\mathbf{t}-\mathbf{k})\}_{\mathbf{k}\in\mathbb{Z}^d}$. Then, the orthogonal decomposition of the space $L^2(\mathbb{R}^d)$ is

$$L^2(\mathbb{R}^d) = \bigoplus_{l\in\mathbb{Z}} U_l,$$

where $U_l = U_{\mu_1,\dots,\mu_j}$ $(l = \sigma(\mu_1) + 2^d\sigma(\mu_2) + \cdots + 2^{(j-1)d}\sigma(\mu_j))$. Denote

$$U_l^{(m)} = U_{\mu_1,\dots,\mu_j}^{(m)} = \mathrm{span}\{\psi_{\mu_1,\dots,\mu_j}(2^m\mathbf{t} - \mathbf{k})\}_{\mathbf{k}\in\mathbb{Z}^d}.$$

Let $f \in U_l^{(1)}$. Then,

$$\begin{cases} f(\mathbf{t}) = \sum\limits_{\mathbf{k}\in\mathbb{Z}^d} a_{\mu_1,\dots,\mu_j}(\mathbf{k})\, \psi_{\mu_1,\dots,\mu_j}(2\mathbf{t} - \mathbf{k}), \\ f(\mathbf{t}) = \sum\limits_{\mathbf{k}\in\mathbb{Z}^d} \sum\limits_{\mu_{j+1}} b_{\mu_1,\dots,\mu_{j+1}}(\mathbf{k})\, \psi_{\mu_1,\dots,\mu_{j+1}}(\mathbf{t} - \mathbf{k}), \end{cases} \tag{3.6.2}$$

where

$$a_{\mu_1,\dots,\mu_j}(\mathbf{k}) = 2^d \int_{\mathbb{R}^d} f(\mathbf{t})\, \overline{\psi}_{\mu_1,\dots,\mu_j}(2\mathbf{t} - \mathbf{k})\mathrm{d}\mathbf{t},$$
$$b_{\mu_1,\dots,\mu_{j+1}}(\mathbf{k}) = \int_{\mathbb{R}^d} f(\mathbf{t})\, \overline{\psi}_{\mu_1,\dots,\mu_{j+1}}(\mathbf{t} - \mathbf{k})\mathrm{d}\mathbf{t}$$

are called *wavelet packet coefficients*. By (3.6.1), the *decomposition formula of wavelet packet coefficients* is

$$\begin{aligned} b_{\mu_1,\dots,\mu_{j+1}}(\mathbf{k}) &= 2^d \sum_{\mathbf{l}\in\mathbb{Z}^d} u_{\mu_{j+1},\mathbf{l}} \int_{\mathbb{R}^d} f(\mathbf{t})\, \overline{\psi}_{\mu_1,\dots,\mu_j}(2\mathbf{t} - 2\mathbf{k} - \mathbf{l})\mathrm{d}\mathbf{t} \\ &= 2^d \sum_{\mathbf{l}\in\mathbb{Z}^d} u_{\mu_{j+1},\mathbf{l}-2\mathbf{k}} \int_{\mathbb{R}^d} f(\mathbf{t})\, \overline{\psi}_{\mu_1,\dots,\mu_j}(2\mathbf{t} - \mathbf{l})\mathrm{d}\mathbf{t} \\ &= 2^d \sum_{\mathbf{l}\in\mathbb{Z}^d} u_{\mu_{j+1},\mathbf{l}-2\mathbf{k}} a_{\mu_1,\dots,\mu_j}(\mathbf{l}). \end{aligned}$$

Substituting (3.6.1) into the second formula in (3.6.2), we get

$$\begin{aligned} f(\mathbf{t}) &= 2^d \sum_{\mathbf{k}\in\mathbb{Z}^d} \sum_{\mu_{j+1}} b_{\mu_1,\dots,\mu_{j+1}}(\mathbf{k}) \sum_{\mathbf{l}\in\mathbb{Z}^d} u_{\mu_{j+1},\mathbf{l}} \psi_{\mu_1,\dots,\mu_j}(2\mathbf{t} - 2\mathbf{k} - \mathbf{l}) \\ &= 2^d \sum_{\mathbf{k}\in\mathbb{Z}^d} \left(\sum_{\mu_{j+1}} \sum_{\mathbf{l}\in\mathbb{Z}^d} u_{\mu_{j+1},\mathbf{k}-2\mathbf{l}} b_{\mu_1,\dots,\mu_{j+1}}(\mathbf{k}) \right) \psi_{\mu_1,\dots,\mu_j}(2\mathbf{t} - \mathbf{k}). \end{aligned}$$

Since $\psi_{\mu_1,\dots,\mu_j}(2\mathbf{t} - \mathbf{k})(\mathbf{k} \in \mathbb{Z}^d)$ are orthogonal, comparing it with the first formula (3.6.2), the *reconstruction formula of wavelet packet coefficients* is

$$a_{\mu_1,\dots,\mu_j}(\mathbf{k}) = 2^d \sum_{\mu_{j+1}} \sum_{\mathbf{l}\in\mathbb{Z}^d} u_{\mu_{j+1},\mathbf{k}-2\mathbf{l}} b_{\mu_1,\dots,\mu_{j+1}}(\mathbf{k}).$$

3.7 Wavelet Variance

Wavelets can decompose a stochastic process into subprocesses, each of which is associated with a particular scale. A wavelet variance is the variance of a subprocess at a given scale and quantifies the amount of variation presented at that particular scale, so it has been used widely to analyze various multiscale processes in atmospheric and oceanic dynamics.

3.7.1 Generalized Wavelet Decomposition

Given the time series $\mathbf{S} = (S_0, \ldots, S_{N-1})^T$ $(N = 2^M)$, the pyramid algorithm for generalized wavelet decomposition is as follows:

Step 1. Define a (wavelet) filter $\{b_l\}_{l=0,\ldots,L-1}$ $(b_0, b_{L-1} \neq 0)$ with even length L satisfying the conditions:

$$\sum_{l=0}^{L-1} b_l = 0, \qquad 2\sum_{l=0}^{L-1} b_l b_{l+2n} = \delta_{0,n} \qquad (n \in \mathbb{Z},\ b_l = 0\ (l < 0,\ l > L)).$$

Define a corresponding (scaling) filter $\{a_l\}_{l=0,\ldots,L-1}$ as $a_l = (-1)^{l+1} b_{L-1-l}$. This implies that

$$2\sum_{l=0}^{L-1} a_l a_{l+2n} = \delta_{0,n}, \qquad \sum_{l=0}^{L-1} a_l b_{l+2n} = 0 \qquad (n \in \mathbb{Z}).$$

Without loss of generality, assume that $\sum_{l=0}^{L-1} a_l = 1$. Let $A(\omega)$ and $B(\omega)$ be transfer functions for $\{a_l\}_{l=0,\ldots,L-1}$ and $\{b_l\}_{l=0,\ldots,L-1}$, respectively,

$$A(\omega) = \sum_{l=0}^{L-1} a_l\, e^{-il\omega}, \qquad B(\omega) = \sum_{l=0}^{L-1} b_l\, e^{-il\omega}.$$

Then, the matrix $\begin{pmatrix} B(\omega) & B(\omega+\pi) \\ A(\omega) & A(\omega+\pi) \end{pmatrix}$ is a unitary matrix.

Step 2. Let the most coarse scaling coefficients be

$$\mathbf{V}_0 = (V_{0,0}, V_{0,1}, \ldots, V_{0,N-1})^T = (S_0, S_1, \ldots, S_{N-1})^T.$$

This yields wavelet coefficients $\mathbf{W}_{-1}$ and scaling coefficients $\mathbf{V}_{-1}$:

$$\mathbf{W}_{-1} = \left(W_{-1,0}, W_{-1,1}, \ldots, W_{-1,\frac{N}{2}-1}\right)^T,$$
$$\mathbf{V}_{-1} = \left(V_{-1,0}, V_{-1,1}, \ldots, V_{-1,\frac{N}{2}-1}\right)^T,$$

where

$$W_{-1,k} = \sum_{l=0}^{L-1} b_l V_{0,2k+1-l \, mod \, \frac{N}{2}},$$
$$V_{-1,k} = \sum_{l=0}^{L-1} a_l V_{0,2k+1-l \, mod \, \frac{N}{2}} \; (k = 0, \ldots, \tfrac{N}{2} - 1).$$

Step 3. Continuing the same procedure as Step 2, we start from $\mathbf{V}_{-1}$ to yield wavelet coefficients and scaling coefficients:

$$\mathbf{W}_{-2} = \left(W_{-2,0}, W_{-2,1}, \ldots, W_{-2,\frac{N}{2^2}-1}\right)^T,$$
$$\mathbf{V}_{-2} = \left(V_{-2,0}, V_{-2,1}, \ldots, V_{-2,\frac{N}{2^2}-1}\right)^T.$$

Repeating the above process again and again, we can start from scaling coefficients

$$\mathbf{V}_{-m+1} = \left(V_{-m+1,0}, V_{-m+1,1}, \ldots, V_{-m+1,\frac{N}{2^{m-1}}-1}\right)^T$$

to yield wavelet coefficients and scaling coefficients:

$$\mathbf{W}_{-m} = \left(W_{-m,0}, W_{-m,1}, \ldots, W_{-m,\frac{N}{2^m}-1}\right)^T,$$
$$\mathbf{V}_{-m} = \left(V_{-m,0}, V_{-m,1}, \ldots, V_{-m,\frac{N}{2^m}-1}\right)^T,$$

where

$$W_{-m,k} = \sum_{l=0}^{L-1} b_l V_{m-1,2k+1-l \, mod \, \frac{N}{2^m}},$$
$$V_{-m,k} = \sum_{l=0}^{L-1} a_l V_{m-1,2k+1-l \, mod \, \frac{N}{2^m}} \; (k = 0, 1, \ldots, \tfrac{N}{2^m} - 1).$$

Finally, we get M_0 vectors of wavelet coefficients $\mathbf{W}_{-1}, \ldots, \mathbf{W}_{-M_0}$ and a vector $\mathbf{V}_{-M_0}$ of scaling coefficients, and the following formula:

$$\| \mathbf{S} \|^2 = \sum_{m=1}^{M_0} \| \mathbf{W}_{-m} \|^2 + \| \mathbf{V}_{-M_0} \|^2,$$

3.7.2 Maximal Overlap Discrete Wavelet Transform

With the help of this pyramid algorithm, we can obtain $\mathbf{W}_{-m}$ and $\mathbf{V}_{-m}$ directly from the time series $\mathbf{S}$ as

$$\begin{aligned} W_{-m,k} &= \sum_{l=0}^{L_m-1} b_{m,l} S_{2^m(k+1)-1-l \bmod N}, \\ V_{-m,k} &= \sum_{l=0}^{L_m-1} a_{m,l} S_{2^m(k+1)-1-l \bmod N}, \end{aligned} \quad (3.7.1)$$

where $\{a_{m,l}\}_{l=0,\ldots,L_m-1}$ and $\{b_{m,l}\}_{l=0,\ldots,L_m-1}$ are the m-order scaling and wavelet filters for the time series $\mathbf{S}$, respectively, and $L_m = (2^m - 1)(L-1) + 1$, where $\|\cdot\|$ is the Euclidean norm.

The maximal overlap discrete wavelet transform (MODWT) is a modified version of generalized wavelet decomposition. MODWT coefficients $\widetilde{\mathbf{W}}_m$ and $\widetilde{\mathbf{V}}_m$ are defined as

$$\widetilde{\mathbf{W}}_m = \left(\widetilde{W}_{m,0}, \widetilde{W}_{m,1}, \ldots, \widetilde{W}_{m,N-1}\right)^T,$$

$$\widetilde{\mathbf{V}}_m = \left(\widetilde{V}_{m,0}, \widetilde{V}_{m,1}, \ldots, \widetilde{V}_{m,N-1}\right)^T,$$

where

$$\widetilde{W}_{m,k} = \sum_{l=0}^{L_m-1} \widetilde{b}_{m,l} S_{k-l \bmod N}, \quad \widetilde{b}_{m,l} = 2^{-\frac{m}{2}} b_{m,l},$$

$$\widetilde{V}_{m,k} = \sum_{l=0}^{L_m-1} \widetilde{a}_{m,l} S_{k-l \bmod N}, \quad \widetilde{a}_{m,l} = 2^{-\frac{m}{2}} a_{m,l}.$$

Here, $\{b_{m,l}\}$ and $\{a_{m,l}\}$ are stated in (3.7.1) and $L_m = (2^m - 1)(L-1) + 1$. The corresponding formula becomes

$$\| \mathbf{S} \|^2 = \sum_{m=1}^{M_0} \| \widetilde{\mathbf{W}}_m \|^2 + \| \widetilde{\mathbf{V}}_{M_0} \|^2 .$$

3.7.3 Wavelet Variance

Assume that $\mathbf{S} = \{S_t\}_{t\in\mathbb{Z}}$ be a real-valued stationary stochastic process with mean 0. Let $\{\widetilde{b}_{m,l}\}_{l=0,\ldots,L_m-1}$ be the mth level MODWT wavelet filter. Define the mth level wavelet coefficient process by

$$\overline{W}_{m,t} = \sum_{l=0}^{L_m-1} \widetilde{b}_{m,l} S_{t-l} \qquad (t \in \mathbb{Z}),$$

where $L_m = (2^m - 1)(L-1) + 1$, and let $\widetilde{H}_m^{(r)}(\omega) = \sum_{l=0}^{L_m-1} \widetilde{b}_{m,l}\, e^{-il\omega}$. Then, the spectral density function is

$$P_m(\omega) = |\widetilde{H}_m^{(r)}(\omega)|^2 P_S(\omega). \tag{3.7.2}$$

The wavelet variance of stochastic process $\{S_t\}_{t\in\mathbb{Z}}$ at scale 2^{m-1} is defined as the variance $\nu_S^2(2^{m-1})$ of wavelet coefficients $\overline{W}_{m,t}$, i.e., $\nu_S^2(2^{m-1}) = \text{Var}\,(\overline{W}_{m,t})$. Since $\{\overline{W}_{m,t}\}_{t\in\mathbb{Z}}$ is stationary, $\text{Var}(\overline{W}_{m,t})$ is independent of t. Since $E[S_t] = 0$, we have $E[\overline{W}_{m,t}] = 0$. By (3.7.2),

$$\text{Var}(\overline{W}_{m,t}) = E[|\overline{W}_{m,t}|^2] = \frac{1}{2\pi}\int_{-\pi}^{\pi} P_m(\omega)\mathrm{d}\omega = \frac{1}{2\pi}\int_{-\pi}^{\pi} |\widetilde{H}_m^{(r)}(\omega)|^2 P_S(\omega)\mathrm{d}\omega,$$

so $\sum_{m\in\mathbb{Z}_+} \nu_S^2(2^{m-1}) = \text{Var}(S_t)$, i.e., $\nu_S^2(2^{m-1})$ is the contribution to the total variability in $\{S_t\}$ at scale 2^{m-1}.

Assuming that only the first N terms of S_t are known, an unbiased estimator of $\nu_S^2(2^{m-1})$ is

$$\hat{\nu}_S^2(2^{m-1}) = \frac{1}{R_m}\sum_{t=L_m-1}^{N-1} \widetilde{W}_{m,t}^2,$$

where $R_m = N - L_m + 1$ and $\{\widetilde{W}_{m,t}\}$ is the mth level MODWT wavelet coefficients for the time series $\{S_t\}$:

$$\widetilde{W}_{m,t} = \sum_{l=0}^{L_m-1} \widetilde{b}_{m,l} S_{t-l \bmod N} \qquad (t = 0, \ldots, N-1).$$

From $\widetilde{W}_{m,t} = \overline{W}_{m,t}$ $(t = L_m - 1, \ldots, N-1)$, it follows that $E[\widetilde{W}_{m,t}^2] = \nu_S^2(2^{m-1})$. When R_m is sufficiently large, the stochastic variable $\hat{\nu}_S^2(2^{m-1})$ is approximately a Gaussian distributed with mean $\nu_S^2(2^{m-1})$ and variance:

$$\text{Var}(\nu_S^2(2^{m-1})) = \frac{1}{\pi R_m}\int_{-\pi}^{\pi} P_m(\omega)\mathrm{d}\omega.$$

From this, the confidence interval for the estimator can be given.

3.8 Significant Tests

Significance tests are to distinguish statistically significant results from pure randomness when only one or few samples are given. The significance test for the wavelet analysis is undoubtedly important since it can extract the real features of climatic and environmental time series from background noise. If a peak in the wavelet power spectrum of time series is significantly well above that of background white noise or red noise (AR(1) time series), then it is a true feature of time series.

A wavelet is a function $\psi \in L^2(\mathbb{R})$ with $\int_{\mathbb{R}} \psi(t)dt = 0$. Examples of popular wavelets include the following:

(a) Morlet wavelet is $\psi^M(t) = \pi^{-\frac{1}{4}} e^{it\omega_0} e^{-\frac{t^2}{2}}$ $(\omega_0 \geq 6)$.

(b) Mexican hat is $\psi^{MH}(t) = -\frac{1}{\sqrt{\Gamma(2.5)}}(1-t^2)\, e^{-\frac{t^2}{2}}$, where Γ is the Gamma function.

(c) Haar wavelet is $H(t) = \begin{cases} 1, & 0 < t \leq \frac{1}{2}, \\ -1, & -\frac{1}{2} < t < 0, \\ 0, & \text{otherwise}. \end{cases}$

Define a *continuous wavelet transform* of $f \in L^2(\mathbb{R})$ as

$$(W_\psi f)(b,a) = \frac{1}{\sqrt{|a|}} \int_{\mathbb{R}} f(t) \overline{\psi}\left(\frac{t-b}{a}\right) dt \qquad (a \neq 0,\ b \in \mathbb{R}), \tag{3.8.1}$$

where a is called the *dilation parameter* and b is called the *translation parameter*.

The wavelet transform $(W_\psi f)(b,a)$ has the capacity of representing local characteristics in the time and frequency domains. The time-frequency window of wavelet transform is:

$$[b + at^* - |a|\Delta_\psi,\ b + at^* + |a|\Delta_\psi] \times \left[\frac{\omega^*}{a} - \frac{\Delta_{\widehat{\psi}}}{|a|},\ \frac{\omega^*}{a} + \frac{\Delta_{\widehat{\psi}}}{|a|}\right],$$

where t^* and ω^* are the centers of ψ and its Fourier transform $\widehat{\psi}$ respectively, and Δ_ψ and $\Delta_{\widehat{\psi}}$ are radii of ψ and $\widehat{\psi}$, respectively. From this, we can see that the low frequency part of the wavelet transform has a lower time resolution and high frequency resolution, while the high frequency part has the high time resolution and lower frequency resolution; therefore, the wavelet transform is suitable for detection of the time series which contains transient anomalies.

If the wavelet ψ satisfies the admissibility condition $C_\psi = \int_{\mathbb{R}} \frac{|\widehat{\psi}(\omega)|^2}{|\omega|} d\omega < \infty$, the *inverse wavelet transform* is

$$f(t) = \frac{1}{C_\psi} \int_{\mathbb{R}^2} (W_\psi f)(b,a) \frac{1}{|a|^{\frac{5}{2}}} \psi\left(\frac{t-b}{a}\right) da db.$$

The *d−variate wavelet transform* of $f \in L^2(\mathbb{R}^d)$ is defined as

$$(W_\psi f)(\mathbf{b}, a) = a^{-\frac{d}{2}} \int_{\mathbb{R}^d} f(\mathbf{t}) \overline{\psi}\left(\frac{\mathbf{t}-\mathbf{b}}{a}\right) d\mathbf{t} \qquad (a > 0,\ \mathbf{b} \in \mathbb{R}^d),$$

where $\psi \in L^2(\mathbb{R}^d)$ is the wavelet function. If ψ is a radial function, then Fourier transform of ψ is also a radial function, denoted by $\eta(|\omega|) = \widehat{\psi}(\omega)$. Under the admissibility condition $C_\psi^{(d)} = (2\pi)^d \int_0^\infty \frac{|\eta(t)|^2}{t} dt < \infty$, the *inverse d-variate wavelet transform* is

$$f(\mathbf{t}) = \frac{1}{C_\psi^{(d)}} \int_0^\infty \frac{da}{a^{\frac{3d}{2}+1}} \int_{\mathbb{R}^d} (W_\psi f)(\mathbf{b}, a)\, \psi\left(\frac{\mathbf{t}-\mathbf{b}}{a}\right) d\mathbf{b} \qquad (a > 0,\ \mathbf{b} \in \mathbb{R}^d).$$

For a stochastic process $S(\mathbf{t})$, the *wavelet power spectrum* $(W_\psi S)(\mathbf{b}, a)$ is defined as $|(W_\psi S)(\mathbf{b}, a)|^2$ and the *wavelet phase* is $\varphi = \tan^{-1} \frac{\mathrm{Im}(W_\psi S)(\mathbf{b},a)}{\mathrm{Re}(W_\psi S)(\mathbf{b},a)}$.

3.8.1 Haar Wavelet Analysis

In climatic or environmental time series, AR(1) time series is used often as the background noise model for climatic and environmental time series. A discrete stochastic process $\{x_n\}_{n=0,\dots,N-1}$ is called an AR(1) time series with parameter α $(0 \le \alpha \le 1)$ if

$$x_n = \alpha x_{n-1} + z_n \quad (n = 1, \dots, N-1), \qquad x_0 = 0, \tag{3.8.2}$$

where $\{z_n\}_{n\in\mathbb{Z}_+}$ is the white noise and satisfies

$$E[z_n] = 0,$$
$$E[z_k z_l] = \delta_{k,l}\sigma^2,$$

and $\{z_n\}_{n\in\mathbb{Z}_+}$ is an independent normal distribution. Using (3.8.1) successively implies that $x_n = \sum_{k=1}^n \alpha^{n-k} z_k$, so each x_n is a Gaussian stochastic variable with mean 0.

Based on the discrete version of (3.8.1), the *wavelet transform* $W_\nu(s)$ *of the time series* $\{x_n\}_{n=0,\dots,N-1}$ associated with Haar function $H(x)$ is defined as

$$W_\nu(s) = \sum_{n=0}^{N-1} x_n H\left(\frac{(n-\nu)\delta t}{s}\right)\sqrt{\frac{\delta t}{s}}, \tag{3.8.3}$$

where δt is a sample period. It is a Gaussian stochastic variable with mean 0. Choose ν, s such that $\frac{s}{2\delta t} \le \nu \le N - 1 - \frac{s}{2\delta t}$. Let $s^* = \frac{s}{2\delta t}$. This implies that

$$\begin{aligned} 2s^*\mathrm{Var}(W_\nu(s)) &= E\left[\left(\sum_{n,m=\nu-s^*}^{\nu+s^*} x_n H\left(\tfrac{(n-\nu)\delta t}{s}\right)\right)^2\right] \\ &= \sum_{n,m=\nu-s^*}^{\nu+s^*} E[x_n x_m] H\left(\tfrac{m-\nu}{2s^*}\right) H\left(\tfrac{n-\nu}{2s^*}\right). \end{aligned} \tag{3.8.4}$$

Since $E[z_k z_l] = \delta_{k,l}\sigma^2$, by (3.8.2), it is clear that

$$E[x_n x_m] = \sum_{k=1}^{\lambda_{mn}} \alpha^{n-k}\alpha^{m-k} E[z_k^2] = \frac{\sigma^2}{\alpha^2-1}\alpha^{m+n}(1-\alpha^{-2\lambda_{mn}}),$$

where $\lambda_{mn} = \min\{m, n\}$. It can be rewritten into

$$E[x_n x_m] = \frac{\sigma^2}{\alpha^2-1}(\alpha^{m+n} - \alpha^{|m-n|}).$$

By the definition of Haar function,

$$H\left(\frac{n-\nu}{2s^*}\right) = \begin{cases} 1, & \nu < n \le \nu + s^*, \\ -1, & \nu - s^* < n < \nu, \\ 0, & \text{otherwise.} \end{cases}$$

From this and (3.8.4), a direct computation shows that

$$2s^*\left(\frac{1-\alpha^2}{\sigma^2}\right)\mathrm{Var}(W_\nu(s)) = J_1 - J_2,$$

where

$$J_1 = \frac{2s^*(1-\alpha^2)-4\alpha(1-\alpha^{s^*})-2\alpha^2(1-\alpha^{s^*})^2}{(1-\alpha)^2},$$

$$J_2 = \frac{\alpha^{2\nu}(1-\alpha^{s^*})^2(\alpha^{-s^*}-\alpha)^2}{(1-\alpha)^2}.$$

Then, the Haar wavelet power spectrum of an AR(1) time series with parameters α and σ^2 is

$$|W_\nu(s)|^2 \Rightarrow \frac{\sigma^2(J_1-J_2)}{2s^*(1-\alpha^2)}\chi_1^2 \qquad (s^* = \frac{s}{2\delta t}),$$

where "$\Rightarrow$" means "is distributed as" and χ_1^2 is the chi-square distribution with one degree of freedom, and δt is the sample period.

The *modulated Haar wavelet* is defined as $\psi^{(k)}(t) = e^{i4k\pi t}H(t)$, where $k \in \mathbb{Z}$ and $H(t)$ is the Haar wavelet.

For a white noise $\{z_n\}_{n=0,1,\dots}$ with variance σ^2, denote by $\widetilde{W}_\nu^{(k)}(s)(\nu \ge s^*)$ its wavelet transform associated with $\psi^{(k)}$. Then,

$$|\widetilde{W}_\nu^{k)}(s)|^2 \Rightarrow \frac{\sigma^2}{2}\chi_2^2.$$

and the tangent of its wavelet phase is distributed as the standard Cauchy distribution, i.e.,

$$\tan\left(\arg \widetilde{W}_\nu^{(k)}(s)\right) = \frac{\mathrm{Re}\widetilde{W}_\nu^{(k)}(s)}{\mathrm{Im}\widetilde{W}_\nu^{(k)}(s)}.$$

For the case $\nu < s^*$, $\widetilde{W}_\nu^{(k)}(s)$ is distorted due to the boundary effect of AR(1) time series at the point 0, so one often ignores the value of $\widetilde{W}_\nu^{(k)}(s)$ for $\nu < s^*$.

Let $\{x_n\}_{n=0,1,..}$ be an AR(1) process with parameters α and σ^2. Its wavelet transform associated with the wavelet $\psi^{(k)}$ is

$$W_\nu^{(k)}(s) = \frac{1}{\sqrt{2s^*}} \sum_{n=0}^{\infty} x_n \overline{\psi^{(k)}}\left(\frac{n-\nu}{2s^*}\right) \qquad (s^* = \frac{s}{2\delta t}).$$

When $\nu \geq s^*$, $\alpha^\nu \approx 0$, we have the following:

(a) $\mathrm{Re}W_\nu^{(k)}(s)$ is distributed as $\sigma_1 X_1$, where X_1 is a Gaussian stochastic variable with mean 0 and variance 1 and $\sigma_1^2 = \frac{\sigma^2\sigma_R^2}{2s^*(1-\alpha^2)}$, where

$$\sigma_R^2 = s^* P_k(\alpha) + \frac{4\alpha(1-\alpha^{s^*})}{(1-\alpha^2)^2}\left(\alpha - \cos\frac{2k\pi}{s^*}\right)\left(1 - \alpha\cos\frac{2k\pi}{s^*}\right) P_k^2(\alpha)$$
$$- \frac{2\alpha^2(1-\alpha^{s^*})^2}{(1-\alpha^2)^2}\left(\alpha - \cos\frac{2k\pi}{s^*}\right)^2 P_k^2(\alpha)$$
$$P_k(\alpha) = \frac{1-\alpha^2}{1 - 2\alpha\cos\frac{2k\pi}{s^*} + \alpha^2};$$

(b) $\mathrm{Im}W_\nu^{(k)}(s)$ is distributed as $\sigma_2 X_2$, where X_2 is a Gaussian stochastic variable with mean 0 and variance 1 and $\sigma_2^2 = \frac{\sigma^2\sigma_I^2}{2s^*(1-\alpha^2)}$, where

$$\sigma_I^2 = s^* P_k(\alpha) + \frac{4\alpha^2(1-\alpha^{s^*})}{(1-\alpha^2)^2}\sin^2\frac{2k\pi}{s^*} P_k^2(\alpha) - \frac{2\alpha^2(1-\alpha^{s^*})^2}{(1-\alpha^2)^2}\sin^2\frac{2k\pi}{s^*} P_k^2(\alpha)$$

and $P_k(\alpha)$ is stated as above;

(c) $\mathrm{Re}W_\nu^{(k)}(s)$ and $\mathrm{Im}W_\nu^{(k)}(s)$ are independent. The wavelet power spectrum:

$$|W_\nu^{(k)}(s)|^2 = \mathrm{Re}^2 W_\nu^{(k)}(s) + \mathrm{Im}^2 W_\nu^{(k)}(s)$$

is distributed as $\frac{\sigma^2}{s^*(1-\alpha^2)}(\sigma_R^2 X_1^2 + \sigma_I^2 X_2^2)$, where X_1 and X_2 are two independent Gaussian stochastic variables with mean 0 and variance 1 and σ_I and σ_R are stated as above, and the tangent of the wavelet phase:

$$\tan\varphi = \frac{\mathrm{Im}W_\nu(s)}{\mathrm{Re}W_\nu(s)}$$

is distributed as $\frac{\sigma_I}{\sigma_R}Y$, where Y is a stochastic variable with standard Cauchy distribution.

Compared with other wavelet transforms, it can be proved that significant tests on Haar wavelet transform can extract more information in low frequency domain.

3.8.2 Morlet Wavelet Analysis

Let $x(t)$ be a continuous white noise, i.e.,

$$E[x(t)] = 0,$$
$$E[x(t_1)x(t_2)] = \sigma^2 \delta(t_1 - t_2),$$

where δ is the Dirac function. Using Morlet wavelet $\psi^M(t) = \pi^{-\frac{1}{4}} e^{i\omega_0 t} e^{-\frac{t^2}{2}}$, where the parameter $\omega_0 \geq 6$, the Morlet wavelet transform of $x(t)$ is

$$(W_{\psi^M} x)(b, a) = \frac{1}{\sqrt{|a|}} \int_{\mathbb{R}} x(t) \, \overline{\psi^M} \left(\frac{t-b}{a} \right) \mathrm{d}t \qquad (a \neq 0, \ b \in \mathbb{R}).$$

In a rigorous statistical framework, Zhang and Moore (2012) showed that for any $a > 0$ and $b \in \mathbb{R}$, $\mathrm{Re}(W_{\psi^M} x)(b, a)$ and $\mathrm{Im}(W_{\psi^M} x)(b, a)$ are independent and the wavelet power of $x(t)$ is distributed as

$$|(W_{\psi^M} x)(b, a)|^2 \Rightarrow \frac{\sigma^2}{2} \left(1 + e^{-\omega_0^2}\right) X_1^2 + \frac{\sigma^2}{2} \left(1 - e^{-\omega_0^2}\right) X_2^2,$$

where X_1 and X_2 are independent Gaussian stochastic variables with mean 0 and variance 1. For $\omega_0 \geq 6$, $e^{-\omega_0^2} \approx 0$, and so

$$|(W_{\psi^M} x)(b, a)|^2 \Rightarrow \frac{\sigma^2}{2}(X_1^2 + X_2^2) \quad \text{or} \quad |(W_{\psi^M} x)(b, a)|^2 = \frac{\sigma^2}{2} \chi_2^2.$$

For the Morlet wavelet power spectra of an AR(1) time series, Torrence and Compo (Torrence and Compo (1998)) used the Monte Carlo method to give the following empirical formula:

$$\frac{|W_\nu(s)|^2}{\widetilde{\sigma}^2} \quad \text{is distributed as} \quad \frac{1 - \alpha^2}{2(1 - 2\alpha \cos(\frac{2\pi k}{N}) + \alpha^2)} \chi_2^2$$

where $\widetilde{\sigma}^2$ is the variance of AR(1) time series and the Fourier frequency k corresponds to the wavelet scale s.

3.9 Wavelet Threshold and Shrinkage

A signal $f = \{f_n\}_{n=0,\dots,N-1}$ is contaminated by the addition of a noise $W = \{W_n\}_{n=0,\dots,N-1}$:

$$X_n = f_n + W_n \qquad (n = 0, \dots, N-1).$$

Assuming that W_n is a zero-mean white noise with variance σ^2, the support of the signal f_n is normalized to $[0, 1]$ and $N = 2^L$ samples spaced N^{-1}. Let $\psi_{j,m}(n)$ ($J < j < L$, $0 \le m < 2^j$) be a discrete orthogonal wavelet basis, which is translated modulo modifications near the boundaries, and φ be an orthogonal scaling function. Then, the wavelet expansion of the signal $f = \{f_n\}_{n=0,\dots,N-1}$ is

$$f_n = \sum_{j=J+1}^{L} \sum_{m=0}^{2^j} (f, \psi_{j,m})\psi_{j,m}(n) + \sum_{m=0}^{2^J} (f, \varphi_{J,m})\varphi_{J,m}(n);$$

the wavelet expansion of the white noise $W = \{W_n\}_{n=0,\dots,N-1}$ is

$$W_n = \sum_{j=J+1}^{L} \sum_{m=0}^{2^j} (W, \psi_{j,m})\psi_{j,m}(n) + \sum_{m=0}^{2^J} (W, \varphi_{J,m})\varphi_{J,m}(n);$$

and the wavelet expansion of the noisy signal $X = \{X_n\}_{n=0,\dots,N-1}$ is

$$X_n = \sum_{j=J+1}^{L} \sum_{m=0}^{2^j} (X, \psi_{j,m})\psi_{j,m}(n) + \sum_{m=0}^{2^J} (X, \varphi_{J,m})\varphi_{J,m}(n),$$

3.9.1 *Wavelet Threshold*

A hard thresholding estimator is defined as

$$\rho_\delta^H(t) = \begin{cases} t, & |t| > \delta, \\ 0, & |t| \le \delta, \end{cases}$$

and then the corresponding thresholding operator acting on X is

$$F_{\delta,n}^H(X) = \sum_{j=J+1}^{L} \sum_{m=0}^{2^j} \rho_\delta^H((X, \psi_{j,m}))\, \psi_{j,m}(n) + \sum_{m=0}^{2^J} \rho_\delta^H((X, \varphi_{J,m}))\, \varphi_{J,m}(n) \qquad (n = 0, \dots, N-1).$$

The thresholding risk is

$$r(F_\delta^H(X),\ f) = E[\|\ F_\delta^H(X) - f\ \|^2]$$
$$= \sum_{j=J+1}^{L}\sum_{m=0}^{2^j} E[\,|(f,\ \psi_{j,m}) - \rho_\delta^H((X,\ \psi_{j,m}))|^2] + \sum_{m=0}^{2^J} E[\,|(f,\ \varphi_{J,m}) - \rho_\delta^H((X,\ \varphi_{J,m}))|^2],$$

where $F_\delta^H(X) = \{F_{\delta,n}^H(X)\}_{n=0,\dots,N-1}$ and $\|\cdot\|$ is the Euclidean norm. Note that

$$|(f,\ \psi_{j,m}) - \rho_\delta^H((X,\ \psi_{j,m}))|^2 = |(W,\ \psi_{j,m})|^2 \quad (|(X,\ \psi_{j,m})| > \delta),$$
$$|(f,\ \psi_{j,m}) - \rho_\delta^H((X,\ \psi_{j,m}))|^2 = |(f,\ \psi_{j,m})|^2 \quad (|(X,\ \psi_{j,m})| \le \delta),$$
$$|(f,\ \varphi_{J,m}) - \rho_\delta^H((X,\ \varphi_{J,m}))|^2 = |(W,\ \varphi_{J,m})|^2 \quad (|(X,\ \varphi_{J,m})| > \delta),$$
$$|(f,\ \varphi_{J,m}) - \rho_\delta^H((X,\ \varphi_{J,m}))|^2 = |(f,\ \varphi_{J,m})|^2 \quad (|(X,\ \varphi_{J,m})| \le \delta).$$

The threshold risk satisfies

$$r(F_\delta^H(X),\ f) \ge \sum_{j=J+1}^{L}\sum_{m=0}^{2^j} \min(|(f,\ \psi_{j,m})|^2,\ \sigma^2) + \sum_{m=0}^{2^J} \min(|(f,\ \varphi_{J,m})|^2,\ \sigma^2) =: r_p(f). \tag{3.9.1}$$

When $\delta = \sigma\sqrt{2\log N}$, Donoho and Johnstone (1998) gave the upper bound of the thresholding risk:

$$r(F_\delta^H(X),\ f) \le 2(\log N + 1)(\sigma^2 + r_p(f)), \tag{3.9.2}$$

3.9.2 *Wavelet Shrinkage*

The hard thresholding operator decreases the amplitude of small coefficients in order to reduce the added noise, while the following soft thresholding operator decreases the amplitude of all noisy coefficients:

$$\rho_\delta^s(t) = \begin{cases} t-\delta,\ t \ge \delta, \\ t+\delta,\ t \le -\delta, \\ 0, \qquad |t| \le \delta. \end{cases} \tag{3.9.3}$$

The soft thresholding of wavelet coefficients is called *wavelet shrinkage*. The corresponding soft thresholding operator is

$$F_{\delta,n}^{s}(X) = \sum_{j=J+1}^{L} \sum_{m=0}^{2^j} \rho_\delta^s((X,\ \psi_{j,m}))\,\psi_{j,m}(n)$$
$$+ \sum_{m=0}^{2^J} \rho_\delta^s((X,\ \varphi_{J,m}))\,\varphi_{J,m}(n) \qquad (n = 0, \ldots, N-1).$$

The upper bound of the thresholding risk is similar to hard thresholding in (3.9.2), and the factor $2 \log N$ cannot be improved. When $r_p(f)$ in (3.9.1) is small, f can be well reconstructed by a few wavelet coefficients. The threshold δ must be chosen such that the absolute values of wavelet coefficients of noise W have a high probability of being $\leq \delta$. At the same time, δ cannot be chosen too large such that there are not too many coefficients that their absolute values are greater than δ. Since W is a vector of N-independent Gaussian stochastic variables with variance σ^2, the maximum amplitude of wavelet coefficients of W has a very high probability of being $\leq \sigma\sqrt{2\log N}$. Reducing by δ the amplitude of all noisy wavelet coefficients, this ensures that the amplitude of an estimated wavelet coefficient is smaller than that of the original wavelet coefficient, i.e.,

$$|\rho_\delta^s((X,\ \psi_{j,m}))| \leq |(f,\ \psi_{j,m})|,$$

$$|\rho_\delta^s((X,\ \varphi_{J,m}))| \leq |(f,\ \varphi_{J,m})|.$$

Both wavelet threshold and shrinkage are such that all wavelet coefficients satisfying $|(X,\ \psi_{j,m})| \leq \delta$ are replaced by zero. This performs an adaptive smoothing that depends on the regularity of signal f. Note that noise coefficients $(W,\ \psi_{j,m})$ must have a high probability of being $\leq \delta$. Hence, if

$$|(X,\ \psi_{j,m})| = |(f,\ \psi_{j,m}) + (W,\ \psi_{j,m})| \geq \delta,$$

then $|(f,\ \psi_{j,m})|$ has a higher probability of being $\geq \delta$. At fine scale, these coefficients are the neighborhood of sharp signal transitions. By keeping them, we avoid smoothing these sharp variations. When $|(X,\ \psi_{j,m})| < \delta$, the original wavelet coefficients $(f,\ \psi_{j,m})$ are often small, and this means f is locally smooth. In this case, let $(f,\ \psi_{j,m}) = 0$ be equivalent to locally averaging the noisy data X.

A compromise between hard and soft threshold is mid-threshold as follows:

$$\rho_\delta^m(t) = \begin{cases} t & \text{if } \ |t| > 2\delta, \\ 2(t-\delta) & \text{if } \ \delta \leq t \leq 2\delta, \\ 2(t-\delta) & \text{if } \ -2\delta \leq t \leq -\delta, \\ 0 & \text{if } \ |t| < \delta. \end{cases}$$

The mid-threshold is also a special case of firm threshold:

$$\rho^f(t) = \begin{cases} 0 & \text{if} \quad |t| < \delta, \\ (\operatorname{sgn} t)\frac{\delta'(|t|-\delta)}{\delta'-\delta} & \text{if} \quad \delta \le |t| \le \delta', \\ t & \text{if} \quad |t| > \delta'. \end{cases}$$

3.9.3 Minimax Estimation

Let $\Theta_{k,s}$ be the set of piecewise polynomial signals on $[0, N-1]$ with at most k polynomial components of degree s. To estimate $f \in \Theta_{k,s}$ from the noisy signal $X_n = f_n + W_n$, by applying a wavelet thresholding estimator in $\widetilde{F} = DX$, the risk of $\widetilde{F}$ is $r(D, f) = E[\| DX - f \|^2]$. The maximum risk for any $f \in \Theta_{k,s}$ is

$$r(D, \Theta_{k,s}) = \sup_{f \in \Theta_{k,s}} E[\| DX - f \|^2],$$

and the minimax risk is the lower bound computed over all thresholding estimators O:

$$r(\Theta_{k,s}) = \inf_{D \in O} r(D, \Theta_{k,s}).$$

The lower bound of minimax risk is $r(\Theta_{k,s}) \ge k(s+1)\sigma^2$. If the Daubechies wavelet basis with $s+1$ vanishing moments is used and $\delta = \sigma\sqrt{2\log N}$, the hard or soft thresholding risk is

$$r_\delta(\Theta_{k,s}) \le 4\sigma^2 k(s+1)\frac{\log^2 N}{\log 2}(1 + o(1)).$$

3.9.4 Adaptive Denoising Algorithm

Suppose that the noisy signal $x_i = f_i + W_i$ $(i = 1, \ldots, n)$, where $n = 2^{J+1}$ and each W_i is white noise with mean 0 and variance σ. Adaptive denoising algorithm consists of three steps:

Step 1. Perform discrete wavelet transform for the signal $\{\frac{x_k}{\sqrt{n}}\}$ yielding noisy wavelet coefficients:

$$\{\alpha_{j,m}\}_{j=j_0,\ldots,J;m=0,\ldots,2^m-1}.$$

Step 2. Apply (3.9.3) to $\alpha_{j,n}$ and yield the estimate of the wavelet coefficients of f_i, where the threshold

$$\delta = \frac{\sqrt{2\log n}\sigma}{\sqrt{n}}.$$

Step 3. Use the estimated wavelet coefficients to yield the estimate of f_i.

3.10 Shearlets, Bandelets, and Curvelets

Wavelet representations are optimal for one-dimensional data with pointwise singularities, but they cannot handle well two-dimensional data with singularities along curves because wavelets are generated by isotropically dilation of a single or finite generator. To overcome these limitations of wavelets, shearlets, bandelets, and curvelets are introduced and have a wide application in the analysis of atmospheric and oceanic circulation.

3.10.1 Shearlets

For $\psi \in L^2(\mathbb{R}^2)$, the *continuous shearlet system* $\mathrm{SH}(\psi)$ is defined by

$$\mathrm{SH}(\psi) = \{\psi_{\alpha,\beta,\mathbf{l}} = T_{\mathbf{l}} D_{A_\alpha} D_{s_\beta} \psi : \quad \alpha > 0, \ \beta \in \mathbb{R}, \ \mathbf{l} \in \mathbb{R}^2\},$$

where $T_{\mathbf{l}}$ ($\mathbf{l} \in \mathbb{R}^2$) is the translation operator, $T_{\mathbf{l}}\psi(t) = \psi(\mathbf{t} - \mathbf{l})$ ($\mathbf{t} \in \mathbb{R}^2$), and $\{D_{A_\alpha}\}_{\alpha>0}$ is the family of dilation operations based on parabolic scaling matrices A_α of the form:

$$A_\alpha = \begin{pmatrix} \alpha & 0 \\ 0 & \alpha^{\frac{1}{2}} \end{pmatrix},$$

and D_{s_β} ($\beta \in \mathbb{R}$) is the shearing operator, where the shearing matrix has the form:

$$s_\beta = \begin{pmatrix} 1 & \beta \\ 0 & 1 \end{pmatrix}.$$

The *shearlet transform* is defined as follows: For $f \in L^2(\mathbb{R}^2)$,

$$\mathrm{SH}_\psi f(\alpha, \beta, \mathbf{l}) = \int_{\mathbb{R}^2} f(\mathbf{t}) \overline{\psi}_{\alpha,\beta,\mathbf{l}}(\mathbf{t}) d\mathbf{t} \qquad (\alpha > 0, \ \beta \in \mathbb{R}, \ \mathbf{l} \in \mathbb{R}^2).$$

To give the inverse shearlet transform, we need the notion of an admissible shearlet. If $\psi \in L^2(\mathbb{R}^2)$ satisfies

$$\int_{\mathbb{R}^2} \frac{|\widehat{\psi}(\omega_1, \omega_2)|^2}{\omega_1^2} d\omega_1 d\omega_2 < \infty,$$

then ψ is called an *admissible shearlet*. It is seen easily that if ψ is compactly supported away from the origin, then ψ is an admissible shearlet. If

$$\int_0^\infty \int_{\mathbb{R}} \frac{|\widehat{\psi}(\omega_1, \omega_2)|^2}{\omega_1^2} d\omega_2 d\omega_1 = \int_{-\infty}^0 \int_{\mathbb{R}} \frac{|\widehat{\psi}(\omega_1, \omega_2)|^2}{\omega_1^2} d\omega_2 d\omega_1 = 1, \qquad (3.10.1)$$

then the continuous shearlet transform is one-to-one correspondence.

Let $\psi \in L^2(\mathbb{R}^2)$ be defined by

$$\widehat{\psi}(\boldsymbol{\omega}) = \widehat{\psi}(\omega_1, \omega_2) = \widehat{\psi}_1(\omega_1)\widehat{\psi}_2\left(\frac{\omega_2}{\omega_1}\right),$$

where ψ_1 satisfies discrete Calderon conditions $\sum_{\nu\in\mathbb{Z}} |\widehat{\psi}_1(2^{-\nu}\omega)|^2 = 1\ (\omega \in \mathbb{R})$ and $\widehat{\psi}_1 \in C^\infty(\mathbb{R})$, and ψ_2 satisfies the conditions $\sum_{k=-1}^{1} |\widehat{\psi}_2(\omega + k)|^2 = 1\ (\omega \in [-1, 1])$, $\widehat{\psi}_2 \in C^\infty(\mathbb{R})$, and $\text{supp}\widehat{\psi}_2 \subset [-1, 1]$. Then, ψ is called a *classical shearlet*. Classical shearlets satisfy the condition (3.10.1).

A *discrete shearlet system* associated with ψ, denoted by SHψ, is defined as

$$(\text{SH}\psi)(\mathbf{t}) = \{\, \psi_{m,k,\mathbf{n}}(\mathbf{t}) = 2^{\frac{4}{3}m}\psi(s_k A_{2^m}\mathbf{t} - \mathbf{n}),\ \ m, k \in \mathbb{Z},\ \ \mathbf{n} \in \mathbb{Z}^2 \,\} \qquad (\mathbf{t} \in \mathbb{R}^2),$$

where

$$s_k = \begin{pmatrix} 1 & k \\ 0 & 1 \end{pmatrix}, \qquad A_{2^m} = \begin{pmatrix} 2^m & 0 \\ 0 & 2^{\frac{m}{2}} \end{pmatrix}.$$

If $\psi \in L^2(\mathbb{R}^2)$ is a classical shearlet, then the *discrete shearlet transform* is a one-to-one map from f to $(f,\ \psi_{m,k,\mathbf{n}})$, and the *reconstruction formula* is

$$f(\mathbf{t}) = \sum_{m\in\mathbb{Z}}\sum_{k\in\mathbb{Z}}\sum_{\mathbf{n}\in\mathbb{Z}^2}(f,\ \psi_{m,k,\mathbf{n}})\psi_{m,k,\mathbf{n}}(\mathbf{t}) \qquad (\mathbf{t} \in \mathbb{R}^2).$$

3.10.2 Bandelets

The two-dimensional wavelet basis can be constructed by one-dimensional wavelet $\psi(t)$ and a scaling function $\varphi(t)$ as follows: For $m \in \mathbb{Z}$ and $\mathbf{n} = (n_1, n_2) \in \mathbb{Z}^2$,

$$\begin{aligned}
\psi^{(1)}_{m,\mathbf{n}}(\mathbf{t}) &= \varphi_{m,\mathbf{n}}(t_1)\psi_{m,\mathbf{n}}(t_2),\\
\psi^{(2)}_{m,\mathbf{n}}(\mathbf{t}) &= \psi_{m,\mathbf{n}}(t_1)\varphi_{m,\mathbf{n}}(t_2),\\
\psi^{(3)}_{m,\mathbf{n}}(\mathbf{t}) &= \psi_{m,\mathbf{n}}(t_1)\psi_{m,\mathbf{n}}(t_2).
\end{aligned}$$

Let f be α times continuously differentiable on D, denoted by $f \in C^\alpha(D)$, and the wavelet has $p > \alpha$ vanishing moment. Then,

$$\| f - f_N \|^2 = O\left(\frac{1}{N^\alpha}\right),$$

where $\| \cdot \|$ is the norm of $L^2(D)$ and $f_N = \sum_{(i,m,\mathbf{n})\in I_N}(f,\ \psi^{(i)}_{m,\mathbf{n}})\psi^{(i)}_{m,\mathbf{n}}(\mathbf{t})$, and I_N is the index set $\{i, m, \mathbf{n}\}$ of the N largest absolute values of the inner product $(f,\ \psi^{(i)}_{m,\mathbf{n}})$.

However, if $f \in C^\alpha(D \setminus L)$ and $f \notin C^\alpha(D)$, where L is a curve in D, then the approximation error becomes large immediately

$$\| f - f_N \|^2 = O\left(\frac{1}{N}\right).$$

The construction of a bandelet basis from a wavelet basis is warped along the geometric flow to gain advantage of the data regularity along this flow. If the equation of the curve L is $t_2 = c(t_1)$, the *bandelet basis* is defined as

$$\begin{aligned}
\psi^{(1)}_{m,\mathbf{n}}(\mathbf{t}) &= \varphi_{m,\mathbf{n}}(t_1)\psi_{m,\mathbf{n}}(t_2 - c(t_1)),\\
\psi^{(2)}_{m,\mathbf{n}}(\mathbf{t}) &= \psi_{m,\mathbf{n}}(t_1)\varphi_{m,\mathbf{n}}(t_2 - c(t_1)),\\
\psi^{(3)}_{m,\mathbf{n}}(\mathbf{t}) &= \psi_{m,\mathbf{n}}(t_1)\psi_{m,\mathbf{n}}(t_2 - c(t_1)).
\end{aligned}$$

For $f \in C^\alpha(D \setminus L)$ and $f \notin C^\alpha(D)$, the bandelet approximation error is

$$\| f - f_N \|^2 = O(\frac{1}{N^\alpha}).$$

which is the same as the wavelet approximation error in the case of $f \in C^\alpha(D)$.

3.10.3 Curvelets

For $\boldsymbol{\mu} = (m, l, \mathbf{n})$ $(m = 0, 1, \ldots; l = 0, 1, \ldots, 2^m; \mathbf{n} = (n_1, n_2) \in \mathbb{Z}^2)$, define the *curvelet* as a function of $\mathbf{t} \in \mathbb{R}^2$ by

$$\gamma_{\boldsymbol{\mu}}(\mathbf{t}) = 2^{\frac{3}{2}m}\gamma(D_m R_{\theta_{\mathbf{M}}}\mathbf{t} - \mathbf{n}_\delta),$$

where $\theta_{\mathbf{M}} = \frac{2\pi l}{2^{\mathrm{m}}}$ is a *rotation angle* with $\mathbf{M} = (m, l)$, $R_{\theta_{\mathbf{M}}}$ and D_m are, respectively, a *rotation matrix* and a *parabolic scaling matrix*:

$$R_{\theta_{\mathbf{M}}} = \begin{pmatrix} \cos\theta_{\mathbf{M}} & -\sin\theta_{\mathbf{M}} \\ \sin\theta_{\mathbf{M}} & \cos\theta_{\mathbf{M}} \end{pmatrix}, \qquad D_m = \begin{pmatrix} 2^{2m} & 0 \\ 0 & 2^m \end{pmatrix},$$

and $\mathbf{n}_\delta = (n_1\delta_1,\, n_2\delta_2)^T$ is a *translation vector*. The m is called a *scale parameter*, l is called an *orientation parameter*, and $\mathbf{n}_\delta$ is called a *translation parameter*. The bivariate function γ is smooth and oscillatory in the horizontal direction and bell-shaped along the vertical direction.

Each curvelet element is oriented in the direction $\theta_{\mathbf{M}} = \frac{2\pi l}{2^m}$. The parabolic scaling matrix in curvelet is such that the width and length of a curvelet element obey the anisotropy scaling relation:

$$\text{width} \approx \text{length}^2.$$

Under a given parabolic scaling and orientations, the translation vector generates a Cartesian grid with a spacing proportional to the length in the direction $\theta_{\mathbf{M}}$ and width in the normal direction.

One can construct the suitable function $\gamma(\mathbf{t})$ such that the following *reconstruction formula* holds:

$$f(\mathbf{t}) = \sum_{\mu} (f, \gamma_\mu)\gamma_\mu(\mathbf{t}) \qquad (\boldsymbol{\mu} = (m, l, \mathbf{n})).$$

Further Reading

A. Averbuch, M. Israeli, L. Vozovoi, A fast Poisson solver of arbitrary order accuracy in rectangular regions. SIAM J. Sci. Comput. **19**, 933–952 (1998)

F. Bilgili, I. Ozturk, E. Kocakk et al., The influence of biomass energy consumption on CO_2 emissions: a wavelet coherence approach Environ. Sci. Pollut. Res. **23**, 19043–19061 (2016)

A. Biswas, B.C. Si, Application of continuous wavelet transform in examining soil spatial variation: a review. Math. Geosci. **43**, 379–396 (2011)

A.E. Briciu, D. Mihaila, Wavelet analysis of some rivers in SE Europe and selected climate indices. Environ. Monit. Assess. **186**, 6263–6286 (2014)

B. Cazelles, M. Chavez, D. Berteaux et al., Wavelet analysis of ecological time series. Oecologia **156**, 287–304 (2008)

R. Coifman, Y. Meyer, M.V. Wickerhauser, Wavelet analysis and signal processing, in *Wavelets and Their Applications*, ed. by M.B. Ruskai, et al. (Jones and Bartlett, Boston, MA, 1992)

D.L. Donoho, I.M. Johnstone, Minimax estimation via wavelet shrinkage. Ann. Stat. **26**, 879–921 (1998)

H. Guan, X. He, X. Zhang, A comprehensive examination of global atmospheric CO_2 teleconnections using wavelet-based multi-resolution analysis. Environ. Earth. Sci. **74**, 7239–7253 (2015)

I.P. Holman, M. Rivas-Casado, J.P. Bloomfield et al., Identifying non-stationary groundwater level response to North Atlantic ocean-atmosphere teleconnection patterns using wavelet coherence. Hydrogeol. J. **19**, 1269–1278 (2011)

L. Hudgins, J. Huang, Bivariate wavelet analysis of Asia monsoon and ENSO. Adv. Atmos. Sci. **13**, 299 (1996)

I.L. Hudson, M.R. Keatley, I. Kang, Wavelet characterization of eucalypt flowering and the influence of climate. Environ. Ecol. Stat. **18**, 513–533 (2011)

S. Kim, Wavelet analysis of precipitation variability in northern California, USA. KSCE J. Civ. Eng. **8**, 471–477 (2004)

Z.S. Lin, W.L. Bian, W.H. You, The wavelets and hierarchies of the climate system. Meteorol. Atmos. Phys. **61**, 19–26 (1996)

S.G. Mallat, A theory for multiresolution signal decomposition: the wavelet representation, IEEE Trans. Pattern Analysis & Machine Intelligence, **11**, 674–693 (1989)

N. Palizdan, Y. Falamarzi, Y.F. Huang et al., Precipitation trend analysis using discrete wavelet transform at the Langat River Basin, Selangor, Malaysia. Stoch. Environ. Res. Risk Assess. **31**, 853–877 (2017)

A. Prokoph, H. El Bilali, Cross-wavelet analysis: a tool for detection of relationships between paleoclimate proxy records. Math. Geosci. **40**, 575–586 (2008)

N.R. Rigozo, L.E.A. Vieira, E. Echer et al., Wavelet analysis of solar-ENSO imprints in tree ring data from southern Brazil in the last century. Climatic Change **60**, 329–340 (2003)

D. Sonechkin, N. Datsenko, Wavelet analysis of nonstationary and chaotic time series with an application to the climate change problem. Pure Appl. Geophys. **157**, 653–677 (2000)

D. Sonechkin, N. Astafyeva, N. Datsenko et al., Multiscale oscillations of the global climate system as revealed by wavelet transform of observational data time series. Theor. Appl. Climatol. **64**, 131–142 (1999)

C. Torrence, G.P. Compo, A practical guide to wavelet analysis, B. Am. Meteorol. Soc., **79**, 61–78 (1998)

J. Xu, Y. Chen, W. Li et al., Integrating wavelet analysis and BPANN to simulate the annual runoff with regional climate change: a case study of Yarkand River, Northwest China. Water Resour. Manage. **28**, 2523–2537 (2014)

Z. Yan, P.D. Jones, Detecting inhomogeneity in daily climate series using wavelet analysis. Adv. Atmos. Sci. **25**, 157–163 (2008)

P. Yang, J. Xia, C. Zhan et al., Discrete wavelet transform-based investigation into the variability of standardized precipitation index in Northwest China during 1960–2014. Theor. Appl. Climatol. (2017, in press)

Z. Zhang, Measures, densities and diameters of frequency bands of scaling functions and wavelets. J. Approx. Theory **148**, 128–147 (2007)

Z. Zhang, A new method of constructions of non-tensor product wavelets. Acta Appl. Math. **111**, 153–169 (2010)

Z. Zhang, P. Jorgensen, Modulated Haar wavelet analysis of climatic background noise. Acta Appl. Math. **140**, 71–93 (2015)

Z. Zhang, J. Moore, Comment on "Significance tests for the wavelet power and the wavelet power spectrum" by Ge. Ann. Geophys. **30**, 1743–1750 (2012)

Z. Zhang, N. Saito, Harmonic wavelet transform and image approximation. J. Math. Imaging Vis. **38**, 14–34 (2010)

Z. Zhang, J.C. Moore, A. Grinsted, Haar wavelet analysis of climatic time series. Int. J. Wavelets Multiresolut. Inf. Process. **12**, 1450020 (2014)

Chapter 4
Stochastic Representation and Modeling

Stochastic methods are a crucial tool for the analysis of multivariate time series in contemporary climate and environmental research. Since it is impossible to resolve all necessary physical processes and scales, a systematic stochastic approach is often used to represent and model small-scale fluctuations and nonlinear features of the climate and environmental system. In this chapter, we will introduce stationarity and trend tests, principal component analysis, factor analysis, cluster analysis, discriminant analysis, canonical correlation analysis, multidimensional scaling, multivariate ARMA models, Monte Carlo methods, Black–Scholes model, and stochastic optimization.

4.1 Stochastic Processes

If $X(t)$ is a stochastic variable for any fixed t, we say $X(t)$ is a *stochastic process*. The probability of $X(t) \leq x$ for any time t is called a *distribution function* of the stochastic process $X(t)$, denoted by $F(x, t)$, i.e.,

$$F(x, t) = P\{X(t) \leq x\}$$

and the partial derivative:

$$f(x, t) = \frac{\partial F(x, t)}{\partial x}$$

is called the *probability density function* of the stochastic process $X(t)$. The expectation value $E[X(t)]$ is called the *mean function* of the stochastic process $X(t)$, denoted by $\mu_X(t)$ and

$$\mu_X(t) = E[X(t)] = \int_{\mathbb{R}} x f(x, t) \mathrm{d}x.$$

Z. Zhang, *Multivariate Time Series Analysis in Climate and Environmental Research*, https://doi.org/10.1007/978-3-319-67340-0_4

The expectation value of $X(t_1)X(t_2)$ is called the *correlation function*, denoted by $R_X(t_1, t_2)$ and

$$R_X(t_1, t_2) = E[X(t_1)X(t_2)] = \int_{\mathbb{R}}\int_{\mathbb{R}} x_1 x_2 f(x_1, x_2; t_1, t_2)\mathrm{d}x_1\mathrm{d}x_2.$$

The *covariance function* of $X(t)$ is defined as

$$C_X(t_1, t_2) = E[(X(t_1) - \mu_X(t_1))(X(t_2) - \mu_X(t_2))].$$

The *variance function* of $X(t)$ is defined as $\sigma^2(t) = E[(X(t) - \mu_X(t))^2]$.

For a stochastic process $X(t)$, if $t \in \mathbb{R}$, $X(t)$ is called a *continuous stochastic process*; if $t \in \mathbb{Z}$ or $t \in \mathbb{Z}_+$, $X(t)$ is called a *discrete stochastic process*.

For two stochastic processes $X(t)$ and $Y(t)$, the *cross-correlation function* is defined as

$$R_{XY}(t_1, t_2) = E[X(t_1)Y(t_2)].$$

If $R_{XY}(t_1, t_2) = 0$, the processes $X(t)$ and $Y(t)$ are *orthogonal*. The *cross-covariance function* is

$$C_{XY}(t_1, t_2) = E[(X(t_1) - \mu_X(t_1))(Y(t_2) - \mu_Y(t_2))].$$

If $C_{XY}(t_1, t_2) = 0$, the processes $X(t)$ and $Y(t)$ are *uncorrelated*.

If $X(t)$ and $Y(t)$ are real stochastic processes, then $Z(t) = X(t) + iY(t)$ is called a *complex stochastic process*. For a complex stochastic process $Z(t)$, the *correlation function* $R(t_1, t_2)$ is defined as

$$R_Z(t_1, t_2) = E[Z(t_1)\overline{Z}(t_2)].$$

4.1.1 Vector Stochastic Processes

Consider a stochastic vector $\mathbf{X} = (X_1, \ldots, X_n)^T$, where each X_k is a stochastic variable. The vector $E[\mathbf{X}] = (E(X_1), \ldots, E[X_n])^T$ is called the *mean* of the stochastic vector $\mathbf{X}$. The covariance matrix of X is defined as

$$\mathrm{Cov}(\mathbf{X}, \mathbf{X}) = (\mathrm{Cov}(X_i, X_j))_{i,j=1,\ldots,n}.$$

The *joint distribution function* of the components $X_1(t), \ldots, X_n(t)$ is defined as

$$F(x_1, \ldots, x_n) = P(X_1 \le x_1, \ldots, X_n \le x_n)$$

and its *joint probability density function* is defined as

$$f(x_1, \ldots, x_n) = \frac{\partial F(x_1, \ldots, x_n)}{\partial x_1 \cdots \partial x_n}.$$

For a stochastic vector transform:

$$\begin{cases} Y_1 = h_1(X_1, \dots, X_n), \\ Y_2 = h_2(X_1, \dots, X_n), \\ \quad \vdots \\ Y_n = h_n(X_1, \dots, X_n). \end{cases}$$

If its inverse transform exists:

$$\begin{cases} X_1 = g_1(Y_1, \dots, Y_n), \\ X_2 = g_2(Y_1, \dots, Y_n), \\ \quad \vdots \\ X_n = g_n(Y_1, \dots, Y_n), \end{cases}$$

then the joint probability density function of $Y_1, \dots, Y_n$ is

$$f_Y(y_1, \dots, y_n) = \frac{f_X(g_1(y_1, \dots, y_n), \dots, g_n(y_1, \dots, y_n))}{|J(y_1, \dots, y_n)|},$$

where the Jacobian determinant is

$$J(y_1, \dots, y_n) = \begin{vmatrix} \frac{\partial g_1}{\partial y_1} & \cdots & \frac{\partial g_1}{\partial y_n} \\ \vdots & \ddots & \vdots \\ \frac{\partial g_n}{\partial y_1} & \cdots & \frac{\partial g_n}{\partial y_n} \end{vmatrix}.$$

For a vector stochastic processes $\mathbf{X}(t) = (X_1(t), X_2(t), \dots, X_d(t))^T$, one needs to study not only each stochastic process but also the relationship among these d stochastic processes. $\mathbf{X}(t)$ is said to be *stationary* if $\mathbf{X}(t)$ has a constant mean vector $\boldsymbol{\mu} = (\mu_1, \dots, \mu_d)^T$, and the cross-covariance between $X_i(t)$ and $X_j(t+l)$ depends only on the difference l but not on time t, for all $i, j = 1, \dots, d$ and $l \in \mathbb{R}$. Therefore, we may write the corresponding cross-covariance as

$$\gamma_{ij}(l) = \text{Cov}(X_i(t),\ X_j(t+l)) = E[(X_i(t) - \mu_i)(X_j(t+l) - \mu_j)].$$

The $d \times d$ cross-covariance matrix at lag l is

$$\Gamma(l) = E[(\mathbf{X}(t) - \mu)(\mathbf{X}(t+l) - \mu)]^T = \begin{pmatrix} \gamma_{11} & \gamma_{12} & \cdots & \gamma_{1d} \\ \gamma_{21} & \gamma_{22} & \cdots & \gamma_{2d} \\ \vdots & \vdots & \ddots & \vdots \\ \gamma_{d1} & \gamma_{d2} & \cdots & \gamma_{dd} \end{pmatrix} \qquad (l \in \mathbb{Z}).$$

Let $\mathbf{U}(t)$ be a $r-$dimensional input stochastic process, $\mathbf{X}(t)$ be a $d-$dimensional output stochastic process, and

$$\mathbf{X}(t) = \sum_{k \in \mathbb{Z}} \Theta_k \mathbf{U}(t-k) \qquad (t \in \mathbb{R}),$$

where $\mathbf{U}(t) = (U_1(t), \ldots, U_r(t))^T$ and $\mathbf{X}(t) = (X_1(t), \ldots, X_d(t))^T$, and $\Theta_k = (\theta_{ij})_{d \times r}$ is a $d \times r$ matrix. The filter is *causal* if $\Theta_k = O \ (k < 0)$. So

$$\mathbf{X}(t) = \sum_{k=0}^{\infty} \Theta_k \mathbf{U}(t-k) \qquad (t \in \mathbb{R}).$$

The filter is *stable* if $\sum_k \parallel \Theta_k \parallel < \infty$, where $\parallel \cdot \parallel$ is the norm of the matrix. Again, if $\mathbf{U}(t)$ is stationary and has uniformly bounded second moments, then the output $\mathbf{X}(t)$ is also stationary. Denote by $\Gamma_{\mathbf{X}}(l)$ and $\Gamma_{\mathbf{U}}(t)$ the cross-covariance matrices of $\mathbf{X}(t)$ and $\mathbf{U}(t)$, respectively. The following relationship holds:

$$\Gamma_{\mathbf{X}}(l) = \mathrm{Cov}(\mathbf{X}(t),\ \mathbf{X}(t+l)) = \sum_{i \in \mathbb{Z}} \sum_{j \in \mathbb{Z}} \Theta_k \Gamma_{\mathbf{U}}(l+i-j) \Theta_k^T.$$

4.1.2 Gaussian, Markov, and Wiener Processes

For a stochastic vector $\mathbf{x} = (x_1, \ldots, x_m)$, we say it follows a *joint normal distribution* if any linear combination of its components follows a univariate normal distribution, denoted by $\mathbf{x} \sim N(\boldsymbol{\mu}, \Sigma)$, where $\boldsymbol{\mu}$ is the mean vector and Σ is the covariance matrix. Its *probability density function* is

$$f(\mathbf{x}) = \frac{1}{(2\pi)^{\frac{m}{2}}} \exp\left(-\frac{1}{2} (\mathbf{x} - \boldsymbol{\mu})^T \Sigma^{-1} (\mathbf{x} - \boldsymbol{\mu}) \right).$$

For a stochastic process $X(\mathbf{t})$, if, for any $t_1, \ldots, t_n$, the stochastic vector $(X(t_1), \ldots, X(t_n))$ follows a joint normal distribution, then we say $X(\mathbf{t})$ is a *normal stochastic process* or *Gaussian stochastic process.*

Let $X(t)$ be a stochastic process. If the future values $X(s)(s > t)$ are unaffected by the past values $X(s)(s < t)$ for any time t, then $X(t)$ is called a *Markov process.* A discrete Markov process $X(n)(n = 0, 1, \ldots)$ is called a *Markov chain.*

Let $\{X(n)\}$ be a Markov chain. Define the transition probability as

$$P_{k,l}^{n,n+1} = P(X(n+1) = l \mid X(n) = k).$$

If the transition probability is independent of n, we say Markov chain has *stable transition probabilities*, denoted by $P_{k,l}$. Let $\{X(n)\}$ be an Markov chain with stable transition probabilities $(P_{k,l})_{k,l}$. If $P(X(0) = k) = p_k$, then

$$P(X(0) = k_0,\ X(1) = k_1, \ldots, X(n) = k_n) = p_{k_0} P_{k_0,k_1} \cdots P_{k_{n-1},k_n}.$$

If $X(t)$ is independent for different t, then $X(t)$ is called an *independent stochastic process*. If, for any $t_1 < t_2 < \cdots < t_n$, the $n-1$ differences $X(t_2) - X(t_1)$, $X(t_3) - X(t_2)$, ..., and $X(t_n) - X(t_{n-1})$ are independent, then $X(t)$ is called an *independent increment process*.

Let $W(t)(t \geq 0)$ be an independent increment process with $W(0) = 0$ and satisfy the following conditions:

(a) for any $t \geq s > 0$, the difference $W(t) - W(s)$ is a normal stochastic variable $N(0, \sigma^2(t-s))$,

(b) the sample path of $W(t)$ is continuous;

Then $W(t)$ is called a *Wiener process*.

We say a vector stochastic process $\mathbf{W}(t) = (W_1(t), \ldots, W_d(t))^T$ is an *normal Wiener process* if $\mathbf{W}(t)$ satisfies $\mathbf{W}(0) = \mathbf{0}$ and possesses continuous sample paths and independent increments, and for all $0 \leq s < t$,

$$\mathbf{W}(t) - \mathbf{W}(s) \sim N(\mathbf{0}, (t-s)I), \tag{4.1.1}$$

where I is a $d-$dimensional unit matrix. From this, it is clear that each $W_i(t)$ is an one-dimensional standard Wiener process and for $i \neq j$, $W_i(t)$ and $W_j(t)$ are independent. More generally, we say $\mathbf{W}(t)$ is a *vector Wiener process with covariance matrix* Σ if (4.1.1) is replaced by

$$\mathbf{W}(t) - \mathbf{W}(s) \sim N(\mathbf{0},\ (t-s)\Sigma).$$

4.2 Stationarity and Trend Tests

For any time series analysis, one often needs to confirm whether it satisfies the assumptions of stationary and has some trend.

4.2.1 Stationarity Tests

A stochastic process $X(t)$ is *stationary* if stochastic vectors $(X(t_1), \ldots, X(t_N))$ and $(X(t_1+s), \ldots, X(t_N+s))$, for any $t_1, \ldots, t_N$ and s, have the same joint distribution:

$$P(X(t_1) \leq x_1, \ldots, X(t_N) \leq x_N) = P(X(t_1+s) \leq x_1, \ldots, X(t_N+s) \leq x_N) \quad (x_1, \ldots, x_N \in \mathbb{R}).$$

This implies that the statistical properties of a stationary stochastic process are unaffected by a shift in time.

Mean function and correlation function of a stationary stochastic process $X(t)$ have the properties:

$$\begin{aligned} \mu_X(t) &= \mu_X(t+s), \\ R_X(t_1, t_2) &= R_X(t_1+s, t_2+s), \end{aligned} \tag{4.2.1}$$

where $t, t_1, t_2, s \in \mathbb{R}$. If a stochastic process $X(t)$ satisfies (4.2.1), then it is called a *wide-sense stationary*. For a wide-sense stationary stochastic process $X(t)$, the mean function $\mu_X(t)$ is independent of t, i.e., $\mu_X(t)$ is a constant, denoted by μ_X; the correlation function $R_X(t_1, t_2)$ depends only on the difference $\tau = t_1 - t_2$, denoted by $R_X(\tau)$. The *covariance function* is defined as

$$C_X(t_1, t_2) = R_X(t_1, t_2) - \mu_X(t_1)\mu_X(t_2).$$

Then $C_X(t_1, t_2) = R_X(\tau) - \mu_X^2$, where $\tau = t_1 - t_2$, i.e., $C_X(t_1, t_2)$ depends only on τ, denoted by $C_X(\tau)$.

(i) Simple $t-$Test

Assume that the time series $x(1), \ldots, x(n)$ is uncorrelated and normally distributed. Divide the time series $x(1), \ldots, x(n)$ into two subseries of sizes n_1 and n_2 such that $n = n_1 + n_2$, where the first subseries is $x(1), \ldots, x(n_1)$ and the second subseries $x(n_1 + 1), \ldots, x(n)$. The variances of these two subseries can be estimated as

$$S_1^2 = \frac{1}{n_1 - 1} \sum_{j=1}^{n_1} (x(j) - \overline{x}_1)^2, \qquad S_2^2 = \frac{1}{n_2 - 1} \sum_{j=n_1+1}^{n_2} (x(j) - \overline{x}_2)^2$$

while $\overline{x}_1 = \frac{1}{n_1} \sum_{j=1}^{n_1} x(j)$ and $\overline{x}_2 = \frac{1}{n_2} \sum_{j=n_1+1}^{n_2} x(j)$.

The *test-statistic* is

$$\text{ts} = \frac{|\overline{x}_2 - \overline{x}_1|}{S\sqrt{\frac{1}{n_1} + \frac{1}{n_2}}}, \qquad \text{where} \qquad S = \sqrt{\frac{(n_1 - 1)S_1^2 + (n_2 - 1)S_2^2}{n - 2}}.$$

If the mean of these two subseries is same, then ts follows a $t-$distribution with $n - 2$ degrees of freedom. So the test-statistic ts can be used to check whether two subseries differ significantly.

(ii) Mann–Whitney Test

The time series $x_1, \ldots, x_n$ is divided into two subseries of sizes n_1 and n_2:

$$\begin{array}{llll} x(1), & x(2), & \cdots, & x(n_1) \\ x(n_1 + 1), & x(n_1 + 2), & \cdots, & x(n). \end{array}$$

A new series $z(1), \ldots, z(n)$ is a rearrangement of the original data $x(1), \ldots, x(n)$ in increasing order of magnitude. The *test-statistic* is

$$u_c = \frac{\sum_{k=1}^{n_1} R(x(k)) - \frac{1}{2}n_1(n_1 + n_2 + 1)}{\left(\frac{n_1 n_2 (n_1 + n_2 + 1)}{12}\right)^{\frac{1}{2}}},$$

where $R(x(k))$ is the rank of the observation $x(k)$ in the ordered series $z(1), \ldots, z(n)$.

4.2.2 Trend Tests

Methods for detecting trend can be divided into parametric and nonparametric methods. Regression test belongs to parametric method, while turning point test, Kendall's phase test, Mann–Kendall test, and difference sign test belong to nonparametric method.

(i) Regression Test

The most commonly used method for trend detection is a linear model $x(t) = \alpha + \beta t + \epsilon(t)$, where $x(t)$ is the observed value at time t, α and β are regression coefficients, and $\epsilon(t)$ is a stochastic error (white noise). The function

$$Q(\alpha, \beta) = \sum_{t=1}^{n} (x(t) - \alpha - \beta t)^2$$

attains the minimal value when

$$\alpha = \hat{\alpha} = \overline{x} - \hat{\beta}\overline{t},$$
$$\beta = \hat{\beta} = \frac{\sum\limits_{t=1}^{n}(t-\overline{t})(x(t)-\overline{x})}{\sum\limits_{t=1}^{n}(t-\overline{t})^2},$$

where $\overline{x} = \frac{1}{n}\sum_{t=1}^{n} x(t)$ and $\overline{t} = \frac{n+1}{2}$. The sum of squares of the residuals is given by

$$r = \sum_{t=1}^{n}(x(t) - \hat{\alpha} - \hat{\beta}t)^2 = \sum_{t=1}^{n}(x(t) - \overline{x})^2 - \hat{\beta}\sum_{t=1}^{n}(x(t) - \overline{x})(t - \overline{t}).$$

Then the *test-statistic* is

$$RS = \frac{\hat{\beta}}{S_{\hat{\beta}}}, \qquad \text{where} \qquad S_{\hat{\beta}} = \frac{(r/(n-2))^{\frac{1}{2}}}{(\sum_{t=1}^{n}(t-\overline{t})^2)^{\frac{1}{2}}}.$$

When a time series $x(t)$ has no trend, the test-statistic RS follows the t-distribution with $n-2$ degree of freedom.

(ii) Turning Point Test

For a time series $x(t)(t = 1, \ldots, n)$, if, at the time $t^* = 2, \ldots, n-1$,

$$x(t^*) > x(t^*-1) \quad \text{and} \quad x(t^*) > x(t^*+1)$$
$$\text{or} \quad x(t^*) < x(t^*-1) \quad \text{and} \quad x(t^*) < x(t^*+1),$$

we say $x(t^*)$ is a *tuning point*. When a time series $x(t)$ has no trend, the expected number $\overline{p}$ of turning points can be expressed as $\overline{p} = \frac{2}{3}(n-2)$. The variance $\mathrm{Var}(\overline{p}) = \frac{16n-29}{90}$. The *test-statistic* is defined by the standard normal variate z:

$$z = \frac{|p - \overline{p}|}{\sqrt{\mathrm{Var}(\overline{p})}}.$$

(iii) Kendall's Phase Test

For a time series $x(t)(t = 1, \ldots, n)$, the *phase* is defined as the interval between any two successive turning points. When a time series $x(t)$ has no trend, the expected number n_p of phases with length d is

$$n_p = \frac{2(n-d-2)(d^2+3d+1)}{(d+3)!}.$$

Therefore, if the observed number of phases is very different from n_p, the time series must have a trend.

(iv) Mann–Kendall Test

For a time series $x(t)(t = 1, \ldots, n)$, the *test-statistic* is defined as

$$MK = \sum_{i<j} \mathrm{sgn}(x(j) - x(i)),$$

where $\mathrm{sgn}(t) = 1(t > 0)$, $\mathrm{sgn}(t) = -1(t < 0)$, and $\mathrm{sgn}(0) = 0$. When a time series $x(t)$ has no trend, the test-statistic MK follows a normal distribution with mean 0 and variance $\frac{n(n-1)(2n+5)}{18}$.

(v) Difference Sign Test

For a time series $x(t)(t = 1, \ldots, n)$, we construct a sequence of successive differences:

$$x(2) - x(1), \quad x(3) - x(2), \quad \ldots, \quad x(n) - x(n-1).$$

The number of " + " signs is denoted by n_+. When a time series $x(t)$ has no trend, n_+ approximately follows a normal distribution with mean μ_{n_+} and variance $\sigma_{n_+}^2$ given by $\mu_{n_+} = \frac{n-1}{2}$ and $\sigma_{n_+}^2 = \frac{n+1}{12}$. Then the *test-statistic*

$$Z = \frac{|n_+ - \mu_{n_+} - 0.5|}{\sigma_{n_+}}$$

follows an normal distribution. Therefore, when the value of Z is large, then the time series $x(t)$ has a trend.

4.3 Patterns and Classification

In this section, we will discuss main techniques and methods to extract key features, patterns and inherent structure from stochastic variables, including principal component analysis, factor analysis, cluster analysis, discriminant analysis, canonical correlation analysis.

4.3.1 Principal Component Analysis

Principal component analysis is a common technique for emphasizing variations and finding strong patterns in a dataset. It often make the data easy to explore and visualize. Assume that $X_1, \ldots, X_p$ be p stochastic variables with mean 0 and covariance matrix Σ. Let $\lambda_1, \ldots, \lambda_p$ be the eigenvalues and $\lambda_1 \geq \lambda_2 \geq \cdots \geq \lambda_p > 0$, and $\boldsymbol{\alpha}_1, \ldots, \boldsymbol{\alpha}_p$ be the corresponding unit orthogonal eigenvectors of the covariance matrix $\Sigma_{\mathbf{X}}$. Denote $\mathbf{X} = (X_1, \ldots, X_p)^T$ and $\boldsymbol{\alpha}_k = (\alpha_{k1}, \ldots, \alpha_{kp})^T$ $(k = 1, \ldots, p)$. Define

$$\begin{aligned} Y_1 &= \boldsymbol{\alpha}_1^T \mathbf{X} = \alpha_{11} X_1 + \alpha_{12} X_2 + \cdots + \alpha_{1p} X_p, \\ Y_2 &= \boldsymbol{\alpha}_2^T \mathbf{X} = \alpha_{21} X_1 + \alpha_{22} X_2 + \cdots + \alpha_{2p} X_p, \\ &\vdots \\ Y_p &= \boldsymbol{\alpha}_p^T \mathbf{X} = \alpha_{p1} X_1 + \alpha_{p2} X_2 + \cdots + \alpha_{pp} X_p. \end{aligned}$$

Its matrix form is

$$\mathbf{Y} = A\mathbf{X},$$

where $\mathbf{Y} = (Y_1, \ldots, Y_p)$ and $A = (\alpha_{ij})_{p \times p}$. The covariance matrix of $\mathbf{Y}$ is

$$\Sigma_{\mathbf{Y}} = E[\mathbf{Y}\mathbf{Y}^T] = E[A\mathbf{X}\mathbf{X}^T A^T] = A\Sigma_{\mathbf{X}} A^T = \begin{pmatrix} \lambda_1 & 0 & \cdots & 0 \\ 0 & \lambda_2 & \cdots & 0 \\ \vdots & \vdots & \ddots & \vdots \\ 0 & 0 & \cdots & \lambda_p \end{pmatrix},$$

so

$$\begin{aligned} &\mathrm{Cov}(Y_k,\ Y_k) = \lambda_k \ (k = 1, \ldots, p), \\ &\mathrm{Cov}(Y_k,\ Y_l) = 0 \quad (k \neq l,\ k, l = 1, \ldots, p). \end{aligned} \tag{4.3.1}$$

It implies that $Y_1, \ldots, Y_p$ are uncorrelated and $\mathrm{Var} Y_k = \lambda_k$ $(k = 1, \ldots, p)$. $Y_1, \ldots,$ Y_p are called the *principal components*, where Y_1 contains the largest variation, Y_2 contains the second largest variation, in general, Y_k contains the kth largest variation. Note that the total variation of $\mathbf{X}$ is

$$\mathrm{Var}(\mathbf{X}) = \mathrm{tr}(\Sigma_{\mathbf{X}}) = \lambda_1 + \cdots + \lambda_p. \tag{4.3.2}$$

The contribution of the kth principal component can be measured by the ratio $\lambda_k / \mathrm{tr}(\Sigma_{\mathbf{X}})$. If the first few principal components can represent most of the variability of $\mathbf{X}$, then the few principal components can be used to replace the original set of variables approximately.

(i) Sample Principal Components

Let the samples of each stochastic variable X_j be $x_{1j}, \ldots, x_{nj} (j = 1, \ldots, p)$. Then the corresponding sample covariance matrix is $\hat{\Sigma} = (\hat{\sigma}_{ij})_{p\times p}$, where $\hat{\sigma}_{ij} = \frac{1}{n}\sum_{k=1}^{n} x_{ki}x_{kj}$. The larger the sample size n, the better the sample estimates are. Since the principal component analysis depends on the units of variables, one uses the normalized data $c_{ij} = x_{ij}/\sqrt{\hat{\sigma}_{jj}}$ to replace x_{ij}. Denote

$$C = \left(\frac{1}{n}\sum_{k=1}^{n} c_{ki}c_{kj}\right)_{p\times p}.$$

Let $\hat{\lambda}_1, \ldots, \hat{\lambda}_p$ be the eigenvalues and $\hat{\boldsymbol{\alpha}}_1, \ldots, \hat{\boldsymbol{\alpha}}_p$ be the corresponding eigenvectors of the covariance matrix C and $\hat{\lambda}_1 \geq \hat{\lambda}_2 \geq \cdots \geq \hat{\lambda}_p$, and $\| \hat{\boldsymbol{\alpha}}_k \| = 1$ $(k = 1, \ldots, p)$. Define

$$\begin{aligned}
\mathbf{y}_1 &= \hat{\alpha}_{11}\mathbf{x}_1 + \hat{\alpha}_{12}\mathbf{x}_2 + \cdots + \hat{\alpha}_{1p}\mathbf{x}_p,\\
\mathbf{y}_2 &= \hat{\alpha}_{21}\mathbf{x}_1 + \hat{\alpha}_{22}\mathbf{x}_2 + \cdots + \hat{\alpha}_{2p}\mathbf{x}_p,\\
&\vdots\\
\mathbf{y}_p &= \hat{\alpha}_{p1}\mathbf{x}_1 + \hat{\alpha}_{p2}\mathbf{x}_2 + \cdots + \hat{\alpha}_{pp}\mathbf{x}_p,
\end{aligned}$$

where $\mathbf{x}_j = (x_{1j}, \ldots, x_{nj})^T$, $\boldsymbol{\alpha}_j = (\alpha_{1j}, \ldots, \alpha_{nj})^T$ $(j = 1, \ldots, p)$. Let $\mathbf{y}_i = (y_{i1}, \ldots, y_{in})^T$ $(i = 1, \ldots, p)$. Then the matrix $(y_{ij})_{p\times n}$ are called *principal component scores*.

(ii) Rotation Principal Components

If we have found L principal components $Y_1, \ldots, Y_L$, but each Y_j does not always have practical physical meaning. To solve this problem, one makes some orthogonal transform for $Y_1, \ldots, Y_L$, i.e., take an $L \times L$ orthogonal matrix H. Let $\mathbf{R} = H\mathbf{Y}$, where $\mathbf{Y} = (Y_1, \ldots, Y_L)^T$ and $\mathbf{R} = (R_1, \ldots, R_L)^T$, which is called the *rotated principal components*. Moreover, since H is an orthogonal matrix, the total variance of $\mathbf{R}$ is

$$\sum_{k=1}^{L} \mathrm{Var}(R_k) = \sum_{k=1}^{L} \mathrm{Var}(Y_k).$$

However, R_i and R_j are correlated.

4.3.2 Factor Analysis

The factor analysis is used for removing redundancy and revealing similar patterns in a dataset. Let $\mathbf{X} = (X_1, \ldots, X_p)^T$ be a stochastic vector with mean 0. The factor analysis can be modeled as

$$X_k = \sum_{l=1}^{m} \lambda_{k,l} f_l + e_k \qquad (k = 1, \ldots, p;\ m < p),$$

where the stochastic variable f_l is called the *lth factor*, the constant $\lambda_{k,l}$ is called a *loading* of the kth variable on the lth factor, and e_k is the stochastic error. The matrix form of the factor analysis model is

$$\mathbf{X} = \Lambda \mathbf{F} + \mathbf{e}, \tag{4.3.3}$$

where $\mathbf{F} = (f_1, \ldots, f_p)^T$ is the *common factor vector*, $\mathbf{e} = (e_1, \ldots, e_p)^T$ is the *error vector*, and

$$\Lambda = \begin{pmatrix} \lambda_{11} & \lambda_{12} & \cdots & \lambda_{1m} \\ \lambda_{21} & \lambda_{22} & \cdots & \lambda_{2m} \\ \vdots & \vdots & \ddots & \vdots \\ \lambda_{p1} & \lambda_{p2} & \cdots & \lambda_{pn} \end{pmatrix}$$

is the *factor loading matrix*. The assumptions for factor analysis are as follows:

(a) The factors f_l's are independent with mean 0 and variance 1.

(b) The errors e_k's are independent with mean 0 and variance ψ_k.

(c) The factor f_l and the error e_k are independent for any k and l.

By (a),

$$E[\mathbf{F}\mathbf{F}^T] = E\left[\begin{pmatrix} f_1^2 & f_1 f_2 & \cdots & f_1 f_p \\ f_2 f_1 & f_2^2 & \cdots & f_2 f_p \\ \vdots & \vdots & \ddots & \vdots \\ f_p f_1 & f_p f_2 & \cdots & f_p^2 \end{pmatrix}\right] = \begin{pmatrix} E[f_1^2] & E[f_1 f_2] & \cdots & E[f_1 f_p] \\ E[f_2 f_1] & E[f_2^2] & \cdots & E[f_2 f_p] \\ \vdots & \vdots & \ddots & \vdots \\ E[f_p f_1] & E[f_p f_2] & \cdots & E[f_p^2] \end{pmatrix} =: I_p,$$

where I_p is the $p-$order unit matrix. By (b),

$$E[\mathbf{e}\mathbf{e}^T] = \begin{pmatrix} \psi_1 & 0 & \cdots & 0 \\ 0 & \psi_2 & \cdots & 0 \\ \vdots & \vdots & \ddots & \vdots \\ 0 & 0 & \cdots & \psi_p \end{pmatrix} = \mathrm{diag}(\psi_1 \cdots \psi_p) = \Psi.$$

By (c),

$$E[\,\mathbf{e}\mathbf{F}^T\,] = O, \qquad E[\,\mathbf{F}\mathbf{e}^T\,] = O,$$

where O is the $p-$order zero matrix. From this and (4.3.3), the covariance matrix of $\mathbf{X}$ is

$$\begin{aligned}\Sigma &= E[\mathbf{X}\mathbf{X}^T] = E[(\Lambda\mathbf{F}+\mathbf{e})(\Lambda\mathbf{F}+\mathbf{e})^T] \\ &= E[(\Lambda\mathbf{F})(\Lambda\mathbf{F})^T] + E[\mathbf{e}(\Lambda\mathbf{F})^T] + E[(\Lambda\mathbf{F})\mathbf{e}^T] + E[\mathbf{e}\mathbf{e}^T] \\ &= \Sigma_1 + \Sigma_2 + \Sigma_3 + \Sigma_4,\end{aligned} \tag{4.3.4}$$

where

$$\begin{aligned}\Sigma_1 &= E[(\Lambda\mathbf{F})(\Lambda\mathbf{F})^T] = E[\Lambda\mathbf{F}\mathbf{F}^T\Lambda^T] = \Lambda E[\mathbf{F}\mathbf{F}^T]\Lambda^T = \Lambda I_p\Lambda^T = \Lambda\Lambda^T, \\ \Sigma_2 &= E[\mathbf{e}(\Lambda\mathbf{F})^T] = E[\mathbf{e}\mathbf{F}^T\Lambda^T] = E[\mathbf{e}\mathbf{F}^T]\Lambda^T = O, \\ \Sigma_3 &= E[(\Lambda\mathbf{F})\mathbf{e}^T] = \Lambda E[\mathbf{F}\mathbf{e}^T] = O, \\ \Sigma_4 &= E[\mathbf{e}\mathbf{e}^T] = \Psi,\end{aligned}$$

and $\Psi = \text{diag}(\psi_1, \ldots, \psi_p)$.

By (4.3.4),

$$\Sigma = \Lambda\Lambda^T + \Psi. \tag{4.3.5}$$

This equation is called a *factor analysis equation*. Let $\Sigma = (\sigma_{k,l})_{p\times p}$. By $\Sigma = E[\mathbf{X}\mathbf{X}^T]$ and (4.3.5),

$$\begin{aligned}\sigma_{k,k} &= E[X_k^2] = \text{Var}(X_k), \\ \sigma_{k,k} &= \sum_{l=1}^m \lambda_{k,l}^2 + \psi_k \qquad (k = 1, \ldots, p),\end{aligned}$$

and so

$$\text{Var}(X_k) = \sum_{l=1}^m \lambda_{k,l}^2 + \psi_k \qquad (k = 1, \ldots, p).$$

Hence, the proportion of variance of X_k explained by factor f_l is $\sum_{l=1}^m (\lambda_{k,l}^2/\sigma_{kk})$.

The algorithm of factor analysis:

Step 1. Compute the sample covariance matrix S of $\mathbf{X} = (X_1, \ldots, X_p)^T$ and find eigenvalues of S: $\lambda_1 \geq \lambda_2 \geq \cdots \geq \lambda_p > 0$ and the corresponding unit orthogonal eigenvectors: $\boldsymbol{\alpha}_1, \boldsymbol{\alpha}_2, \ldots, \boldsymbol{\alpha}_p$.

Step 2. Determine a minimal integer m such that

$$\frac{\sum\limits_{k=1}^m \lambda_k}{\sum\limits_{k=1}^p \lambda_k} \geq 0.7.$$

Take the first m eigenvectors $\boldsymbol{\alpha}_1, \boldsymbol{\alpha}_2, \ldots, \boldsymbol{\alpha}_m$ as factors.

Step 3. Compute the factor loading matrix,

$$\Lambda = (\sqrt{\lambda_1}\alpha_1, \sqrt{\lambda_2}\alpha_2, \ldots, \sqrt{\lambda_m}\alpha_m) = (\lambda_{k,l})_{p\times m},$$

to obtain $\lambda_{k,l}$ as a loading of X_k on the factor α_l.

Step 4. Compute the variance ψ_k of the error e_k, using the formula:

$$\psi_k = S_{k,k} - \sum_{l=1}^{m} \lambda_{k,l}^2,$$

where $S_{k,k}$ is the $(k,k)-$entry of the matrix S.

4.3.3 Cluster Analysis

Cluster analysis can capture the natural structure of the data. It is to partition all individuals into groups (clusters) such that individuals in the same group have similar characteristic. Let $\mathbf{Y}_1, \ldots, \mathbf{Y}_n$ be n stochastic variables. The samples of each $\mathbf{Y}_i$ are $y_{i1}, \ldots, y_{im}$. Make a normalized transform $x_{ij} = \frac{y_{ij}-\overline{Y}_i}{s_j}$ for these samples, where

$$s_j^2 = \frac{1}{n-1}\sum_{i=1}^{n}(y_{ij} - \overline{Y}_j)^2,$$
$$\overline{Y}_i = \frac{1}{m}\sum_{j=1}^{m} y_{ij}.$$

Let $\mathbf{X}_i = (x_{i1}, \ldots, x_{im})^T (i = 1, \ldots, n)$. Generally, cluster analysis uses the following distance measurement to classify $\mathbf{X}_1, \ldots, \mathbf{X}_n$ into groups.

(i) Various distances between $\mathbf{X}_i$ and $\mathbf{X}_j$

Absolute value distance: $d_{ij} = \sum_{k=1}^{m} |x_{ik} - x_{jk}|$.
Eucliden distance: $d_{ij} = \sum_{k=1}^{m} |x_{ik} - x_{jk}|^2$.
Chebyshev distance: $d_{ij} = \max_{k=1,\ldots,m} |x_{ik} - x_{jk}|$.
Mahalanobis distance:

$$d_{ij} = (\mathbf{X}_i - \mathbf{X}_j)^T \Sigma^{-1} (\mathbf{X}_i - \mathbf{X}_j),$$

where $\mathbf{X}_i$ and $\mathbf{X}_j$ are assumed to have the same covariance matrix Σ.

(ii) Various distances between groups

Let G_k and G_l be two groups arisen from $\mathbf{X}_1, \ldots, \mathbf{X}_n$. The distance $D_{k,l}$ between G_k and G_l can be measured by

$$D_{kl} = \min_{\mathbf{X}_i \in G_k;\, \mathbf{X}_j \in G_l} d_{ij},$$
$$D_{kl} = \max_{\mathbf{X}_i \in G_k,\, \mathbf{X}_j \in G^l} d_{ij},$$
$$D_{kl} = \frac{1}{n_k n_l} \sum_{\mathbf{X}_i \in G_k,\, \mathbf{X}_j \in G_l} d_{ij}^2,$$

where n_k and n_l are a cardinal number of G_k and G_l, respectively.

4.3.4 Discriminant Analysis

Suppose that $G_1, G_2, \ldots, G_N$ are N groups. For a given sample $\mathbf{X} = (x_1, \ldots, x_m)^T$, the discriminant analysis is used to decide which group $\mathbf{X}$ belongs to. The estimates of the mean μ and covariance matrix Σ_l for the group G_l are, respectively,

$$\overline{\mathbf{X}}_l = \frac{1}{n_l} \sum_{k=1}^{n_l} \mathbf{X}_k^{(l)},$$
$$S_l = \frac{1}{n_l - 1} \sum_{k=1}^{n_l} (\mathbf{X}_k^{(l)} - \overline{\mathbf{X}}_l)(\mathbf{X}_k^{(l)} - \overline{\mathbf{X}}_l)^T,$$

where $\mathbf{X}_1^{(l)}, \ldots, \mathbf{X}_{n_l}^{(l)}$ are samples of G_l. The Mahalanobis distance between an individual $\mathbf{X} = (x_1, \ldots, x_m)^T$ and a group G_l is

$$d(\mathbf{X}, G_l) = (\mathbf{X} - \mu_l)^T \Sigma_l^{-1} (\mathbf{X} - \mu_l),$$

which can be estimated as

$$\hat{d}(\mathbf{X}, G_l) = (\mathbf{X} - \overline{\mathbf{X}}_l)^T S_l^{-1} (\mathbf{X} - \overline{X}_l).$$

If $\hat{d}(\mathbf{X}, G_j) = \min_{l=1,\ldots,N} \hat{d}(\mathbf{X}, G_l)$, then $\mathbf{X} \in G_j$.

In the previous process of discriminant analysis, one does not consider the prior probability. Suppose that $G_1, \ldots, G_N$ have the known prior probabilities $p_1, \ldots, p_N$. A generalized distance between an individual $\mathbf{X} = (x_1, \ldots, x_m)^T$ and a group G_l becomes

$$D(\mathbf{X}, G_l) = d(\mathbf{X}, G_l) + \varphi_1(l) + \varphi_2(l) \qquad (l = 1, \ldots, k),$$

where $d(\mathbf{X}, G_l)$ is the Mahalanobis distance between $\mathbf{X}$ and G_l and

$$\varphi_1(l) = \begin{cases} 0, & \Sigma_1 = \cdots = \Sigma_N, \\ \log |S_l|, & \text{otherwise}, \end{cases}$$
$$\varphi_2(l) = \begin{cases} 0, & p_1 = \cdots = p_N, \\ -2 \log |p_l|, & \text{otherwise}, \end{cases}$$

Here Σ_l is the covariance matrix of G_l and $|S_l|$ is the determinant of the sample covariance matrix S_l. If $D(\mathbf{X}, G_j) = \min_{l=1,\ldots,N} D(\mathbf{X}, G_l)$, then $\mathbf{X} \in G_j$.

4.3.5 Canonical Correlation Analysis

Canonical correlation analysis is to study the relationship between the $p-$dimensional stochastic vector $\mathbf{X} = (X_1, \ldots, X_p)^T$ with mean 0 and covariance matrix $\Sigma_{\mathbf{X}}$ and the $q-$dimensional stochastic vector $\mathbf{Y} = (Y_1, \ldots, Y_q)^T$ with mean 0 and covariance matrix $\Sigma_{\mathbf{Y}}$. Denote by $\Sigma_{\mathbf{XY}}$ the cross-covariance matrix of $\mathbf{X}$ and $\mathbf{Y}$. Let $S = \Sigma_{\mathbf{X}}^{-\frac{1}{2}} \Sigma_{\mathbf{XY}} \Sigma_{\mathbf{Y}}^{-\frac{1}{2}}$ have r nonzero singular values $\mu_1 \geq \mu_2 \geq \cdots \geq \mu_r > 0$, and $\mathbf{C}_k$, $\mathbf{D}_k$ be its left and right singular vectors corresponding to $\mu_k (k = 1, \ldots, r)$. Let

$$
\begin{aligned}
\mathbf{f}_k &= \Sigma_{\mathbf{X}}^{-\frac{1}{2}} \mathbf{C}_k, \\
\mathbf{g}_k &= \Sigma_{\mathbf{Y}}^{-\frac{1}{2}} \mathbf{D}_k.
\end{aligned}
$$

Then

$$
\begin{aligned}
&(\mathbf{f}_k, \ \mathbf{f}_l) = (\mathbf{g}_k, \ \mathbf{g}_l) = 0 \qquad (k \neq l), \\
&(\mathbf{f}_k, \ \mathbf{g}_l) = \delta_{kl},
\end{aligned}
$$

where $(\cdot, \cdot)$ is the inner products of vectors and δ_{kl} is the Kroneccker Delta. Define

$$
\begin{aligned}
\alpha_k &= (\mathbf{X}, \mathbf{f}_k) = \sum_{k=1}^{p} f_{1k} X_k, \\
\beta_k &= (\mathbf{Y}, \mathbf{g}_k) = \sum_{k=1}^{q} g_{1k} Y_k \qquad (k = 1, \ldots, r).
\end{aligned}
$$

Then $\operatorname{Var}(\alpha_k) = \operatorname{Var}(\beta_k) = 1$ $(k = 1, \ldots, r)$. The correlation coefficient of α_1 and β_1 attains the maximal value. In general, the correlation coefficient of α_k and β_k attains the kth maximal value.

4.4 Multidimensional Scaling

Let $\mathbf{x}_1, \ldots, \mathbf{x}_p$ be p points in the $n-$dimensional space. Assume that $p > n$ and

$$
\sum_{k=1}^{p} \mathbf{x}_k = \mathbf{0}. \tag{4.4.1}
$$

Each point $\mathbf{x}_k$ has the form:

$$\begin{aligned}
\mathbf{x}_1 &= (x_{1,1}, x_{1,2}, \ldots, x_{1,n})^T, \\
\mathbf{x}_2 &= (x_{2,1}, x_{2,2}, \ldots, x_{2,n})^T, \\
&\vdots \\
\mathbf{x}_p &= (x_{p,1}, x_{p,2}, \ldots, x_{p,n})^T.
\end{aligned} \tag{4.4.2}$$

The Euclidean distance between the ith point $\mathbf{x}_i$ and the jth point $\mathbf{x}_j$ is

$$d_{i,j} = \sum_{k=1}^{p} (x_{k,i} - x_{k,j})^2 \qquad (i, j = 1, \ldots, p). \tag{4.4.3}$$

Multidimensional scaling aims to find the original Euclidean coordinates $x_{i,j}$ ($i = 1, \ldots, p$; $j = 1, \ldots, n$) from their distance matrix $D = (d_{i,j})_{p\times p}$. Define

$$\alpha_{i,j} = -\frac{1}{2} d_{i,j}^2, \qquad A = (\alpha_{i,j})_{p\times p}.$$

Let $\alpha_{i,\cdot}$ ($i = 1, \ldots, p$) be the mean of the ith row and $\alpha_{\cdot,j}$ ($j = 1, \ldots, p$) be the mean of the jth column, and $\alpha_{\cdot,\cdot}$ be the mean of all $\alpha_{i,j}$ ($i, j = 1, \ldots, p$), i.e.,

$$\begin{aligned}
\alpha_{i,\cdot} &= \tfrac{1}{p} \sum_{j=1}^{p} \alpha_{i,j}, \\
\alpha_{\cdot,j} &= \tfrac{1}{p} \sum_{i=1}^{p} \alpha_{i,j}, \\
\alpha_{\cdot,\cdot} &= \tfrac{1}{p^2} \sum_{i=1}^{p} \sum_{j=1}^{p} \alpha_{i,j}.
\end{aligned}$$

Let $\beta_{i,j} = \alpha_{i,j} - \alpha_{i,\cdot} - \alpha_{\cdot,j} + \alpha_{\cdot,\cdot}$. For $i, j = 1, \ldots, n$, by (4.4.3), we get

$$d_{i,j}^2 = \sum_{k=1}^{p} x_{k,i}^2 + \sum_{k=1}^{p} x_{k,j}^2 - 2\sum_{k=1}^{p} x_{k,i}x_{k,j} = \mathbf{x}_i^T\mathbf{x}_i + \mathbf{x}_j^T\mathbf{x}_j - 2\mathbf{x}_i^T\mathbf{x}_j.$$

This implies that

$$\begin{aligned}
-2\beta_{i,j} &= -2(\alpha_{i,j} - \alpha_{i,\cdot} - \alpha_{\cdot,j} + \alpha_{\cdot,\cdot}) = d_{i,j}^2 - \tfrac{1}{p}\sum_{k=1}^{p} d_{i,k}^2 - \tfrac{1}{p}\sum_{k=1}^{p} d_{k,j}^2 + \tfrac{1}{p^2}\sum_{k=1}^{p}\sum_{l=1}^{n} d_{k,l}^2 \\
&= (\mathbf{x}_i^T\mathbf{x}_i + \mathbf{x}_j^T\mathbf{x}_j - 2\mathbf{x}_i^T\mathbf{x}_j) - \tfrac{1}{p}\sum_{k=1}^{p}(\mathbf{x}_i^T\mathbf{x}_i + \mathbf{x}_k^T\mathbf{x}_k - 2\mathbf{x}_i^T\mathbf{x}_k) \\
&\quad - \tfrac{1}{p}\sum_{k=1}^{p}(\mathbf{x}_k^T\mathbf{x}_k + \mathbf{x}_j^T\mathbf{x}_j - 2\mathbf{x}_k^T\mathbf{x}_j) + \tfrac{1}{p^2}\sum_{k=1}^{p}\sum_{l=1}^{p}(\mathbf{x}_k^T\mathbf{x}_k + \mathbf{x}_l^T\mathbf{x}_l - 2\mathbf{x}_k^T\mathbf{x}_l).
\end{aligned}$$

From this and (4.4.1), it follows that $\beta_{ij} = \mathbf{x}_i^T\mathbf{x}_j$. Define the matrix $Q = (\beta_{ij})_{p\times p}$. Then $Q = (\mathbf{x}_i^T\mathbf{x}_j)_{p\times p}$.

Let

$$X = (x_{kl})_{p\times n}.$$

It is clear that $Q = XX^T$ and $Q^T = (XX^T)^T = X^T X = Q$. For any vector $\mathbf{a} = (a_1, \ldots, a_p)^T$,

$$\mathbf{a}^T Q\mathbf{a} = (\mathbf{a}^T X)(X^T \mathbf{a}) = (\mathbf{a}^T X)(\mathbf{a}^T X)^T \geq 0.$$

Since $p > n$, the matrix X has rank n. Therefore, Q is a $p \times p$ symmetric positive semi-definite matrix with rank n. This implies that the matrix Q has n positive eigenvalues $\lambda_1 \geq \lambda_2 \geq \cdots \geq \lambda_n > 0$ and $p - n$ zero eigenvalues $\lambda_{n+1} = \cdots = \lambda_p = 0$. The corresponding eigenvectors $\boldsymbol{\tau}_j = (\tau_{1j}, \ldots, \tau_{pj})^T$ $(j = 1, \ldots, n)$ form an orthogonal matrix $P = (\boldsymbol{\tau}_1, \ldots, \boldsymbol{\tau}_p) = (\tau_{ij})_{p\times p}$. This gives a spectral decomposition formula $Q = P\Lambda P^T$, where

$$\Lambda = \begin{pmatrix} \lambda_1 & & 0 \\ & \ddots & \\ 0 & & \lambda_p \end{pmatrix}$$

Let

$$\Lambda^{\frac{1}{2}} = \begin{pmatrix} \lambda_1^{\frac{1}{2}} & & 0 \\ & \ddots & \\ 0 & & \lambda_p^{\frac{1}{2}} \end{pmatrix}.$$

Then $Q = (P\Lambda^{\frac{1}{2}})(P\Lambda^{\frac{1}{2}})^T$. From this and $Q = XX^T$, we finally obtain

$$X = P_n \begin{pmatrix} \lambda_1^{\frac{1}{2}} & & 0 \\ & \ddots & \\ 0 & & \lambda_n^{\frac{1}{2}} \end{pmatrix},$$

where $P_n = (\tau_{i,j})_{p\times n}$.

The algorithm of multidimensional scaling:

Step 1. Compute $\alpha_{i,j}$, $\alpha_{i,\cdot}$, $\alpha_{\cdot,j}$, and $\alpha_{\cdot,\cdot}$ by the following formulas

$$\alpha_{i,j} = -\tfrac{1}{2}d_{i,j}^2, \qquad \alpha_{i,\cdot} = \tfrac{1}{p}\sum_{j=1}^{p}\alpha_{i,j},$$
$$\alpha_{\cdot,j} = \tfrac{1}{p}\sum_{i=1}^{p}\alpha_{i,j}, \qquad \alpha_{\cdot,\cdot} = \tfrac{1}{p^2}\sum_{i=1}^{p}\sum_{j=1}^{p}\alpha_{i,j}.$$

Step 2. Compute $p \times p$ matrix $Q = (\beta_{i,j})_{p\times p}$, where $\beta_{i,j} = \alpha_{i,j} - \alpha_{i,\cdot} - \alpha_{\cdot,j} + \alpha_{\cdot,\cdot}$.

Step 3. Compute all the n nonzeros eigenvalues $\lambda_1 \geq \lambda_2 \geq \cdots \geq \lambda_n > 0$ of the matrix Q and the corresponding eigenvectors $\boldsymbol{\tau}_j = (\tau_{1,j}, \ldots, \tau_{n,j})^T$ $(j = 1, \ldots, n)$, and $P_n = (\tau_{i,j})_{p\times n}$.

Step 4. Compute the matrix:

$$X = P_n \begin{pmatrix} \lambda_1^{\frac{1}{2}} & & 0 \\ & \ddots & \\ 0 & & \lambda_n^{\frac{1}{2}} \end{pmatrix}.$$

4.5 Vector ARMA Processes

The simplest model for vector stochastic processes is the white noise $\mathbf{Z}(t) = (Z_1(t), \ldots, Z_d(t))^T$ satisfying the following conditions: for all $t \in \mathbb{R}$,

(a) $E[\mathbf{Z}(t)] = (E[Z_1(t)], \ldots, E[Z_d(t)])^T = \mathbf{0}$;

(b) $\text{Cov}(\mathbf{Z}(t),\ \mathbf{Z}^T(t+l)) = E[\mathbf{Z}(t), \mathbf{Z}^T(t+l)] = O\ (l \neq 0)$, where O is a $d \times d$ matrix with all its entry being 0, i.e., $E[Z_i(t)Z_j(t+l)] = 0$;

(c) $\text{Cov}(\mathbf{Z}(t),\ \mathbf{Z}^T(t)) = E[\mathbf{Z}(t)\mathbf{Z}^T(t)] = V$, where $V = (v_{ij})_{d\times d}$ is a constant matrix and $v_{ij} = E[Z_i(t)Z_j(t)]$. Denote $\mathbf{Z}(t) \sim WN(\mathbf{0}, V)$, where WN represents the white noise.

A vector white noise $\mathbf{Z}(t)$ is stationary vector stochastic process with mean $\mathbf{0}$ and cross-covariance:

$$\Gamma(l) = \text{Cov}(\mathbf{Z}(t),\ \mathbf{Z}^T(t+l)) = \begin{cases} O,\ l \neq 0, \\ V,\ l = 0. \end{cases}$$

Wold's theorem shows that any vector stochastic process $\mathbf{X}(t) = (X_1(t), \ldots, X_d(t))^T$ with mean $\mathbf{0}$ can be represented by an infinite moving average (MA) process:

$$\mathbf{X} = \sum_{k=0}^{\infty} \Theta_k \mathbf{Z}(t-k),$$

where $\mathbf{Z}(t)$ is a vector white noise and each Θ_k is a $d \times d$ constant matrix.

Vector autoregressive moving average (ARMA) processes are fundamental tools in modeling and predicting unknown stochastic process with the minimum number of parameters. As special cases, pure autoregressive (AR) or moving average (MA) processes are also widely used.

4.5.1 Vector MA(q) Processes

Let $\mathbf{X}(t)$ be a vector stochastic process and

$$\mathbf{X}(t) = \mathbf{Z}(t) - \sum_{k=1}^{q} \Theta_k \mathbf{Z}(t-k), \qquad (4.5.1)$$

i.e.,

$$\begin{pmatrix} X_1(t) \\ X_2(t) \\ \vdots \\ X_d(t) \end{pmatrix} = \begin{pmatrix} Z_1(t) \\ Z_2(t) \\ \vdots \\ Z_d(t) \end{pmatrix} - \sum_{k=1}^{q} \begin{pmatrix} \theta_{11}^{(k)} & \theta_{12}^{(k)} & \cdots & \theta_{1d}^{(k)} \\ \theta_{21}^{(k)} & \theta_{22}^{(k)} & \cdots & \theta_{2d}^{(k)} \\ \vdots & \vdots & \ddots & \vdots \\ \theta_{d1}^{(k)} & \theta_{d2}^{(k)} & \cdots & \theta_{dd}^{(k)} \end{pmatrix} \begin{pmatrix} Z_1(t-k) \\ Z_2(t-k) \\ \vdots \\ Z_d(t-k) \end{pmatrix},$$

where $\mathbf{Z}(t) \sim WN(O, V)$ and Θ_k is a constant matrix. Then $\mathbf{X}(t)$ is called a *vector* MA(q) *process*. Define a *time shift operator*:

$$B: \qquad B\mathbf{Y}(t) = \mathbf{Y}(t-1),$$

where $\mathbf{Y}(t)$ is a stationary vector stochastic process. By (4.5.1), it follows that

$$\mathbf{X}(t) = \mathbf{Z}(t) - \sum_{k=1}^{q} \Theta_k \mathbf{Z}(t) = \left(I - \sum_{k=1}^{q} \Theta_k B^k \right) \mathbf{Z}(t),$$

where $I - \sum_{k=1}^{q} \Theta_k B^k$ is an operator polynomial with matrix coefficients. When $\det(I - \sum_{k=1}^{q} \Theta_k z^k) \neq 0$ for all $|z| \leq 1$, the inverse operator $(I - \sum_{k=1}^{q} \Theta_k B^k)^{-1}$ exists. So

$$\mathbf{Z}(t) = \left(I - \sum_{k=1}^{q} \Theta_k B^k \right)^{-1} \mathbf{X}(t).$$

Note that the following power series expansion holds:

$$\left(I - \sum_{k=1}^{q} \Theta_k B^k \right)^{-1} = I - \sum_{k=1}^{\infty} R_k B^k,$$

i.e.,

$$\left(I - \sum_{k=1}^{q} \Theta_k B^k \right) \left(I - \sum_{k=1}^{\infty} R_k B^k \right) = I.$$

Equating coefficients matrices of various powers of B^k on both sides, we can solve out R_k as follows:

$$R_k = \Theta_1 R_{k-1} + \Theta_2 R_{k-2} + \cdots + \Theta_q R_{k-q}, \qquad (R_0 = -I, \quad R_k = \mathrm{O} \ (k < 0)).$$

So $\mathbf{Z}(t) = \mathbf{X}(t) - \sum_{k=1}^{\infty} R_k \mathbf{Z}(t-k)$.

By (4.5.1), we know that MA(q) processes are both stationary and causal. Clearly,

$$E[\mathbf{X}(t)] = E[\mathbf{Z}(t)] - \sum_{k=1}^{q} \Theta_k E[\mathbf{Z}(t-k)] = \mathbf{0} \qquad (t \in \mathbb{R}).$$

Let $\Theta_0 = I$ and $\Theta_j = \mathrm{O}\,(j < 0,\ j > q)$. Then

$$\begin{aligned}
&\mathrm{Cov}(\mathbf{X}(t),\ \mathbf{X}(t+l)) = E[\mathbf{X}(t)\mathbf{X}^T(t+l)] \\
&= E\left[\left(\sum_{k=0}^{q} \Theta_k \mathbf{Z}_{t-k}\right)\left(\sum_{k=0}^{q} \Theta_k \mathbf{Z}_{t+l-k}\right)^T\right] \\
&= \sum_{k=0}^{q}\sum_{j=0}^{q} \Theta_j E[\mathbf{Z}_{t-j}\mathbf{Z}_{t+l-k}^T]\Theta_k^T \\
&= \sum_{k=0}^{q}\sum_{j=0}^{q} \Theta_j \delta_{j,k-l} V \Theta_k^T = \sum_{k=0}^{q-l} \Theta_k V \Theta_{k+l}^T \qquad (l = 0, 1, \ldots, q),
\end{aligned}$$

i.e., $\mathrm{Cov}(\mathbf{X}(t),\ \mathbf{X}(t+l))$ depends only on l. So the cross-covariance matrix of $\mathbf{X}(t)$ is

$$\Gamma(l) = \sum_{k=0}^{q-l} \Theta_k V \Theta_{k+l}^T \qquad (l \in \mathbb{Z}),$$

where V is the covariance matrix of the white noise $\mathbf{Z}(t)$.

For $d = 1$ and $q = 1$, the white noise $Z(t)$ satisfies $E[Z(t)] = 0$, $E[Z^2(t)] = \sigma^2$, and $E[Z(t)Z(t-1)] = 0$, the Eq. (4.5.1) is reduced to $X(t) = Z(t) - \theta Z(t-1)$, and so

$$\begin{aligned}
&E[X(t)] = E[Z(t)] - \theta E[Z(t-1)] = 0, \\
&\mathrm{Var}(X(t)) = E[X^2(t)] = E[Z^2(t)] + \theta^2 E[Z^2(t-1)] - 2\theta E[Z(t)Z(t-1)] = (1+\theta^2)\sigma^2,
\end{aligned}$$

$$\mathrm{Cov}(X(t),\ X(t-l)) = E[X(t)X(t-l)] = \begin{cases} -\theta\sigma^2, & l = \pm 1, \\ 0, & l = \pm 2, \pm 3, \ldots \end{cases}$$

For $d = 2$ and $q = 1$,

$$\begin{pmatrix} X_1(t) \\ X_2(t) \end{pmatrix} = \begin{pmatrix} Z_1(t) \\ Z_2(t) \end{pmatrix} - \begin{pmatrix} \theta_{11} & \theta_{12} \\ \theta_{21} & \theta_{22} \end{pmatrix}\begin{pmatrix} Z_1(t-1) \\ Z_2(t-1) \end{pmatrix},$$

i.e.,

$$\begin{aligned}
X_1(t) &= Z_1(t) - \theta_{11}Z_1(t-1) - \theta_{12}Z_2(t-1), \\
X_2(t) &= Z_2(t) - \theta_{21}Z_1(t-1) - \theta_{22}Z_2(t-1).
\end{aligned} \qquad (4.5.2)$$

Since $E[Z_i(t)Z_j(t)] = v_{ij}\ (t \in \mathbb{R})$ and $E[Z_i(t-1)Z_j(t)] = 0\,(t \in \mathbb{R})$, the covariance of $X_i(t)$ and $X_j(t)$ is

$$\begin{aligned}
\mathrm{Cov}(X_1(t),\ X_2(t)) &= v_{12} + \theta_{11}\theta_{21}v_{11} + \theta_{12}\theta_{22}v_{22} + \theta_{11}\theta_{22}v_{12} + \theta_{12}\theta_{21}v_{12} \\
&= (1 + \theta_{11}\theta_{22} + \theta_{12}\theta_{21})v_{12} + \theta_{11}\theta_{21}v_{11} + \theta_{12}\theta_{22}v_{22}
\end{aligned}$$

and

$$\begin{aligned}\mathrm{Cov}(X_1(t),\ Z_1(t)) &= E[X_1(t)Z_1(t)] = v_{11},\\ \mathrm{Cov}(X_1(t),\ Z_2(t)) &= E[X_1(t)Z_2(t)] = v_{12},\\ \mathrm{Cov}(X_2(t),\ Z_1(t)) &= E[X_2(t)Z_1(t)] = v_{12},\\ \mathrm{Cov}(X_2(t),\ Z_2(t)) &= E[X_1(t)Z_2(t)] = v_{22}.\end{aligned}$$

Let

$$\mathbf{X}(t) = \begin{pmatrix} X_1(t) \\ X_2(t) \end{pmatrix}, \quad \mathbf{Z}(t) = \begin{pmatrix} Z_1(t) \\ Z_2(t) \end{pmatrix}, \quad \Theta = \begin{pmatrix} \theta_{11} & \theta_{12} \\ \theta_{21} & \theta_{22} \end{pmatrix}.$$

Then the matrix form of (4.5.2) is $\mathbf{X}(t) = \mathbf{Z}(t) - \Theta\mathbf{Z}(t-1)$. Using the time shift operator B, it is rewritten in the form:

$$\mathbf{X}(t) = I\mathbf{Z}(t) - \Theta B\mathbf{Z}(t) = (I - \Theta B)\mathbf{Z}(t), \tag{4.5.3}$$

where

$$(I - \Theta B) = \begin{pmatrix} 1 - \theta_{11}B & -\theta_{12}B \\ -\theta_{21}B & 1 - \theta_{22}B \end{pmatrix}.$$

When $\det(I - \Theta z) = (1 - \theta_{11}z)(1 - \theta_{22}z) - \theta_{12}\theta_{21}z^2 \neq 0\,(|z| \leq 1)$, the inverse operator $(1 - \Theta B)^{-1}$ exists and

$$(I - \Theta B)^{-1} = \sum_{j=0}^{\infty} \Theta^j B^j.$$

From this and (4.5.3),

$$\mathbf{Z}(t) = (I - \Theta B)^{-1}\mathbf{X}(t) = \sum_{j=0}^{\infty} \Theta^j B^j \mathbf{X}(t).$$

By $B^j\mathbf{Y}(t) = \mathbf{Y}(t-j)$, the *inversion formula* is

$$\mathbf{Z}(t) = \mathbf{X}(t) + \sum_{j=1}^{\infty} \Theta^j \mathbf{X}(t-j).$$

4.5.2 Vector AR(p) Processes

Let $\mathbf{X}(t)$ be a vector stochastic processes and

$$\mathbf{X}(t) - \sum_{k=1}^{p} \Phi_k \mathbf{X}(t-k) = \mathbf{Z}(t), \tag{4.5.4}$$

i.e.,

$$\begin{pmatrix} X_1(t) \\ \vdots \\ X_d(t) \end{pmatrix} - \sum_{k=1}^{p} \begin{pmatrix} \varphi_{11}^{(k)} & \cdots & \varphi_{1d}^{(k)} \\ \vdots & \ddots & \vdots \\ \varphi_{d1}^{(k)} & \cdots & \varphi_{dd}^{(k)} \end{pmatrix} \begin{pmatrix} X_1(t-k) \\ \vdots \\ X_d(t-k) \end{pmatrix} = \begin{pmatrix} Z_1(t) \\ \vdots \\ Z_d(t) \end{pmatrix},$$

where $\mathbf{Z}(t) \sim WN(\mathbf{0}, V)$ and Φ_k is a constant matrix. Then $\mathbf{X}(t)$ is called a *vector* AR(p) *process.*

When $d = 1$, the Eq. (4.5.4) is reduced to

$$X(t) - \sum_{k=1}^{p} \varphi_k X(t-k) = Z(t),$$

$$\text{or} \quad \left(I - \sum_{k=1}^{p} \varphi_k B^k\right) X(t) = Z(t).$$

If $1 - \sum_{k=1}^{p} \varphi_k z^k \neq 0\,(|z| \leq 1)$, then the inverse operator $(I - \sum_{k=1}^{p} \varphi_k B^k)^{-1}$ exists. Again, if

$$\left(I - \sum_{k=1}^{p} \varphi_k z^k\right)^{-1} = 1 - \sum_{j=1}^{\infty} \psi_j z^j \; (|z| \leq 1),$$

then the inverse operator $\left(I - \sum_{k=1}^{p} \varphi_k B^k\right)^{-1} = 1 - \sum_{j=1}^{\infty} \psi_j B^j$ and

$$X(t) = \left(I - \sum_{k=1}^{p} \varphi_k B^k\right)^{-1} Z(t) = Z(t) - \sum_{j=1}^{\infty} \psi_j Z(t-j) \qquad (t \in \mathbb{R}).$$

For a vector AR(1) process,

$$\mathbf{X}(t) - \Phi\mathbf{X}(t-1) = \mathbf{Z}(t), \tag{4.5.5}$$

where Φ is a $d \times d$ matrix. If the determinant $\det(I - \Phi z) \neq 0\,(|z| \leq 1)$, then the series $\sum_{k=0}^{\infty} \Phi^k Z(t-k)$ converges and

$$\mathbf{X}(t) = \sum_{k=0}^{\infty} \Phi^k Z(t-k).$$

In general, for a vector AR(p) process in (4.5.4), if the determinant:

$$\det\left(I - \sum_{k=1}^{p} \Phi_k z^k\right) \neq 0 \qquad (|z| \leq 1),$$

then the inverse operator $(I - \sum_{k=1}^{p} \Phi_k B^k)^{-1}$ exists. Again, if

$$\left(I - \sum_{k=1}^{p} \Phi_k z^k\right)^{-1} = I - \sum_{j=1}^{\infty} \Psi_j z^j,$$

then

$$\mathbf{X}(t) = \left(I - \sum_{k=1}^{p} \Phi_k B^k\right)^{-1} \mathbf{Z}(t) = \mathbf{Z}(t) - \sum_{j=1}^{\infty} \Psi_j \mathbf{Z}(t-j) \qquad (t \in \mathbb{R}),$$

where Ψ_j is a $d \times d$ matrix. Note that $E[\mathbf{Z}(t-k)\mathbf{Z}^T(t)] = \delta_{k0} V$. Then

$$E[\mathbf{X}(t-l)\mathbf{Z}^T(t)] = E[\mathbf{Z}(t-l)\mathbf{Z}^T(t)] - \sum_{j=1}^{\infty} \Psi_j E[\mathbf{Z}(t-l-j)\mathbf{Z}^T(t)] = \delta_{l0} V \quad (l \geq 0).$$

By (4.5.4), the cross-covariance matrix is equal to

$$\begin{aligned}\Gamma(l) &= E[\mathbf{X}(t-l)\mathbf{X}^T(t)] \\ &= E[\mathbf{X}(t-l)\mathbf{Z}^T(t)] + \sum_{j=1}^{p} E[\mathbf{X}(t-l)\mathbf{X}^T(t-j)]\Phi_j^T.\end{aligned}$$

This implies the following *Yule–Walker equation*:

$$\begin{aligned}\Gamma(0) &= V + \sum_{j=1}^{p} \Gamma(j)\Phi_j^T, \\ \Gamma(l) &= \sum_{j=1}^{p} \Gamma(l-j)\Phi_j^T \qquad (l = 1, \ldots, p).\end{aligned}$$

For $d = 1$, the AR(p) process becomes

$$X(t) - \sum_{k=1}^{p} \varphi_k X(t-k) = Z(t), \tag{4.5.6}$$

where $Z(t) \sim WN(0, \sigma^2)$. The Yule–Walker equation is

$$\begin{aligned}\gamma(0) &= \sigma^2 + \sum_{j=1}^{p} \varphi_j \gamma(j), \\ \gamma(l) &= \sum_{j=1}^{p} \varphi_j \gamma(l-j) \qquad (l = 1, \ldots, p),\end{aligned} \tag{4.5.7}$$

where $\gamma(l) = E[X(t)X(t+l)]$. From (4.5.7), the covariance function $\gamma(l)$ of $X(t)$ can be computed by coefficients $\varphi_1, \ldots, \varphi_p$ and σ^2. Conversely, the coefficients $\varphi_1, \ldots, \varphi_p$ and σ^2 can be computed by the covariance function of $X(t)$. In practice, one often replaces $\gamma(l)$ by the sample covariance function $\hat{\gamma}(l)$ and then find the coefficient estimates $\hat{\varphi}_1, \dots, \hat{\varphi}_p$ using the system of linear equations:

$$\begin{pmatrix} \hat{\gamma}(0) & \hat{\gamma}(1) & \cdots & \hat{\gamma}(l-1) \\ \hat{\gamma}(1) & \hat{\gamma}(0) & \cdots & \hat{\gamma}(l-2) \\ \vdots & \vdots & \ddots & \vdots \\ \hat{\gamma}(l-1) & \hat{\gamma}(l-2) & \cdots & \hat{\gamma}(0) \end{pmatrix} \begin{pmatrix} \varphi_1 \\ \varphi_2 \\ \vdots \\ \varphi_l \end{pmatrix} = \begin{pmatrix} \hat{\gamma}(1) \\ \hat{\gamma}(2) \\ \vdots \\ \hat{\gamma}(l) \end{pmatrix}.$$

The variance σ^2 can be estimated by

$$\hat{\sigma}^2 = \hat{\gamma}(0) - \sum_{k=1}^{l} \hat{\varphi}_k \hat{\gamma}(k).$$

These estimates give an approach of how to fit the stochastic process by the AR (p) process.

Let $\tau(t) = X(t) - \sum_{k=1}^{p} \varphi_k X(t-k)$. Then

$$\begin{aligned} \frac{\partial E[\tau^2(t)]}{\partial \varphi_k} &= \frac{\partial}{\partial \varphi_k}\left(\gamma(0) - 2\sum_{k=1}^{p} \varphi_k \gamma(k) + \sum_{k=1}^{p}\sum_{j=1}^{p} \varphi_k \varphi_j \gamma(k-j) \right) \\ &= -2\gamma(k) + 2\sum_{j=1}^{p} \varphi_j \gamma(k-j). \end{aligned}$$

From this and (4.5.7),

$$\frac{\partial E[\tau^2(t)]}{\partial \varphi_k} = 0.$$

Therefore, the linear combination $\sum_{k=1}^{p} \varphi_k X(t-k)$ is the best prediction of $X(t)$. By (4.5.6),

$$E[\tau^2(t)] = E[Z^2(t)] = \sigma^2.$$

Similarly, for a vector AR(p) process, by (4.5.4), $\sum_{k=1}^{p} \Phi_k \mathbf{X}(t-k)$ is the best linear prediction of $\mathbf{X}(t)$ and the cross-covariance matrix of the error vector $\boldsymbol{\tau}(t) = \mathbf{X}(t) - \sum_{k=1}^{p} \Phi_k \mathbf{X}(t-k)$ satisfies

$$E[\boldsymbol{\tau}(t)\boldsymbol{\tau}^T(t)] = E[\mathbf{Z}(t)\mathbf{Z}^T(t)] = V.$$

4.5.3 Vector ARMA(p, q) Processes

Let $\mathbf{X}(t)$ be a vector stochastic process and

$$\mathbf{X}(t) - \sum_{k=1}^{p} \Phi_k \mathbf{X}(t-k) = \mathbf{Z}(t) - \sum_{k=1}^{q} \Theta_k \mathbf{Z}(t-k), \tag{4.5.8}$$

i.e.,

$$\begin{pmatrix} X_1(t) \\ \vdots \\ X_d(t) \end{pmatrix} - \sum_{k=1}^{p} \begin{pmatrix} \varphi_{11}^{(k)} & \cdots & \varphi_{1d}^{(k)} \\ \vdots & \ddots & \vdots \\ \varphi_{d1}^{(k)} & \cdots & \varphi_{dd}^{(k)} \end{pmatrix} \begin{pmatrix} X_1(t-k) \\ \vdots \\ X_d(t-k) \end{pmatrix}$$

$$= \begin{pmatrix} Z_1(t) \\ \vdots \\ Z_d(t) \end{pmatrix} - \sum_{k=1}^{q} \begin{pmatrix} \theta_{11}^{(k)} & \cdots & \theta_{1d}^{(k)} \\ \vdots & \ddots & \vdots \\ \theta_{d1}^{(k)} & \cdots & \theta_{dd}^{(k)} \end{pmatrix} \begin{pmatrix} Z_1(t-k) \\ \vdots \\ Z_d(t-k) \end{pmatrix},$$

where $\mathbf{Z}(t) \sim WN(\mathbf{0}, V)$ and Φ_k, Θ_k are constant matrices. Then $\mathbf{X}(t)$ is called an ARMA(p, q) *process*.

The Eq. (4.5.8) can be written in the operator form:

$$\Phi(B)\mathbf{X}(t) = \Theta(B)\mathbf{Z}(t), \tag{4.5.9}$$

where

$$\Phi(B) = I - \sum_{k=1}^{p} \Phi_k B^k,$$
$$\Theta(B) = I - \sum_{k=1}^{q} \Theta_k B^k.$$

If $\det(\Phi(z)) \neq 0$ $(|z| \le 1)$, then $\Psi(z) = \Theta(z)\Phi^{-1}(z)$ can be expanded into a power series of the following form:

$$\Psi(z) = I - \sum_{k=1}^{\infty} \Psi_k z^k.$$

where the coefficient Ψ_k is a $d \times d$ matrix. Note that $\Phi(z)\Psi(z) = \Theta(z)$. Then

$$\left(I - \sum_{k=1}^{\infty} \Phi_k z^k\right)\left(I - \sum_{k=1}^{p} \Psi_k z^k\right) = I - \sum_{k=1}^{q} \Theta_k z^k.$$

Comparing the coefficient matrices of power z^k, we get

$$\Psi_k = \Phi_1 \Psi_{k-1} + \Psi_2 \Psi_{k-2} + \cdots + \Phi_p \Psi_{k-p} + \Theta_k \qquad (k \in \mathbb{Z}_+),$$

where $\Psi_0 = I$, $\Psi_k = O$ $(j < 0)$, and $\Theta_k = O$ $(k > q)$. From this, we solve out the coefficient Ψ_k $(k \in \mathbb{Z}_+)$ successively.

When $d = 1$, the Eq. (4.5.9) is reduced to

$$X(t) - \sum_{k=1}^{p} \varphi_k X(t-k) = Z(t) - \sum_{k=1}^{q} \theta_k Z(t-k),$$

where $Z(t) \sim WN(0, \sigma^2)$ and φ_k, θ_k are constants.

4.6 Monte Carlo Methods

Monte Carlo methods are a class of numerical methods that rely on a lot of stochastic samples. It has a wide application in climate and environmental research. In 1998, in order to extract significant feature of climatic time series, Torrence and Compo used the Monte Carlo method to estimate the distribution of wavelet power spectrum of AR(1) process (see Sect. 3.8).

Now we compute the high-dimensional integral using Monte Carlo method.

The high-dimensional integral $I = \int_G f(\mathbf{t})\mathrm{d}\mathbf{t}$, where G is a bounded domain in $\mathbb{R}^d$, can be viewed as the expectation $E[f(U]$, where U is the uniform distribution on G. Precisely,

$$I = vol(G)E[f(U)] = \int_G f(\mathbf{t})\mathrm{d}\mathbf{t}. \tag{4.6.1}$$

where $vol\, G$ is the volume of G in $\mathbb{R}^d$

Using a uniform stochastic number generator, we produce the samples $U_1, U_2, \ldots,$ $U_n, \ldots$ of the stochastic variable U. Then we get the Monte Carlo estimate:

$$I_n = \frac{vol\,(G)}{n} \sum_{l=1}^{n} f(U_l) \qquad (n = 1, 2, \ldots). \tag{4.6.2}$$

By the *Law of Large Numbers*, it is clear that

$$\lim_{n\to\infty} \frac{1}{n} \sum_{l=1}^{n} f(U_l) = E[f(U)]$$

in the probability sense. From this and (4.6.1), and (4.6.2), it follows that $\lim_{n\to\infty} I_n = I$.

By the *Central Limit Theorem*, when n is large, the error term $I_n - I$ approximates to a normal distribution with mean 0 and standard variance $\sqrt{E[(I_n - I)^2]} = \frac{\sigma(f)}{\sqrt{n}}$, where

$$\sigma^2(f) = \int_G \left(f(\mathbf{t}) - \frac{I}{vol\, G} \right)^2 \mathrm{d}\mathbf{t}.$$

The parameter $\sigma^2(f)$ can be estimated by sample variance:

$$S^2(f) = \frac{1}{n-1}\sum_{l=1}^{n}\left(f(U_l) - \frac{I_n}{vol\, G}\right)^2.$$

Since the standard error $\sqrt{E[(I_n - I)^2]} = O(\frac{1}{\sqrt{n}})$ is independent of the dimensionality d. When dimensionality d increases, the error order is invariable. However, for classic numerical methods, when the dimensionality d increases, the convergence speed of the numerical integral always becomes slowly. Therefore, the numerical integral by Monte Carlo method has obvious advantages over classic numerical methods.

4.7 Black–Scholes Models

Black–Scholes model plays a key role in monitoring and predicting carbon prices in carbon trade market. It is described by the following stochastic differentiable equation:

$$dS(t) = \mu S(t)dt + \sigma S(t)dW(t), \tag{4.7.1}$$

where μ and σ are two real parameters, and $W(t)$ is a Wiener process. Below we find the solution of the stochastic differentiable Eq. (4.7.1).

Let $f(S) = \log S$. Then $df(S) = f'(S)dS + \frac{1}{2}f''(S)(dS)^2 = \frac{1}{S}dS - \frac{1}{2S^2}(dS)^2$, i.e.,

$$d\log S(t) = \frac{1}{S}dS - \frac{1}{2S^2}(dS)^2. \tag{4.7.2}$$

By (4.7.1),

$$(dS)^2 = S^2(\mu^2(dt)^2 + 2\sigma\mu dW dt + \sigma^2(dW)^2). \tag{4.7.3}$$

Since $W(t)$ is a Wiener process,

$$W(t) - W(S) \sim N(0, t - S).$$

This implies $(\Delta W)^2 \approx \Delta t$, and so $(dW)^2 = dt$. By (4.7.3), $(dS)^2 = \sigma^2 S^2 dt$. Again, by (4.7.1) and (4.7.2),

$$d\log S(t) = \sigma dW(t) + \left(\mu - \frac{\sigma^2}{2}\right)dt.$$

Integrating both sides from 0 to T,

$$\int_0^T d\log S(t) = \sigma\int_0^T dW(t) + \int_0^T \left(\mu - \frac{\sigma^2}{2}\right)dt.$$

Since $W(0) = 0$, it is clear that

$$\log S(T) = \log S(0) + \sigma W(T) + \left(\mu - \frac{\sigma^2}{2}\right) T.$$

Hence, the stochastic differentiable Eq. (4.7.1) has a solution:

$$S(T) = S(0)\, exp \left\{\sigma W(T) + \left(\mu - \frac{\sigma^2}{2})T\right)\right\}.$$

Since $W(T)$ is a Wiener process, $W(T) \sim N(0, T)$. Let $Z \sim N(0, 1)$. Then $W(T)$ and $\sqrt{T}Z$ are same distributed. Hence, the solution can be written into

$$S(T) = S(0)\, exp \left\{\sigma\sqrt{T}Z + \left(\mu - \frac{\sigma^2}{2}\right) T\right\}.$$

Clearly,

$$\log S(T) = \log S(0) + \left(\mu - \frac{\sigma^2}{2}\right) T + \sigma\sqrt{T}Z.$$

So $\log S(T) \sim N(\lambda, \sigma^2 T)$, where $\lambda = \log S(0) + (\mu - \frac{\sigma^2}{2})T$, i.e., $S(T)$ is a logarithm normal distribution.

4.8 Stochastic Optimization

Let $f(\mathbf{x}) = E[g(\mathbf{x}, \xi)]$ be a $d-$variate target function on the bounded convex domain $\Omega \subset \mathbb{R}^d$, where ξ is a stochastic variable. The stochastic optimization is to find the minimal value of $f(\mathbf{x})$ on Ω.

First of all, we need to estimate the expectation $E[g(\mathbf{x}, \xi)]$ by various methods:

The first method is based on the Monte Carlo method. Let $\xi_1, \ldots, \xi_N$ be the samples of ξ. Then

$$f(\mathbf{x}) \approx \frac{1}{N} \sum_{l=1}^{N} g(\mathbf{x}, \xi_l).$$

The second method is based on the triangular distribution method. If $a \le \xi \le b$ and c is the most likely value of the stochastic variable ξ. We will approximate the probability density function $p(t)$ of ξ by the triangular distribution method. From $\frac{1}{2}(b - a)p(c) = 1$, it follows that $p(c) = \frac{2}{b-a}$. So the probability density function of ξ is

$$p(t) \approx \begin{cases} 0, & t < a, \\ \frac{2(t-a)}{(b-a)(c-a)}, & a \le t \le c, \\ \frac{2(b-t)}{(b-a)(b-c)}, & c < t \le b, \\ 0, & t > b \end{cases}.$$

Hence,

$$f(\mathbf{x}) = E[g(\mathbf{x}, \xi)] = \int_a^b g(\mathbf{x}, t)p(t)\mathrm{d}t \approx \int_a^c g(\mathbf{x}, t)\frac{2(t-a)}{(b-a)(c-a)}\mathrm{d}t + \int_c^b g(\mathbf{x}, t)\frac{2(b-t)}{(b-a)(b-c)}\mathrm{d}t$$

The third method is based on interpolation method. Assume that the values of probability density function $p(x)$ at $a, x_0, \ldots, x_n, b$ can be estimated, where $a < x_0 < \cdots < x_n < b$, and $p(x) = 0\,(x < a \text{ or } x > b)$. Let $L_n(x)$ be an interpolation polynomial of degree n satisfying the condition $L_n(x_k) = p(x_k)\ (k = 0, \ldots, n)$. Then the expectation can be estimated by

$$f(\mathbf{x}) = E[g(\mathbf{x}, \xi)] = \int_a^b g(\mathbf{x}, t)p(t)\mathrm{d}t \approx \int_a^b g(\mathbf{x}, t)L_n(t)\mathrm{d}t.$$

After $f(\mathbf{x})$ is well estimated, we use the steepest descent algorithm to find the minimal value of $f(\mathbf{x})$ on the bounded convex domain Ω.

The steepest descent algorithm:

Step 1. Start at an initial point $\mathbf{x}_0 \in \Omega$.

Step 2. Stop if $\nabla f(\mathbf{x}_0) = \mathbf{0}$. If $\nabla f(\mathbf{x}_0) \neq \mathbf{0}$, the descent direction is $\mathbf{d}_0 = -\nabla f(\mathbf{x}_0)$, where

$$\nabla f(\mathbf{t}) = \left(\frac{\partial f}{\partial t_1}, \ldots, \frac{\partial f}{\partial t_d}\right)^T \qquad (\mathbf{t} = (t_1, \ldots, t_d)).$$

Step 3. Let $\mathbf{x}_1 = \mathbf{x}_0 + \mu_0\mathbf{d}_0$, where $\mu_0 > 0$ is such that $\mathbf{x}_1 \in \Omega$ and $f(\mathbf{x}_0 + \mu_0\mathbf{d}_0)$ is the minimal value.

Step 4. Stop if $\nabla f(\mathbf{x}_1) = 0$. If $\nabla f(\mathbf{x}_1) \neq 0$, the decent direction is $\mathbf{d}_1 = -\nabla f(\mathbf{x}_1)$. Let $\mathbf{x}_2 = \mathbf{x}_1 + \mu_1\mathbf{d}_0$, where $\mu_1 > 0$ is such that $\mathbf{x}_2 \in \Omega$ and $f(\mathbf{x}_1 + \mu_1\mathbf{d}_1)$ is the minimal value.

Step 5. Continuing the procedure in Step 4. The algorithm will terminate when $\nabla f(\mathbf{x}_k) = \mathbf{0}$. At that time, $f(\mathbf{x})$ attains the minimal value at $\mathbf{x}_k \in \Omega$.

Further Reading

A.G. Barnston, C.F. Ropelewski, Prediction of ENSO episodes using canonical correlation analysis. J. Clim. **5**, 1316–1345 (1992)

T.A. Buishand, Some methods for testing the homogeneity of rainfall records. J. Hydrol. **58**, 11–27 (1982)

M. Casas-Prat, X.L. Wang, J.P. Sierra, A physical-based statistical method for modeling ocean wave heights. Ocean Model. **73**, 59–75 (2014)

M. Ceschia, R. Garfagnini, A. Lavenia, E. Toppano, An analysis by ARMA stochastic processes of the daily average temperatures collected during, in Udine and Lignano Sabbiadoro. Il Nuovo Cimento C **9**(1986), 859–884 (1984)

K. Cheng, J. Hou, Y. Wu, J. Liou, Assessing the impact of climate change on annual typhoon rainfall- a stochastic simulation approach. Paddy Water Environ. **7**, 333–340 (2009)

S. Cherry, Singular value analysis and canonical correlation analysis. J. Clim. **9**, 2003–2009 (1996)
U. Cubasch, B.D. Santer, A. Hellbach, G. Hegerl et al., Monte Carlo climate change forecasts with a global coupled ocean-atmosphere model. Clim. Dyn. **10**, 1–19 (1994)
P. Fernandez, S. Mourato, M. Moreira, L. Pereira, A new approach for computing a flood vulnerability index using cluster analysis. Phys. Chem. Earth **94**, 47–55 (2016)
J.M. Gibbons, S.J. Ramsden, Robustness of recommended farm plans in england under climate change: a Monte Carlo simulation. Clim. Change **68**, 113–133 (2005)
M. Gocic, S. Trajkovic, Analysis of changes in meteorological variables using Mann-Kendall and Sens slope estimator statistical tests in Serbia. Global Planet. Change **100**, 172–182 (2013)
K.H. Hamed, Trend detection in hydrologic data: the Mann-Kendall trend test under the scaling hypothesis. J. Hydrol. **349**, 350–363 (2008)
J.D. Horel, Complex principal component analysis: theory and examples. J. Clim. Appl. Meteorol. **23**, 1660–1673 (1984)
G.K. Kanji, *100 Statistical Tests* (Sage Publication, New Delhi, 2001)
M.G. Kendall, *Time Series* (Charles Griffin and Co., Ltd., London, UK, 1973)
N.C. Matalas, Stochastic hydrology in the context of climate change. Clim. Change **37**, 89–101 (1997)
K. Mohammadi, H.R. Eslami, R. Kahawita, Parameter estimation of an ARMA model for river flow forecasting using goal programming. J. Hydrol. **331**, 293–299 (2006)
J. von Neumann, Distribution of the ratio of the mean square successive difference to the variance. Ann. Math. Statist. **12**, 367–395 (1941)
M. New, M. Hulme, Representing uncertainty in climate change scenarios: a Monte-Carlo approach. Integr. Assess. **1**, 203–213 (2000)
X. Niu, M. Stein, Space-Time ARMA Models for Satellite Ozone Data, in *Computing Science and Statistics*, ed. by C. Page, R. LePage (Springer, New York, 1990)
D.B. Owen, *Handbook of Statistical Table* (Addison-Wesley, Reading, Mass, 1962)
S. Patidar, D. Jenkins, P. Banfill, G. Gibson, Simple statistical model for complex probabilistic climate projections: overheating risk and extreme events. Renew. Energy **61**, 23–28 (2014)
J. Parente, M.G. Pereira, M. Tonini, Space-time clustering analysis of wildfires: the influence of dataset characteristics, fire prevention policy decisions, weather and climate. Sci. Total Environ. **559**, 151C165 (2016)
A.O. Pektas, H.K. Cigizoglu, ANN hybrid model versus ARIMA and ARIMAX models of runoff coefficient. J. Hydrol. **500**, 21–36 (2013)
A. Sarhadi, D.H. Burn, G. Yang, A. Ghodsi, Advances in projection of climate change impacts using supervised nonlinear dimensionality reduction techniques. Clim. Dyn. **48**, 1329–1351 (2017)
M. Valipour, M.E. Banihabib, S.M.R. Behbahani, Comparison of the ARMA, ARIMA, and the autoregressive artificial neural network models in forecasting the monthly inflow of Dez dam reservoir. J. Hydrol. **476**, 433–441 (2013)

Chapter 5
Multivariate Spectral Analysis

The complexity of climatic and environmental variability on all timescales requires the use of advanced methods to unravel its primary dynamics from observations, so various spectral methods for multivariate stochastic processes are developed and applied to unravel amplitudes and phases of periodic components embedded in climatic and environmental data. In this chapter, we introduce various spectral analysis and estimation, including periodogram method, Blackman–Tukey method, Welch method, maximum entropy method, multitaper method, ARMA spectrum, and multichannel SSA.

5.1 Power Spectral Density

Let $x(\mathbf{t})$ $(\mathbf{t} = (t_1, \ldots, t_d) \in \mathbb{Z}^d)$ be a stationary multivariate stochastic process with mean $\mathbf{0}$. Its covariance function is

$$\gamma(\mathbf{k}) = E[x(\mathbf{t})\overline{x}(\mathbf{t} - \mathbf{k})] \qquad (\mathbf{k} = (k_1, \ldots, k_d) \in \mathbb{Z}^d).$$

It is clear that the covariance function $\gamma(\mathbf{k})$ satisfies $\gamma(\mathbf{k}) = \overline{\gamma}(-\mathbf{k})$ and $\gamma(\mathbf{0}) \geq |\gamma(\mathbf{k})|$. The *power spectral density* (PSD) of $x(\mathbf{t})$ is defined by the Fourier transform of its covariance function $\gamma(\mathbf{k})$ as follows:

$$S(\omega) = \sum_{\mathbf{k}\in\mathbb{Z}^d} \gamma(\mathbf{k})\, e^{-i(\mathbf{k}\cdot\omega)} \qquad (\mathbf{k} = (k_1, \ldots, k_d),\ \ \omega = (\omega_1, \ldots, \omega_d)).$$

Its inverse transform is

$$\gamma(\mathbf{k}) = \frac{1}{(2\pi)^d} \int_{[-\pi,\pi]^d} S(\omega)\, e^{i(\mathbf{k}\cdot\omega)} \mathrm{d}\omega \qquad (\mathbf{k} \in \mathbb{Z}^d). \tag{5.1.1}$$

Z. Zhang, *Multivariate Time Series Analysis in Climate and Environmental Research*, https://doi.org/10.1007/978-3-319-67340-0_5

Especially,

$$E[|x(\mathbf{t})|^2] = \gamma(\mathbf{0}) = \frac{1}{(2\pi)^d}\int_{-[\pi,\pi]^d} S(\boldsymbol{\omega})d\boldsymbol{\omega}.$$

Another definition of PSD of $x(\mathbf{t})$ is

$$S(\boldsymbol{\omega}) = \lim_{N\to\infty} E\left[\frac{1}{N^d}\left|\sum_{t\in([1,N]^d\bigcap\mathbb{Z}^d)} x(\mathbf{t})\, e^{-i(\mathbf{t}\cdot\boldsymbol{\omega})}\right|^2\right].$$

Since the covariance function $\gamma(\mathbf{k})$ for stationary multivariate stochastic process often decays, under some mild condition, both definitions of PSD are equivalent. Without loss of generality, we only prove this result for the case $d = 1$. Note that

$$\begin{aligned}
E\left[\tfrac{1}{N}\left|\sum_{t=1}^{N} x(t)\, e^{-it\omega}\right|^2\right] &= \tfrac{1}{N}\sum_{t=1}^{N}\sum_{s=1}^{N} E[x(t)\overline{x}(s)]\, e^{-i(t-s)\omega}\\
&= \tfrac{1}{N}\sum_{t=1}^{N}\sum_{s=1}^{N} \gamma(t-s)\, e^{-i(t-s)\omega} = \sum_{k=-N+1}^{N-1}\left(1-\tfrac{k}{N}\right)\gamma(k)\, e^{-ik\omega}\\
&= \sum_{k=-N+1}^{N-1}\gamma(k)\, e^{-ik\omega} - \tfrac{1}{N}\sum_{k=-N+1}^{N-1} k\gamma(k)\, e^{-ik\omega}.
\end{aligned}$$

Under the mild condition $\gamma(k) = o\left(\frac{1}{k}\right)$, the second term of the last equality tends to zero as $N \to \infty$. So

$$\lim_{N\to\infty} E\left[\frac{1}{N}\left|\sum_{t=1}^{N} x(t)\, e^{-i\omega t}\right|^2\right] = \sum_{k\in\mathbb{Z}}\gamma(k)\, e^{-ik\omega} = S(\omega),$$

i.e., both definitions of PSD are equivalent.

For a linear time-invariant system with the input $u(\mathbf{t})$, the filter $h(\mathbf{t})$, and the output $x(\mathbf{t})$, if the input $u(\mathbf{t})$ is stationary, then the output $x(\mathbf{t})$ must be stationary and satisfy the following:

$$x(\mathbf{t}) = \sum_{\mathbf{m}\in\mathbb{Z}^d} h(\mathbf{m})u(\mathbf{t}-\mathbf{m}).$$

Denote the covariance function of $x(\mathbf{t})$ and $u(\mathbf{t})$ by $\gamma_x(\mathbf{k})$ and $\gamma_u(\mathbf{k})$, respectively. It is clear that

$$\gamma_x(\mathbf{k}) = E[x(\mathbf{t})\overline{x}(\mathbf{t}-\mathbf{k})] = \sum_{\mathbf{l}\in\mathbb{Z}^d}\sum_{\mathbf{m}\in\mathbb{Z}^d}\overline{h}(\mathbf{l})h(\mathbf{m})\gamma_u(\mathbf{l}+\mathbf{k}-\mathbf{m})$$

and the PSD of $x(\mathbf{t})$ is

$$\begin{aligned}S_x(\omega) &= \sum_{\mathbf{k}\in\mathbb{Z}^d} \gamma_x(\mathbf{k})\, e^{-i(\mathbf{k}\cdot\omega)}\\ &= \sum_{\mathbf{k}\in\mathbb{Z}^d}\sum_{\mathbf{l}\in\mathbb{Z}}\sum_{\mathbf{m}\in\mathbb{Z}^d} \overline{h}(\mathbf{l})h(\mathbf{m})\gamma_u(\mathbf{l}+\mathbf{k}-\mathbf{m})\, e^{-i(\mathbf{k}\cdot\omega)}\\ &= \left(\sum_{\mathbf{m}\in\mathbb{Z}^d} h(\mathbf{m})\, e^{-i(\mathbf{m}\cdot\omega)}\right)\overline{\left(\sum_{\mathbf{l}\in\mathbb{Z}^d} h(\mathbf{l})\, e^{-i(\mathbf{l}\cdot\omega)}\right)}\left(\sum_{\mu\in\mathbb{Z}^d} \gamma_u(\mu)\, e^{-i(\mu\cdot\omega)}\right).\end{aligned}$$

Note that the transfer function of the system is $H(\omega) = \sum_{\mathbf{k}\in\mathbb{Z}^d} h(\mathbf{k})\, e^{-i(\mathbf{k}\cdot\omega)}$. Then the transfer of PSD is

$$S_x(\omega) = |H(\omega)|^2 S_u(\omega). \tag{5.1.2}$$

For two stationary stochastic processes $x(\mathbf{t}),\ u(\mathbf{t})$ with cross-covariance function $\gamma_{x,u}(\mathbf{k}) = E[x(\mathbf{t})\overline{u}(\mathbf{t}-\mathbf{k})]$, the *cross-spectrum* of $x(\mathbf{t})$ and $u(\mathbf{t})$ is defined as

$$S_{x,u}(\omega) = \sum_{\mathbf{k}\in\mathbb{Z}^d} \gamma_{x,u}(\mathbf{k})\, e^{-i(\mathbf{k}\cdot\omega)}.$$

For a linear system $x(\mathbf{t}) = \sum_{\mathbf{m}\in\mathbb{Z}^d} h(\mathbf{m})u(\mathbf{t}-\mathbf{m})$, the cross-covariance function of $x(\mathbf{t})$ and $u(\mathbf{t})$ becomes

$$\begin{aligned}\gamma_{x,u}(\mathbf{k}) &= E[x(\mathbf{t})\overline{u}(\mathbf{t}-\mathbf{k})] = E\left[\sum_{\mathbf{m}\in\mathbb{Z}^d} h(\mathbf{m})u(\mathbf{t}-\mathbf{m})\overline{u}(\mathbf{t}-\mathbf{k})\right]\\ &= \sum_{\mathbf{m}\in\mathbb{Z}^d} h(\mathbf{m})\gamma_u(\mathbf{m}-\mathbf{k}).\end{aligned}$$

Therefore, the cross-spectrum is

$$\begin{aligned}S_{x,u}(\omega) &= \sum_{\mathbf{k}\in\mathbb{Z}^d} \gamma_{x,u}(\mathbf{k})\, e^{-i(\mathbf{k}\cdot\omega)} = \left(\sum_{\mathbf{m}\in\mathbb{Z}^d} h(\mathbf{m})\, e^{-i(\mathbf{m}\cdot\omega)}\right)\left(\sum_{\mathbf{k}\in\mathbb{Z}^d} \gamma_u(\mathbf{k})\, e^{-i(\mathbf{k}\cdot\omega)}\right)\\ &= H(\omega)S_u(\omega),\end{aligned}$$

which is called the *Wiener–Hopf equation*, where $H(\omega) = \sum_{\mathbf{m}\in\mathbb{Z}^d} h(\mathbf{m})\, e^{-i(\mathbf{m}\cdot\omega)}$.

Let $x(\mathbf{t})$ and $u(\mathbf{t})$ be two stationary stochastic processes. Consider the linear square error $\epsilon(\mathbf{t})$ between $x(\mathbf{t})$ and the linear combination of $\{u(\mathbf{t}-\mathbf{k})\}_{\mathbf{k}\in\mathbb{Z}^d}$, i.e.,

$$\epsilon(\mathbf{t}) = x(\mathbf{t}) - \sum_{\mathbf{k}\in\mathbb{Z}^d} h(\mathbf{k})\, u(\mathbf{t}-\mathbf{k}).$$

If $H(\omega) = \sum_{\mathbf{k}\in\mathbb{Z}^d} h(\mathbf{k})\, e^{-i(\mathbf{k}\cdot\omega)} = \frac{S_{x,u}(\omega)}{S_u(\omega)}$, then the error $E[|\epsilon(\mathbf{t})|^2]$ attains the minimal value.

Let $y(\mathbf{t}) = \sum_{\mathbf{k}\in\mathbb{Z}^d} h(\mathbf{k})u(\mathbf{t}-\mathbf{k})$. By $x(\mathbf{t}) = \epsilon(\mathbf{t}) + y(\mathbf{t})$, it can be proved that

$$S_x(\omega) = S_\epsilon(\omega) + S_y(\omega).$$

Note that $S_y(\omega) = |H(\omega)|^2 S_u(\omega)$. By Wiener–Hopf equation, we get

$$\begin{aligned} E[|\epsilon(\mathbf{t})|^2] = \gamma_\epsilon(\mathbf{0}) &= \tfrac{1}{(2\pi)^d}\int_{[-\pi,\pi]^d} S_\epsilon(\omega)\mathrm{d}\omega \\ &= \tfrac{1}{(2\pi)^d}\int_{[-\pi,\pi]^d} (S_x(\omega) - S_y(\omega))\mathrm{d}\omega \\ &= \tfrac{1}{(2\pi)^d}\int_{[-\pi,\pi]^d} \left(1 - \tfrac{|H(\omega)|^2 S_u(\omega)}{S_x(\omega)}\right) S_x(\omega)\mathrm{d}\omega \\ &= \tfrac{1}{(2\pi)^d}\int_{[-\pi,\pi]^d} \left(1 - \tfrac{|S_{x,u}(\omega)|^2}{S_x(\omega)S_u(\omega)}\right) S_x(\omega)\mathrm{d}\omega. \end{aligned}$$

Let $C_{x,u}(\omega) = \frac{S_{x,u}(\omega)}{(S_x(\omega)S_u(\omega))^{\frac{1}{2}}}$, which is called *complex coherency*. Finally,

$$E[|\epsilon(\mathbf{t})|^2] = \frac{1}{(2\pi)^d}\int_{[-\pi,\pi]^d} (1 - |C_{x,u}(\omega)|^2) S_x(\omega)\mathrm{d}\omega.$$

5.2 Periodogram and Correlogram

Periodogram and correlogram methods are the most classical nonparametric methods of spectrum estimation. Both of them can be computed fast by fast Fourier transform.

5.2.1 Algorithms

Let $\{x(\mathbf{t})\}_{t\in[1,N]^d\bigcap\mathbb{Z}^d}$ be the sample of a multivariate stochastic process. The *periodogram estimator* of PSD is defined as

$$\hat{S}_p(\omega) = \frac{1}{N^d}\left|\sum_{t\in([1,N]^d\bigcap\mathbb{Z}^d)} x(\mathbf{t})\, e^{-i(\mathbf{t}\cdot\omega)}\right|^2.$$

The *correlogram estimator* of PSD is defined as

$$\hat{S}_c(\omega) = \sum_{\mathbf{k}\in([-N+1,N-1]^d\bigcap\mathbb{Z}^d)} \hat{\gamma}(\mathbf{k})\, e^{-i(\mathbf{k}\cdot\omega)},$$

where $\hat{\gamma}(\mathbf{k})$ is an estimate of the covariance $\gamma(\mathbf{k})$,

$$\hat{\gamma}(\mathbf{k}) = \frac{1}{N^d} \sum_{\mathbf{t} \in (\prod_{i=1}^{d} [k_i+1, N]^d \bigcap \mathbb{Z}^d)} x(\mathbf{t})\overline{x}(\mathbf{t}-\mathbf{k}) \qquad (\mathbf{k} = (k_1, \ldots, k_d) \in [0, N-1]^d)$$

and $\hat{\gamma}(-\mathbf{k}) = \overline{\hat{\gamma}}(\mathbf{k})$. Since the first factor in the right side of the above formula is $\frac{1}{N^d}$, $\hat{\gamma}(\mathbf{k})$ is a *biased estimator* of the covariance $\gamma(\mathbf{k})$.

Now we prove that the correlogram estimator is equivalent to the periodogram estimator of PSD. Let $e(\mathbf{t})$ be a white noise with unit variance, i.e.,

$$E[e(\mathbf{t})\overline{e}(\mathbf{s})] = \delta_{\mathbf{t},\mathbf{s}}.$$

Consider a linear system with the filter $\frac{x(\mathbf{k})}{\sqrt{N^d}}$ $(\mathbf{k} \in ([1, N]^d \bigcap \mathbb{Z}^d))$, i.e.,

$$y(\mathbf{t}) = \frac{1}{\sqrt{N^d}} \sum_{\mathbf{k} \in ([1,N]^d \bigcap \mathbb{Z}^d)} x(\mathbf{k})e(\mathbf{t}-\mathbf{k}).$$

By (5.1.2), $S_y(\omega) = |H(\omega)|^2 S_e(\omega)$, where H is the transfer function and

$$H(\omega) = \frac{1}{\sqrt{N^d}} \sum_{\mathbf{k} \in ([1,N]^d \bigcap \mathbb{Z}^d)} x(\mathbf{k})\, e^{-i(\mathbf{k}\cdot\omega)}.$$

Since the PSD of $e(\mathbf{t})$ is 1, it is clear that

$$S_y(\omega) = \frac{1}{N^d} \left| \sum_{\mathbf{k} \in ([1,N]^d \bigcap \mathbb{Z}^d)} x(\mathbf{k})\, e^{-i(\mathbf{k}\cdot\omega)} \right|^2 = \hat{S}_p(\omega). \tag{5.2.1}$$

On the other hand, the autocorrelation function is

$$\begin{aligned}
\gamma_y(\mathbf{k}) &= E[y(\mathbf{t})\overline{y}(\mathbf{t}-\mathbf{k})] \\
&= \tfrac{1}{N^d} \sum_{\mathbf{l},\mathbf{m} \in ([1,N]^d \bigcap \mathbb{Z}^d)} x(\mathbf{l})\overline{x}(\mathbf{m}) E[e(\mathbf{t}-l)\overline{e}(\mathbf{t}-\mathbf{k}-\mathbf{m})] \\
&= \tfrac{1}{N^d} \sum_{\mathbf{l} \in (\prod_{i=1}^{d} [k_i+1,N]^d \bigcap \mathbb{Z}^d)} x(\mathbf{l})\overline{x}(\mathbf{l}-\mathbf{k}) = \hat{\gamma}(\mathbf{k}) \qquad (\mathbf{k} \in ([-N+1, N-1]^d \bigcap \mathbb{Z}^d)), \\
\gamma_y(\mathbf{k}) = 0 & \qquad (\mathbf{k} \notin [-N+1, N-1]^d \bigcap \mathbb{Z}^d).
\end{aligned}$$

By the definition of PSD, we get

$$S_y(\omega) = \sum_{\mathbf{k} \in ([-N+1,N-1]^d \bigcap \mathbb{Z}^d)} \hat{\gamma}(\mathbf{k})\, e^{-i(\mathbf{k}\cdot\omega)} = \hat{S}_c(\omega).$$

Combining this with (5.2.1), we have $\hat{S}_c(\omega) = \hat{S}_p(\omega)$.

5.2.2 Bias Analysis

By the definition of the correlogram,

$$\begin{aligned} E[\hat{S}_c(\omega)] &= \sum_{\mathbf{k}\in([-N+1,N-1]^d \bigcap \mathbb{Z}^d)} E[\hat{\gamma}(\mathbf{k})]\, e^{-i(\mathbf{k}\cdot\omega)} \\ &= \sum_{\mathbf{k}\in([-N+1,N-1]^d \bigcap \mathbb{Z}^d)} \prod_{j=1}^{d} \left(1 - \frac{k_j}{N}\right) \gamma(\mathbf{k})\, e^{-i(\mathbf{k}\cdot\omega)} \\ &= \sum_{\mathbf{k}\in\mathbb{Z}^d} w_B(\mathbf{k})\gamma(\mathbf{k})\, e^{-i(\mathbf{k}\cdot\omega)}, \end{aligned}$$

where $\mathbf{k} = (k_1, \ldots, k_d)$ and

$$w_B(\mathbf{k}) = \begin{cases} \prod_{j=1}^{d} \left(1 - \frac{k_j}{N}\right) & (\mathbf{k} \in ([-N+1, N-1]^d \bigcap \mathbb{Z}^d)), \\ 0 & \text{otherwise} \end{cases} \tag{5.2.2}$$

which is called the *Bartlett window*. By (5.1.1),

$$\gamma(\mathbf{k}) = \frac{1}{(2\pi)^d} \int_{[-\pi,\pi]^d} S(\boldsymbol{\theta})\, e^{i(\mathbf{k}\cdot\boldsymbol{\theta})} \mathrm{d}\boldsymbol{\theta}.$$

It follows that

$$\begin{aligned} E[\hat{S}_c(\omega)] &= \frac{1}{(2\pi)^d} \sum_{\mathbf{k}\in\mathbb{Z}^d} w_B(\mathbf{k}) \left(\int_{[-\pi,\pi]^d} S(\boldsymbol{\theta})\, e^{i(\mathbf{k}\cdot\boldsymbol{\theta})} \mathrm{d}\boldsymbol{\theta}\right) e^{-i(\mathbf{k}\cdot\omega)} \\ &= \frac{1}{(2\pi)^d} \int_{[-\pi,\pi]} S(\boldsymbol{\theta}) W_B(\omega - \boldsymbol{\theta}) \mathrm{d}\boldsymbol{\theta}, \end{aligned} \tag{5.2.3}$$

where $W_B(\omega)$ is the Fourier transform of $\omega_B(\mathbf{k})$. By (5.2.2),

$$W_B(\omega) = \sum_{\mathbf{k}\in([-N+1,N-1]^d \bigcap \mathbb{Z}^d)} \prod_{j=1}^{d} \left(1 - \frac{k_j}{N}\right) e^{-i(\mathbf{k}\cdot\omega)} = \frac{1}{N^d} \prod_{j=1}^{d} \left(\frac{\sin(\frac{\omega_j N}{2})}{\sin(\frac{N}{2})}\right)^2.$$

This is just the Fejér kernel of Fourier series. Therefore, the right-hand side of (5.2.3) is the Fejér sums of Fourier series of $S(\boldsymbol{\theta})$. By the approximation theory of Fourier series, we deduce that

$$\lim_{N\to\infty} E[\hat{S}_p(\omega)] = \lim_{N\to\infty} E[\hat{S}_c(\omega)] = S(\omega). \tag{5.2.4}$$

Therefore, the periodogram is an asymptotically unbiased spectral estimator.

5.2.3 Variance Analysis

Assume that $e(\mathbf{t})$ ($\mathbf{t} \in \mathbb{Z}^d$) is a Gaussian complex white noise with variance σ^2, i.e.,

$$E[e(\mathbf{t})\overline{e}(\mathbf{s})] = \sigma^2 \delta_{\mathbf{t},\mathbf{s}}.$$

Using the expectation formula of Gaussian stochastic variable, we have

$$E[e(\mathbf{t})\overline{e}(\mathbf{s})e(\mathbf{l})\overline{e}(\mathbf{m})] = \sigma^4(\delta_{\mathbf{t},\mathbf{s}}\delta_{\mathbf{l},\mathbf{m}} + \delta_{\mathbf{t},\mathbf{m}}\delta_{\mathbf{s},\mathbf{l}}).$$

Note that $\hat{S}_p(\boldsymbol{\omega}) = \frac{1}{N^d} | \sum_{\mathbf{t}\in([1,N]^d \bigcap \mathbb{Z}^d)} e(\mathbf{t})\, e^{-i(\mathbf{t}\cdot\boldsymbol{\omega})} |^2$. It follows that

$$\begin{aligned} E[\hat{S}_p(\boldsymbol{\omega}^{(1)})\hat{S}_p(\boldsymbol{\omega}^{(2)})] &= \frac{\sigma^4}{N^{2d}} \sum_{\mathbf{t},\mathbf{s},\mathbf{l},\mathbf{m}\in([1,N]^d \bigcap \mathbb{Z}^d)} (\delta_{\mathbf{t},\mathbf{s}}\delta_{\mathbf{l},\mathbf{m}} + \delta_{\mathbf{t},\mathbf{m}}\delta_{\mathbf{s},\mathbf{l}})\, e^{-i((\mathbf{t}-\mathbf{s})\cdot\boldsymbol{\omega}^{(1)}) - i((\mathbf{l}-\mathbf{m})\cdot\boldsymbol{\omega}^{(2)})} \\ &= \sigma^4 + \frac{\sigma^4}{N^{2d}} \sum_{\mathbf{t},\mathbf{s}\in([1,N]^d \bigcap \mathbb{Z}^d)} e^{-i((\mathbf{t}-\mathbf{s})\cdot(\boldsymbol{\omega}^{(1)}-\boldsymbol{\omega}^{(2)}))} \\ &= \sigma^4 + \frac{\sigma^4}{N^{2d}} \left| \sum_{\mathbf{t}\in([1,N]^d \bigcap \mathbb{Z}^d)} e^{-i(\mathbf{t}\cdot(\boldsymbol{\omega}^{(1)}-\boldsymbol{\omega}^{(2)}))} \right|^2 \\ &= \sigma^4 + \frac{\sigma^4}{N^{2d}} \prod_{i=1}^{d} \left(\frac{\sin(\omega_i^{(1)}-\omega_i^{(2)})N/2}{\sin(\omega_i^{(1)}-\omega_i^{(2)})/2} \right)^2 . \end{aligned}$$

Where $\boldsymbol{\omega}^{(1)} = (\omega_1^{(1)}, \ldots, \omega_d^{(1)})$ and $\boldsymbol{\omega}^{(2)} = (\omega_1^{(2)}, \ldots, \omega_d^{(2)})$. Further,

$$\lim_{N\to\infty} E[\hat{S}_p(\boldsymbol{\omega}^{(1)})\hat{S}_p(\boldsymbol{\omega}^{(2)})] = \begin{cases} 2\sigma^4 & (\boldsymbol{\omega}^{(1)} = \boldsymbol{\omega}^{(2)}), \\ \sigma^4 & (\boldsymbol{\omega}^{(1)} \neq \boldsymbol{\omega}^{(2)}). \end{cases}$$

From this and (5.2.4), it follows that

$$\begin{aligned} &\lim_{N\to\infty} E[(\hat{S}_p(\boldsymbol{\omega}^{(1)}) - S(\boldsymbol{\omega}^{(1)}))(\hat{S}_p(\boldsymbol{\omega}^{(2)}) - S(\boldsymbol{\omega}^{(2)}))] \\ &\quad = \lim_{N\to\infty} E[\hat{S}_p(\boldsymbol{\omega}^{(1)})\hat{S}_p(\boldsymbol{\omega}^{(2)})] - S(\boldsymbol{\omega}^{(1)})S(\boldsymbol{\omega}^{(2)}) \end{aligned}$$

Note that the PSD of the white noise $e(\mathbf{t})$ is $S(\boldsymbol{\omega}) = \sigma^2$ for all $\boldsymbol{\omega}$. So,

$$\lim_{N\to\infty} E[(\hat{S}_p(\boldsymbol{\omega}^1) - S(\boldsymbol{\omega}^{(1)}))(\hat{S}_p(\boldsymbol{\omega}^2) - S(\boldsymbol{\omega}^{(2)}))] = \sigma^4 \delta_{\boldsymbol{\omega}^{(1)},\boldsymbol{\omega}^{(2)}}.$$

This shows that the variance of periodogram and correlogram methods is very large.

5.3 Blackman–Tukey Method

Blackman–Tukey method is to estimate the PSD using a windowed Fourier transform of the autocorrelation function of multivariate time series. It was developed by Blackman and Tukey in order to reduce variance of periodogram and correlogram.

5.3.1 *Blackman–Tukey Estimator*

Let $w(\mathbf{k})$ be a smooth window function satisfying the conditions:

$$w(-\mathbf{k}) = w(\mathbf{k}), \qquad w(\mathbf{0}) = 1,$$

$$w(\mathbf{k}) = 0 \quad \text{for } \mathbf{k} \notin [-M+1, M-1]^d \bigcap \mathbb{Z}^d.$$

The *Blackman–Tukey estimator* is defined as

$$\hat{S}_{BT}(\boldsymbol{\omega}) = \sum_{\mathbf{k}\in([-M+1,M-1]^d \bigcap \mathbb{Z}^d)} w(\mathbf{k})\hat{\gamma}(\mathbf{k})\, e^{-i(\mathbf{k}\cdot\boldsymbol{\omega})}$$
$$\text{or } \hat{S}_{BT}(\boldsymbol{\omega}) = \sum_{\mathbf{k}\in\mathbb{Z}^d} w(\mathbf{k})\hat{\gamma}(\mathbf{k})\, e^{-i(\mathbf{k}\cdot\boldsymbol{\omega})}. \qquad (5.3.1)$$

Denote by $W(\boldsymbol{\omega})$ the Fourier transform of $w(\mathbf{k})$. Note that the Fourier transform of $\hat{\gamma}(\mathbf{k})$ is $\hat{S}_p(\boldsymbol{\omega})$. Using the convolution formula of Fourier transform, the integral representation of the Blackman–Tukey estimator is

$$\hat{S}_{BT}(\boldsymbol{\omega}) = (\hat{S}_p * W)(\boldsymbol{\omega}) = \frac{1}{(2\pi)^d}\int_{[-\pi,\pi]^d} \hat{S}_p(\boldsymbol{\theta})W(\boldsymbol{\omega}-\boldsymbol{\theta})\mathrm{d}\boldsymbol{\theta}. \qquad (5.3.2)$$

(a) If $w(\mathbf{k})$ is the rectangular window:

$$w(\mathbf{k}) = \begin{cases} 1, & \mathbf{k} \in ([-M+1, M-1]^d \bigcap \mathbb{Z}^d), \\ 0, & \text{otherwise}, \end{cases}$$

then DTFT of $w(\mathbf{k})$ is

$$W(\boldsymbol{\omega}) = \sum_{\mathbf{k}\in([-M+1,M-1]^d \bigcap \mathbb{Z}^d)} e^{-i(\mathbf{k}\cdot\boldsymbol{\omega})} = \prod_{j=1}^{d} \frac{\sin(M-\frac{1}{2})\omega_j}{\sin\frac{\omega_j}{2}}.$$

which is called the *Dirichlet kernel*. The integral representation becomes

$$\hat{S}_{BT}(\boldsymbol{\omega}) = \frac{1}{(2\pi)^d} \int_{[-\pi,\pi]^d} \hat{S}_p(\boldsymbol{\theta}) \prod_{j=1}^{d} \frac{\sin(M-\frac{1}{2})(\omega_j-\theta_j)}{\sin\frac{\omega_j-\theta_j}{2}} \mathrm{d}\boldsymbol{\theta}.$$

(b) If $w(\mathbf{k})$ is the triangle window:

$$w(\mathbf{k}) = \begin{cases} \prod_{j=1}^{d} \left(1-\frac{|k_j|}{M}\right), & \mathbf{k} \in ([-M+1, M-1]^d \bigcap \mathbb{Z}^d), \\ 0, & \text{otherwise,} \end{cases}$$

then DTFT of $w(\mathbf{k})$ is

$$W(\boldsymbol{\omega}) = \sum_{\mathbf{k}\in([-M+1,M-1]^d \bigcap \mathbb{Z}^d)} \prod_{j=1}^{d} \left(1-\frac{|k_j|}{M}\right) e^{-i(\mathbf{k}\cdot\boldsymbol{\omega})} = \frac{1}{M^d} \prod_{j=1}^{d} \left(\frac{\sin\frac{M\omega_j}{2}}{\sin\frac{\omega_j}{2}} \right)^2.$$

which is called the *Fejér kernel*. The integral representation becomes

$$\hat{S}_{BT}(\boldsymbol{\omega}) = \frac{1}{(2\pi M)^d} \int_{[-\pi,\pi]^d} \hat{S}_p(\boldsymbol{\theta}) \prod_{j=1}^{d} \left(\frac{\sin\frac{M(\omega_j-\theta_j)}{2}}{\sin\frac{(\omega_j-\theta_j)}{2}} \right)^2 \mathrm{d}\boldsymbol{\theta}.$$

Since the windows $W(\boldsymbol{\omega})$ have often a dominant relatively narrow peak at $\boldsymbol{\omega} = \mathbf{0}$, the Blackman–Tukey spectral estimator $\hat{S}_{BT}(\boldsymbol{\omega})$ can be viewed as the locally weighted average of the periodogram. In the other words, the Blackman–Tukey spectral estimator is to smooth the periodogram and eliminate large fluctuation of the periodogram. Therefore, compared with periodogram method, $\hat{S}_{BT}(\boldsymbol{\omega})$ reduces the variance but also reduces the resolution. The larger the size of the window $W(\boldsymbol{\omega})$ is, the larger the reduction in variance and the lower the resolution.

5.3.2 Several Common Windows

The windows satisfy

$$w(\mathbf{k}) = 0 \qquad (\mathbf{k} \notin [-M+1, M-1]^d \bigcap \mathbb{Z}^d).$$

For $\mathbf{k} \in ([-M+1, M-1]^d \bigcap \mathbb{Z}^d)$,

(a) Rectangular window: $w(\mathbf{k}) = 1$;

(b) Bartlett window: $w(\mathbf{k}) = \prod_{j=1}^{d} \frac{M-|k_j|}{M}$;

(c) Hanning window: $w(\mathbf{k}) = \prod_{j=1}^{d} \left(\frac{1}{2} + \frac{1}{2}\cos\frac{\pi|k_j|}{M}\right);$

(d) Hamming window: $w(\mathbf{k}) = \prod_{j=1}^{d} \left(0.54 + 0.46\cos\frac{\pi k_j}{M-1}\right);$

(e) Blackman window: $w(\mathbf{k}) = \prod_{j=1}^{d} \left(0.42 + 0.5\cos\frac{\pi k_j}{M-1} + 0.08\cos\frac{2\pi k_j}{M-1}\right);$

(f) Kaiser window:

$$w(\mathbf{k}) = \prod_{j=1}^{d} \frac{I_0(\gamma\sqrt{1-(k_j/(M-1))})^2}{I_0(\gamma)},$$

where $I_0(x) = \sum_{k=0}^{\infty} \frac{x^{2k}}{2^{2k}(k!)^2}$ is the zeroth-order modified Bessel function of the first kind. The parameter γ can control the spectral parameters such as the mainlobe width and the ripple ratio. To achieve a peak sidelobe level of B dB below the peak value:

$$\gamma \approx \begin{cases} 0 & (B < 21), \\ 0.584(B-21)^{0.4} + 0.0789(B-21) & (21 \le B \le 50), \\ 0.11(B-8.7) & (B > 50). \end{cases}$$

As special cases, a Kaiser window may reduce to a rectangular window ($\gamma = 0$), Hamming window ($\gamma = 5.4414$), Blackman window ($\gamma = 8.885$).

5.3.3 *Positive Semidefinite Window*

Since the power spectral density $S(\boldsymbol{\omega}) \ge 0$, it is natural to require that the Blackman–Tukey spectral estimator $\hat{S}_{BT}(\boldsymbol{\omega}) \ge 0$. If $\{w(\mathbf{k})\}$ is a window such that its Fourier transform $W(\boldsymbol{\omega}) \ge 0$, by (5.3.2) and $\hat{S}_p(\boldsymbol{\theta}) \ge 0$, it follows immediately that $\hat{S}_{BT}(\boldsymbol{\omega}) \ge 0$. Below, we give the construction of the window $\{w(\mathbf{k})\}$ whose Fourier transform satisfies $W(\boldsymbol{\omega}) \ge 0$.

Given a sequence $\{v(k)\}_{|k|<M}$. Let

$$V(\omega) = \sum_{|k|<M} v(k)\, e^{-ik\omega}. \tag{5.3.3}$$

Define the *spectral window* as $W(\boldsymbol{\omega}) = \prod_{i=1}^{d} |V(\omega_i)|^2$.

To maximize the relative energy in the mainlobe of $W(\boldsymbol{\omega})$, we need to choose $\{v(k)\}$ such that

$$F = \frac{\int_{[-\beta\pi,\beta\pi]^d} W(\boldsymbol{\omega})\mathrm{d}\boldsymbol{\omega}}{\int_{[-\pi,\pi]^d} W(\boldsymbol{\omega})\mathrm{d}\boldsymbol{\omega}} = \prod_{i=1}^{d} \frac{\int_{-\beta\pi}^{\beta\pi} |V(\omega_i)|^2 \mathrm{d}\omega}{\int_{-\pi}^{\pi} |V(\omega_i)|^2 \mathrm{d}\omega_i} \tag{5.3.4}$$

attains the maximal value, where β is a given window design parameter which makes the balance between spectral resolution and statistical variance. Let

$$\mathbf{v} = (v(0), \ldots, v(M-1))^T,$$

$$\mathbf{b}(\omega_i) = (1, e^{-i\omega_i}, \ldots, e^{-i(M-1)\omega_i})^T,$$

where T means transpose. By (5.3.3),

$$V(\omega_i) = \mathbf{v}^T \mathbf{b}(\omega_i), \qquad |V(\omega_i)|^2 = \mathbf{v}^T \mathbf{b}(\omega_i) \overline{\mathbf{b}}^T(\omega_i) \overline{\mathbf{v}}.$$

By Parseval identity, $\frac{1}{2\pi} \int_{-\pi}^{\pi} |V(\omega_i)|^2 \mathrm{d}\omega_i = \overline{\mathbf{v}}^T \mathbf{v}$. From this and (5.3.4),

$$F = \prod_{i=1}^{d} \frac{\mathbf{v}^T \left(\frac{1}{2\pi} \int_{-\beta\pi}^{\beta\pi} \mathbf{b}(\omega_i) \overline{\mathbf{b}}^T(\omega_i) \mathrm{d}\omega_i \right) \overline{\mathbf{v}}}{\mathbf{v}^T \overline{\mathbf{v}}}.$$

Note that

$$\Gamma = \frac{1}{2\pi} \int_{-\beta\pi}^{\beta\pi} \mathbf{b}(\omega_i) \overline{\mathbf{b}}^T(\omega_i) \mathrm{d}\omega_i =: (\gamma(m-n))_{M\times M},$$

where $\gamma(m-n) = \begin{cases} \frac{\sin(m-n)\beta\pi}{(m-n)\pi} & (m \neq n), \\ \beta & (m = n). \end{cases}$ Then,

$$F = \left(\frac{\mathbf{v}^T \Gamma \overline{\mathbf{v}}}{\mathbf{v}^T \overline{\mathbf{v}}} \right)^d.$$

Since the matrix Γ is an $M \times M$ positive semi-definite matrix, all of its eigenvalues are all nonnegative real numbers. Let $\widetilde{\mathbf{v}}$ be the eigenvector corresponding to the maximal eigenvalue $\lambda_{\max}$ of Γ. It can be proved that F attains the maximal value $\lambda_{\max}^d$ when $\mathbf{v} = \widetilde{\mathbf{v}}$. Therefore, the positive semidefinite window which maximizes the relative energy in the mainlobe interval $[-\beta\pi, \beta\pi]^d$ can be designed as

$$w(\mathbf{k}) = \prod_{i=1}^{d} \left(\sum_{l=0}^{M-1} \widetilde{\mathbf{v}}(l) \overline{\widetilde{\mathbf{v}}}(l - k_i) \right).$$

5.4 Welch Method

The Welch method is one of most frequently used PSD estimation methods. To reduce the large fluctuation of the periodogram, the Welch method is carried out by dividing the samples of the multivariate stochastic process into successive blocks, forming the periodogram for each block and averaging.

Let $\{x(\mathbf{t})\}_{t\in[1,\,LM]^d \bigcap \mathbb{Z}^d}$ be the sample of a multivariate stochastic process. We split up it into L^d subsamples with size M^d:

$$x_{\mathbf{j}}(\mathbf{t}) = x(M\mathbf{j}+\mathbf{t}) \qquad (\mathbf{t} \in ([1,\,M]^d \bigcap \mathbb{Z}^d),\ \mathbf{j} \in ([0,\,L-1]^d \bigcap \mathbb{Z}^d)).$$

Define a *windowed periodogram* corresponding to $\{x_{\mathbf{j}}(\mathbf{t})\}_{\mathbf{t}\in([1,M]^d \bigcap \mathbb{Z}^d)}$ with a window $p(\mathbf{t})$ as

$$\hat{S}_{\mathbf{j}}(\omega) = \frac{1}{M^d P}\left| \sum_{\mathbf{t}\in([1,M]^d \bigcap \mathbb{Z}^d)} p(\mathbf{t})x_{\mathbf{j}}(\mathbf{t})\, e^{-i(\mathbf{t}\cdot\omega)} \right|^2,$$

where $P = \frac{1}{M^d}\sum_{\mathbf{t}\in([1,M]^d \bigcap \mathbb{Z}^d)} |p(\mathbf{t})|^2$. Define the *Welch estimate* of PSD as

$$\hat{S}_W(\omega) = \frac{1}{L^d}\sum_{\mathbf{j}\in([0,L-1]^d \bigcap \mathbb{Z}^d)} \hat{S}_{\mathbf{j}}(\omega).$$

It is clear that

$$\begin{aligned}
\hat{S}_W(\omega) &= \frac{1}{L^d}\sum_{\mathbf{j}\in([0,L-1]^d \bigcap \mathbb{Z}^d)} \left(\frac{1}{M^d P}\left| \sum_{\mathbf{t}\in([1,M]^d \bigcap \mathbb{Z}^d)} p(\mathbf{t})x_{\mathbf{j}}(\mathbf{t})\, e^{-i(\mathbf{t}\cdot\omega)} \right|^2 \right) \\
&= \frac{1}{L^d}\sum_{\mathbf{j}\in([0,L-1]^d \bigcap \mathbb{Z}^d)} \left(\frac{1}{M^d P}\left(\sum_{\mathbf{t}\in([1,M]^d \bigcap \mathbb{Z}^d)} p(\mathbf{t})x_{\mathbf{j}}(\mathbf{t})\, e^{-i(\mathbf{t}\cdot\omega)} \right) \right. \\
&\quad \times \left. \overline{\left(\sum_{\mathbf{s}\in([1,M]^d \bigcap \mathbb{Z}^d)}^{M} p(\mathbf{s})x_j(\mathbf{s})\, e^{-i(\mathbf{s}\cdot\omega)} \right)} \right) \\
&= \frac{1}{L^d}\sum_{\mathbf{j}\in([0,L-1]^d \bigcap \mathbb{Z}^d)} \frac{1}{M^d P}\sum_{\mathbf{t}\in([1,M]^d \bigcap \mathbb{Z}^d)}\sum_{\mathbf{s}\in([1,M]^d \bigcap \mathbb{Z}^d)} p(\mathbf{t})\overline{p}(\mathbf{s})x_{\mathbf{j}}(\mathbf{t})\overline{x}_{\mathbf{j}}(\mathbf{s})\, e^{-i((\mathbf{t}-\mathbf{s})\cdot\omega)} \\
&= \frac{1}{M^d P}\sum_{\mathbf{t}\in([1,M]^d \bigcap \mathbb{Z}^d)}\sum_{\mathbf{s}\in([1,M]^d \bigcap \mathbb{Z}^d)} p(\mathbf{t})\overline{p}(\mathbf{s})\left(\frac{1}{L^d}\sum_{\mathbf{j}\in([0,L-1]^d \bigcap \mathbb{Z}^d)} x_{\mathbf{j}}(\mathbf{t})\overline{x}_{\mathbf{j}}(\mathbf{s}) \right) e^{-i((\mathbf{t}-\mathbf{s})\cdot\omega)}.
\end{aligned}$$

Since the average:

$$\frac{1}{L^d}\sum_{\mathbf{j}\in([0,L-1]^d \bigcap \mathbb{Z}^d)} x_{\mathbf{j}}(\mathbf{t})\overline{x}_{\mathbf{j}}(\mathbf{s}) \approx \gamma(\mathbf{t}-\mathbf{s})$$

for large L, it follows that

$$\hat{S}_W(\omega) \approx \frac{1}{M^d P} \sum_{\mathbf{t}\in([1,M]^d \bigcap \mathbb{Z}^d)} \sum_{\mathbf{s}\in([1,M]^d \bigcap \mathbb{Z}^d)} p(\mathbf{t})\overline{p}(\mathbf{s})\gamma(\mathbf{t}-\mathbf{s})\, e^{-i((\mathbf{t}-\mathbf{s})\cdot\omega)}$$

$$= \frac{1}{M^d P} \sum_{\mathbf{t}\in([1,M]^d \bigcap \mathbb{Z}^d)} \sum_{\mathbf{k}\in(\mathbf{t}-[1,M]^d \bigcap \mathbb{Z}^d)} p(\mathbf{t})\overline{p}(\mathbf{t}-\mathbf{k})\gamma(\mathbf{k})\, \mathbf{e}^{-\mathbf{i}(\mathbf{k}\cdot\omega)}.$$

Let $p(\mathbf{t}) = 0\,(\mathbf{t} \notin ([1, M]^d \bigcap \mathbb{Z}^d))$. Then,

$$\hat{S}_W(\omega) \approx \sum_{\mathbf{k}\in[-M+1,M-1]^d} \left(\frac{1}{M^d P} \sum_{\mathbf{t}\in([1,M]^d \bigcap \mathbb{Z}^d)} p(\mathbf{t})\overline{p}(\mathbf{t}-\mathbf{k}) \right) \gamma(\mathbf{k})\, e^{-i(\mathbf{k}\cdot\omega)}$$

$$= \sum_{\mathbf{k}\in[-M+1,M-1]^d} w(\mathbf{k})\gamma(\mathbf{k})\, e^{-i(\mathbf{k}\cdot\omega)},$$

where $w(\mathbf{k}) = \frac{1}{M^d P} \sum_{\mathbf{t}\in([1,M]^d \bigcap \mathbb{Z}^d)} p(\mathbf{t})\overline{p}(\mathbf{t}-\mathbf{k})$.

5.5 Multitaper Method

The multitaper method attempts to reduce the variance of spectral estimates by using a small set of tapers rather than the unique taper or window. Its basic idea is to compute a set of independent estimates of the PSD and then average over these spectral estimates to yield a better and more stable spectral estimate.

Consider the following windowed periodogram estimator of the PSD:

$$\hat{S}_p^{(b)}(\omega) = \left| \sum_{\mathbf{t}\in([0,N-1]^d \bigcap \mathbb{Z}^d)} x(\mathbf{t})b(\mathbf{t})\, e^{-i(\mathbf{t}\cdot\omega)} \right|^2,$$

where the window function $b(\mathbf{t})$ satisfies the normalized condition $\sum_{\mathbf{t}\in([0,N-1]^d \bigcap \mathbb{Z}^d)} |b(\mathbf{t})|^2 = 1$. Denote

$$B(\mathbf{t}) = \sum_{\mathbf{t}\in([0,N-1]^d \bigcap \mathbb{Z}^d)} b(\mathbf{t})\, e^{-i(\mathbf{t}\cdot\omega)}.$$

By the convolution property of Fourier transform, it follows that

$$E[\hat{S}_p^{(b)}(\omega)] = \frac{1}{(2\pi)^d} \int_{[-\pi,\pi]^d} |B(\omega')|^2 S(\omega-\omega')\mathrm{d}\omega',$$

where $S(\omega)$ is the PSD of $x(\mathbf{t})$. If the spectral window $|B(\omega')|^2$ has low amplitudes when $|\omega'|$ is large, then $\hat{S}_p^{(b)}(\omega)$ is close to $S(\omega)$. Given a bandwidth W, where

$\frac{1}{N} < W \le \pi$. Denote

$$\lambda^W = \int_{[-W,W]^d} |B(\boldsymbol{\omega})|^2 \mathrm{d}\boldsymbol{\omega}.$$

By the Parseval identity, it follows from the normalized condition $\sum_{t=0}^{N-1} |b(t)|^2 = 1$ that

$$\int_{[-\pi,\pi]^d} |B(\boldsymbol{\omega})|^2 \mathrm{d}\boldsymbol{\omega} = (2\pi)^d.$$

To reduce the bias of window periodogram estimator, one needs to choose a window function $B(\boldsymbol{\omega})$ such that λ^W are large. Below, we give the construction method of such windows:

Let D is be $N \times N$ matrix whose components are

$$D(t,t') = \frac{\sin W(t-t')}{\pi(t-t')} \qquad (t,t' = 0,1,\ldots,N-1).$$

Since D is an $N \times N$ symmetric matrix, the N eigenvalues of D satisfy $\lambda_0 > \cdots > \lambda_{N-1} > 0$, and the associated eigenvectors $\mathbf{g}_0,\ldots,\mathbf{g}_{N-1}$, where $\mathbf{g}_j = (g_j(0),\ldots,g_j(N-1))^T$ $(j = 0,\ldots,N-1)$, satisfy

$$\sum_{t=0}^{N-1} g_j(t) g_k(t) = \delta_{j,k}.$$

These eigenvectors $\mathbf{g}_0,\ldots,\mathbf{g}_{N-1}$ are called the *slepian sequence*. The Fourier transform of $\mathbf{g}_k$:

$$G_k(\omega) = \sum_{t=0}^{N-1} g_k(t)\, e^{-it\omega}$$

is called the *slepian function* and has the following properties:

(a) $\frac{1}{2\pi}\int_{-\pi}^{\pi} G_j(\omega) G_k^*(\omega)\mathrm{d}\omega = \delta_{j,k}$;
(b) $\frac{1}{2\pi}\int_{-W}^{W} G_k(\omega) G_k^*(\omega)\mathrm{d}\omega = \lambda_k$.

For the first K largest eigenvectors $\mathbf{g}_0,\ldots,\mathbf{g}_{K-1}$, define a series of window functions as follows:

$$B_{\mathbf{m}}(\mathbf{t}) = \prod_{i=1}^{d} g_{m_i}(t_i),$$

where $\mathbf{m} = (m_1,\ldots,m_d) \in ([0,K-1]^d \bigcap \mathbb{Z}^d)$ and $\mathbf{t} = (t_1,\ldots,t_d) \in ([0,N-1]^d \bigcap \mathbb{Z}^d)$. These window functions are also called *orthogonal tapers*. Based on them, the multitaper estimator is

$$\hat{S}(\boldsymbol{\omega}) = \frac{1}{K^d} \sum_{\mathbf{m}\in([0,K-1]^d \bigcap \mathbb{Z}^d)} \hat{S}_{\mathbf{m}}(\boldsymbol{\omega})$$

$$= \frac{1}{K^d} \sum_{\mathbf{m}\in([0,K-1]^d \bigcap \mathbb{Z}^d)} \left| \sum_{\mathbf{t}\in([0,N-1]^d \bigcap \mathbb{Z}^d)} x(\mathbf{t}) B_{\mathbf{m}}(\mathbf{t})\, e^{-i(\mathbf{t}\cdot\boldsymbol{\omega})} \right|^2 .$$

5.6 Maximum Entropy Method

Let $X(\mathbf{t})$ be a stationary multivariate real-valued stochastic process with mean $\mathbf{0}$ and correlation function $\gamma_x(\mathbf{k})(\mathbf{k} \in \mathbb{Z}^d)$. The *entropy* H of $X(\mathbf{t})$ is defined as

$$H = \frac{1}{(2\pi)^d} \int_{[-\pi,\pi]^d} \log S_x(\boldsymbol{\omega}) \mathrm{d}\boldsymbol{\omega}, \tag{5.6.1}$$

where $S_x(\boldsymbol{\omega})$ is the power spectrum density of $X(\mathbf{t})$.

Assume that the correlation function $\gamma_x(\mathbf{k})$ on $[-N,N]^d \bigcap \mathbb{Z}^d$ can be well estimated. In the absence of more prior knowledge about the process that generates the time series $X(\mathbf{t})$, the *maximum entropy spectrum* estimate for $X(\mathbf{t})$ is to choose the spectrum $S_y(\boldsymbol{\omega})$ of the stochastic process $Y(\mathbf{t})$ such that $Y(\mathbf{t})$ have the maximum entropy and its autocorrelation function satisfies $\gamma_y(\mathbf{k}) = \gamma_x(\mathbf{k})$ ($\mathbf{k} \in [-N,N]^d \bigcap \mathbb{Z}^d$). Therefore, the power spectrum $S_y(\boldsymbol{\omega})$ can be expressed as a sum of two terms:

$$S_y(\boldsymbol{\omega}) = \sum_{\mathbf{k}\in([-N,N]^d \bigcap \mathbb{Z}^d)} \gamma_x(\mathbf{k})\, e^{-i(\mathbf{k}\cdot\boldsymbol{\omega})} + R_N(\boldsymbol{\omega}), \tag{5.6.2}$$

where $R_N(\boldsymbol{\omega}) = \sum_{\mathbf{k}\notin([-N,N]^d \bigcap \mathbb{Z}^d)} \gamma_y(\mathbf{k})\, e^{-i(\mathbf{k}\cdot\boldsymbol{\omega})}$. From this and (5.6.1), the entropy H of $Y(\mathbf{t})$ is a function of the unknown correlation coefficients $\gamma_y(\mathbf{k})$ ($\mathbf{k} \notin ([-N, N]^d \bigcap \mathbb{Z}^d)$). Adjust the values of unknown $\gamma_y(\mathbf{k})$'s such that H attains the maximal value. By (5.6.1) and (5.6.2),

$$\frac{\partial H}{\partial \gamma_y(\mathbf{k})} = \frac{1}{(2\pi)^d} \int_{[-\pi,\pi]^d} \frac{1}{S_y(\boldsymbol{\omega})} e^{-i(\mathbf{k}\cdot\boldsymbol{\omega})} \mathrm{d}\boldsymbol{\omega} = 0 \qquad (\mathbf{k} \notin ([-N,N]^d \bigcap \mathbb{Z}^d)).$$

By the orthogonality of the system $\{e^{-i(\mathbf{k}\cdot\boldsymbol{\omega})}\}$, $\frac{1}{S_y(\boldsymbol{\omega})}$ is a trigonometric polynomial:

$$\frac{1}{S_y(\boldsymbol{\omega})} = \sum_{\mathbf{k}\in([-N,N]^d \bigcap \mathbb{Z}^d)} \lambda(\mathbf{k})\, e^{-i(\mathbf{k}\cdot\boldsymbol{\omega})},$$

Since $\frac{1}{S_y(\boldsymbol{\omega})}$ is a real-valued even function, the sequence of Fourier coefficients is an even sequence, i.e.,

$$\lambda(k_1, \ldots, k_{l-1}, -k_l, k_{l+1}, \ldots, k_d)$$
$$= \lambda(k_1, \ldots, k_{l-1}, k_l, k_{l+1}, \ldots, k_d) \quad \text{for } l = 1, \ldots, d.$$

So, $\frac{1}{S_y(\omega)}$ can be expanded into Fourier cosine series:

$$\frac{1}{S_y(\omega)} = \sum_{\mathbf{k}\in([0,N]^d \bigcap \mathbb{Z}^d)} p(\mathbf{k})\lambda(\mathbf{k}) \prod_{j=1}^{d} \cos(k_j\omega_j) = Q_N(\omega), \tag{5.6.3}$$

where $p(\mathbf{k}) = 2^{\sum_{j=1}^{d} u(k_j)}$ and

$$u(k_j) = \begin{cases} 1, \ k_j = 0, \\ 2, \ k_j \neq 0. \end{cases}$$

Similarly, since $S_y(\omega)$ is a real-valued even function, by (5.6.2), it follows that

$$S_y(\omega) = P_N(\omega) + R_N(\omega),$$

where

$$P_N(\omega) = \sum_{\mathbf{k}\in([0,N]^d \bigcap \mathbb{Z}^d)} p(\mathbf{k})\gamma(\mathbf{k}) \prod_{j=1}^{d} \cos(k_j\omega_j),$$

$$R_N(\omega) = \sum_{\mathbf{k}\notin([0,N]^d \bigcap \mathbb{Z}^d)} p(\mathbf{k})\gamma(\mathbf{k}) \prod_{j=1}^{d} \cos(k_j\omega_j).$$

From this and (5.6.3), it is clear that

$$Q_N(\omega)(P_N(\omega) + R_N(\omega)) = 1.$$

For $\mathbf{k} = (k_1, \ldots, k_d) \in ([0, N]^d \bigcap \mathbb{Z}^d)$, the coefficients of $\prod_{j=1}^{d} \cos(k_j\omega_j)$ on the left-hand side of the above equality are determined completely by the term $Q_N(\omega)P_N(\omega)$. Comparing these coefficients, we get a system of equations with unknown $\{\lambda_{\mathbf{k}}\}_{\mathbf{k}\in[0,N]^d}$. After solving $\{\lambda_{\mathbf{k}}\}_{\mathbf{k}\in[0,N]^d}$, we get

$$\hat{S}_x(\omega) = S_y(\omega) = \frac{1}{Q_N(\omega)}.$$

5.7 Rational Spectral Estimation

Consider a multivariate stochastic process $X(\mathbf{t})$ with mean $\mathbf{0}$ and covariance $\gamma_x(\mathbf{k})$, ($\mathbf{k} = (k_1, \ldots, k_d)$). Assume that the correlation function $\gamma_x(\mathbf{k})$ on $[-N, N]^d \bigcap \mathbb{Z}^d$ can be well estimated. We will use the rational spectrum method to estimate the power spectral density of $X(\mathbf{t})$:

$$S_x(\omega) = \sum_{\mathbf{k}\in\mathbb{Z}^d} \gamma_x(\mathbf{k})\, e^{-i(\mathbf{k}\cdot\omega)}.$$

A *rational power spectral density* is a positive rational function of $e^{-i\omega}$ as follows:

$$R(\omega) = \frac{\sum\limits_{\mathbf{k}\in([-q,q]^d \bigcap \mathbb{Z}^d)} \alpha_{\mathbf{k}}\, e^{-i(\mathbf{k}\cdot\omega)}}{\sum\limits_{\mathbf{k}\in([-p,p]^d \bigcap \mathbb{Z}^d)} \beta_{\mathbf{k}}\, e^{-i(\mathbf{k}\cdot\omega)}}.$$

Assume that the numerator and the denominator are both real-valued even functions, then

$$\sum_{\mathbf{k}\in([-q,q]^d \bigcap \mathbb{Z}^d)} \alpha_{\mathbf{k}}\, e^{-i(\mathbf{k}\cdot\omega)} = \sum_{\mathbf{k}\in([0,q]^d \bigcap \mathbb{Z}^d)} p(\mathbf{k})\alpha_{\mathbf{k}} \prod_{j=1}^{d} \cos(k_j\omega_j) =: A(\omega),$$

$$\sum_{\mathbf{k}\in([-p,p]^d \bigcap \mathbb{Z}^d)} \beta_{\mathbf{k}}\, e^{-i(\mathbf{k}\cdot\omega)} = \sum_{\mathbf{k}\in([0,p]^d \bigcap \mathbb{Z}^d)} p(\mathbf{k})\beta_{\mathbf{k}} \prod_{j=1}^{d} \cos(k_j\omega_j) =: B(\omega).$$

Since $S_x(\omega)$ is positive even function,

$$S_x(\omega) = \sum_{\mathbf{k}\in([0,N]^d \bigcap \mathbb{Z}^d)} p(\mathbf{k})\gamma(\mathbf{k}) \prod_{j=1}^{d} \cos(k_j\omega_j) + W_N(\omega),$$

where $W_N(\omega) = \sum_{\mathbf{k}\notin([0,N]^d \bigcap \mathbb{Z}^d)} p(\mathbf{k})\gamma(\mathbf{k}) \prod_{j=1}^{d} \cos(k_j\omega_j)$. We will choose the suitable $\alpha_{\mathbf{k}}$ and $\beta_{\mathbf{k}}$ such that $R(\omega) \approx S_x(\omega)$, then

$$A(\omega) = B(\omega) \sum_{\mathbf{k}\in([0,N]^d \bigcap \mathbb{Z}^d)} p(\mathbf{k})\gamma(\mathbf{k}) \prod_{j=1}^{d} \cos(k_j\omega_j) + B(\omega)W_N(\omega). \qquad (5.7.1)$$

For $\mathbf{k} = (k_1, \ldots, k_d) \in ([0, N]^d \bigcap \mathbb{Z}^d)$, the coefficients of $\prod\limits_{j=1}^{d} \cos(k_j\omega_j)$ on both sides of (5.7.1) are determined completely by two terms:

$$A(\omega) \quad \text{and} \quad B(\omega) \qquad \sum_{\mathbf{k}\in([0,N]^d \bigcap \mathbb{Z}^d)} \quad p(\mathbf{k})\gamma(\mathbf{k}) \prod_{j=1}^{d} \cos(k_j\omega_j).$$

Comparing the coefficients of these two terms, we get a system of equations with unknown $\{\alpha_{\mathbf{k}}\}_{\mathbf{k}\in([0,q]^d \bigcap \mathbb{Z}^d)}$ and $\{\beta_{\mathbf{k}}\}_{\mathbf{k}\in([0,p]^d \bigcap \mathbb{Z}^d)}$. Solving these equations, we get the rational spectral estimation of $X(\mathbf{t})$:

$$\hat{S}_x(\omega) = R(\omega) = \frac{A(\omega)}{B(\omega)}.$$

5.8 Discrete Spectral Estimation

The stochastic processes with discrete spectrum can be described by the following exponential model:

$$x(\mathbf{t}) = u(\mathbf{t}) + e(\mathbf{t}) = \sum_{l=1}^{n} r_l \, e^{i((\mathbf{t}\cdot\omega_l)+\varphi_l)} + e(\mathbf{t}), \tag{5.8.1}$$

where the noise $e(\mathbf{t})$ is complex white noise with mean 0 and variance σ^2. In this section, we will give two methods to estimate $\omega_l, \ldots, \omega_n$.

(i) Least Squared Method

We rewrite (5.8.1) in the form:

$$x(\mathbf{t}) = \sum_{l=1}^{n} \theta_l \, e^{i(\mathbf{t}\cdot\omega_l)} + e(\mathbf{t}) \qquad (\theta_l = r_l \, e^{i\varphi_l}).$$

Now we introduce several methods to estimate parameters $\omega_l (l = 1, \ldots, n)$ and $\theta_k (k = 1, \ldots, n)$ from observation.

Suppose that we have known observation data $\{x(\mathbf{t}_l)\}_{l=1,\ldots,N}$. Let

$$f(\Omega, \boldsymbol{\theta}) = \sum_{k=1}^{N} \left| x(\mathbf{t}_k) - \sum_{l=1}^{n} \theta_l \, e^{i(\mathbf{t}_k\cdot\omega_l)} \right|^2, \tag{5.8.2}$$

where $\Omega = (\omega_1, \ldots, \omega_n)$ and $\boldsymbol{\theta} = (\theta_1, \ldots, \theta_n)^T$. Choose Ω and $\boldsymbol{\theta}$ such that $f(\omega, \boldsymbol{\theta})$ attains the minimal value. Denote $\mathbf{X} = (x(\mathbf{t}_1), \ldots, x(\mathbf{t}_N))^T$ and

$$A = \begin{pmatrix} e^{i(\mathbf{t}_1 \cdot \omega_1)} & \dots & e^{i(\mathbf{t}_1 \cdot \omega_n)} \\ \vdots & \ddots & \vdots \\ e^{i(\mathbf{t}_N \cdot \omega_1)} & \dots & e^{i(\mathbf{t}_N \cdot \omega_n)} \end{pmatrix}.$$

By (5.8.2),

$$f(\Omega, \boldsymbol{\theta}) = (\mathbf{X} - A\boldsymbol{\theta})^*(\mathbf{X} - A\boldsymbol{\theta}), \tag{5.8.3}$$

where $*$ means conjugate and transpose. Since $(A^*A)^{-1}$ exists for $N \geq n$, (5.8.3) can be rewritten as

$$f(\Omega, \boldsymbol{\theta}) = (\boldsymbol{\theta} - (A^*A)^{-1}A^*\mathbf{X})^*(A^*A)(\boldsymbol{\theta} - (A^*A)^{-1}A^*\mathbf{X}) + \mathbf{X}^*\mathbf{X} - \mathbf{X}^*A(A^*A)^{-1}A^*\mathbf{X}. \tag{5.8.4}$$

First, we choose $\Omega = \widetilde{\Omega} = (\widetilde{\omega}_1, \dots, \widetilde{\omega}_n)$ such that $\mathbf{X}^*A(A^*A)^{-1}A^*\mathbf{X}$ attains the maximal value. Denote

$$\widetilde{A} = \begin{pmatrix} e^{i(\mathbf{t}_1 \cdot \widetilde{\omega}_1)} & \dots & e^{i(\mathbf{t}_1 \cdot \widetilde{\omega}_n)} \\ \vdots & \ddots & \vdots \\ e^{i(\mathbf{t}_N \cdot \widetilde{\omega}_1)} & \dots & e^{i(\mathbf{t}_N \cdot \widetilde{\omega}_n)} \end{pmatrix}.$$

Let $\widetilde{\boldsymbol{\theta}} = (\widetilde{A}^*\widetilde{A})^{-1}\widetilde{A}^*\mathbf{X}$. By (5.8.4), we know that $f(\widetilde{\Omega}, \widetilde{\boldsymbol{\theta}})$ attains the minimal value of $f(\Omega, \boldsymbol{\theta})$.

(ii) Covariance Function Method

Suppose that in (5.8.1), the initial phases $\varphi_l (l = 1, \dots, n)$ are independent uniform stochastic variables on $(-\pi, \pi)$, i.e., the probability density function of each φ_l is $\rho(x) = \frac{1}{2\pi}\ (-\pi < x \leq \pi)$.

Let

$$C = \begin{pmatrix} 1 & 1 & \cdots & 1 \\ e^{(i\mathbf{t}_1 \cdot \omega_1)} & e^{i(\mathbf{t}_1 \cdot \omega_2)} & \dots & e^{i(\mathbf{t}_1 \cdot \omega_n)} \\ \vdots & \vdots & \vdots & \vdots \\ e^{i(\mathbf{t}_{m-1} \cdot \omega_1)} & e^{i(\mathbf{t}_{m-1} \cdot \omega_2)} & \dots & e^{i(\mathbf{t}_{m-1} \cdot \omega_n)} \end{pmatrix}$$

be an $m \times n$ matrix. It is clear that (5.8.1) can be rewritten as

$$\mathbf{x}(\mathbf{t}) = C\mathbf{u}(\mathbf{t}) + \mathbf{e}(\mathbf{t}), \tag{5.8.5}$$

where

$$\mathbf{x(t)} = (x(\mathbf{t}), x(\mathbf{t}+\mathbf{t}_1), \ldots, x(\mathbf{t}+\mathbf{t}_{m-1}))^T,$$

$$\mathbf{u(t)} = (u_1(\mathbf{t}), \ldots, u_n(\mathbf{t}))^T, \qquad (u_l(\mathbf{t}) = r_l\, e^{i((\mathbf{t}\cdot\omega_l)+\varphi_l)}),$$

$$\mathbf{e(t)} = (e(\mathbf{t}), \ldots, e(\mathbf{t}+\mathbf{t}_{m-1}))^T$$

Considering the covariance matrix of $\mathbf{x}(t)$, we get

$$R = E[\mathbf{x(t)x^*(t)}] = CQC^* + \sigma^2 I, \tag{5.8.6}$$

where $*$ represents conjugate and transpose, and

$$Q = \begin{pmatrix} r_1^2 & \cdots & 0 \\ & \ddots & \\ 0 & \cdots & r_n^2 \end{pmatrix}.$$

When $m \geq n$, it follows that $\mathrm{rank}(CQC^*) = n$, so the eigenvalues $\eta_1, \ldots, \eta_m$ of CQC^* satisfy $\eta_1 \geq \cdots \geq \eta_n > 0$ and $\eta_{n+1} = \cdots = \eta_m = 0$. Let $\lambda_1 \geq \cdots \geq \lambda_m$ be the eigenvalues of R. By (5.8.6), $\lambda_k = \eta_k + \sigma^2$ $(k = 1, \ldots, m)$, and so

$$\lambda_k > \sigma^2 \quad (k = 1, \ldots, n), \qquad \lambda_k = \sigma^2 \quad (k = n+1, \ldots, m).$$

Denote by $\mathbf{h}_1, \ldots, \mathbf{h}_{m-n}$ the eigenvectors corresponding eigenvalues $\lambda_{n+1}, \ldots, \lambda_m$. Let $H = (\mathbf{h}_1, \ldots, \mathbf{h}_{m-n})$. Then, $RH = \sigma^2 H$. On the other side, from (5.8.6), it follows that

$$RH = CQC^*H + \sigma^2 H.$$

This implies that $CQC^*H = 0$. Since CQ has full column rank, $C^*H = 0$, i.e., $\omega_1, \ldots, \omega_n$ can be estimated from the equation $C^*H = 0$.

5.9 Vector ARMA Spectrum

Since dependence is very common in stochastic processes, the autoregressive moving average (ARMA) processes are fundamental tools in modeling and predicting unknown stochastic process. By allowing the order of an ARMA model to increase, one can approximate any stationary stochastic process with desirable accuracy.

A vector ARMA process satisfies

$$\mathbf{X}(t) - \sum_{k=1}^{p} \Phi_k \mathbf{X}(t-k) = \mathbf{Z}(t) - \sum_{k=1}^{q} \Theta_k \mathbf{Z}(t-k), \tag{5.9.1}$$

i.e.,

$$\begin{pmatrix} X_1(t) \\ \vdots \\ X_d(t) \end{pmatrix} - \sum_{k=1}^{p} \begin{pmatrix} \varphi_{11}^{(k)} & \cdots & \varphi_{1d}^{(k)} \\ \vdots & \ddots & \vdots \\ \varphi_{d1}^{(k)} & \cdots & \varphi_{dd}^{(k)} \end{pmatrix} \begin{pmatrix} X_1(t-k) \\ \vdots \\ X_d(t-k) \end{pmatrix}$$
$$= \begin{pmatrix} Z_1(t) \\ \vdots \\ Z_d(t) \end{pmatrix} - \sum_{k=1}^{q} \begin{pmatrix} \theta_{11}^{(k)} & \cdots & \theta_{1d}^{(k)} \\ \vdots & \ddots & \vdots \\ \theta_{d1}^{(k)} & \cdots & \theta_{dd}^{(k)} \end{pmatrix} \begin{pmatrix} Z_1(t-k) \\ \vdots \\ Z_d(t-k) \end{pmatrix},$$

where $\mathbf{Z}(t) \sim WN(\mathbf{0}, V)$ and Φ_k, Θ_k are constant matrices.

The $d \times d$ cross-covariance matrix at lag l of $\mathbf{X}(t)$ is

$$\Gamma(k) = E[\mathbf{X}(t)\overline{\mathbf{X}}(t-k)] = \begin{pmatrix} \gamma_{11} & \gamma_{12} & \cdots & \gamma_{1d} \\ \gamma_{21} & \gamma_{22} & \cdots & \gamma_{2d} \\ \vdots & \vdots & \ddots & \vdots \\ \gamma_{d1} & \gamma_{d2} & \cdots & \gamma_{dd} \end{pmatrix} \quad (k \in \mathbb{Z}).$$

Then the spectrum matrix of $\mathbf{X}(t)$ is defined as

$$S_{\mathbf{X}}(\omega) = \sum_{k \in \mathbb{Z}} \Gamma(k)\, e^{-ik\omega}$$

Let B be the time shift operator. Then the Eq. (5.9.1) can be written in the operator form:

$$\Phi(B)\mathbf{X}(t) = \Theta(B)\mathbf{Z}(t),$$

where $\Phi(B) = I - \sum_{k=1}^{p} \Phi(B)B^k$ and $\Theta(B) = I - \sum_{k=1}^{q} \Theta_k B^k$. It can be proved that

$$S_{\mathbf{X}}(\omega) = \Phi^{-1}(e^{-i\omega})\Theta(e^{-i\omega})V(\Theta(e^{-i\omega}))^*(\Phi^{-1}(e^{-i\omega}))^*$$

where V is the covariance matrix of the white noise $\mathbf{Z}(t)$ and $*$ represents conjugate and transpose.

As special cases, if $\mathbf{X}(t)$ is a vector MA(q) process: $\mathbf{X}(t) = \mathbf{Z}(t) - \sum_{k=1}^{q} \Theta_k \mathbf{Z}(t-k)$, then

$$S_{\mathbf{X}}(\omega) = \Theta(e^{-i\omega})V(\Theta(e^{-i\omega}))^*;$$

if $\mathbf{X}(t)$ is a vector AR(p) process: $\mathbf{X}(t) - \sum_{k=1}^{p} \Phi_k \mathbf{X}(t-k) = \mathbf{Z}(t)$, then

$$S_{\mathbf{X}}(\omega) = \Phi^{-1}(e^{-i\omega})V(\Phi^{-1}(e^{-i\omega}))^*$$

5.10 Multichannel Singular Spectrum Analysis

The singular spectrum analysis (SSA) technique is a novel and powerful technique for vector stochastic processes. It can decompose the original vector stochastic process into the sum of independent and interpretable components such as trend component, oscillatory component, and noise component.

Let $\mathbf{X}(t) = (X_1(t), \ldots, X_L(t))$ be a stationary vector stochastic process with mean $\mathbf{0}$. The lag cross-covariance between $X_i(j)$ and $X_{i'}(j')$ depends only on $j - j'$, denoted by $\gamma_{ii'}(j - j')$, where

$$\gamma_{ii'}(j - j') = E[X_i(j)X_{i'}(j')].$$

Let $D_{ii'}$ be the cross-covariance matrix between $X_i(t)$ and $X_{i'}(t)$:

$$D_{ii'} = (\gamma_{ii'}(j-j'))_{j,j'=1,\ldots,M} = \begin{pmatrix} \gamma_{ii'}(0) & \gamma_{ii'}(1) & \cdots & \gamma_{ii'}(M-1) \\ \gamma_{ii'}(1) & \gamma_{ii'}(0) & \cdots & \gamma_{ii'}(M-2) \\ \vdots & \vdots & \ddots & \vdots \\ \gamma_{ii'}(M-1) & \gamma_{ii'}(M-2) & \cdots & \gamma_{ii'}(0) \end{pmatrix}.$$

Construct a grand block-matrix $D_{\mathbf{X}} = (D_{ii'})_{L\times L}$ and each entry is a matrix $D_{ii'} = (\gamma_{ii'}(j-j'))_{M\times M}$. The main diagonal of $D_{\mathbf{X}}$ contains an estimate of the lag-zero covariance between $X_i(t)$ and $X_{i'}(t)$.

We will first give the *algorithm to estimate the lagged cross-covariances*:

Step 1. Construct the multichannel trajectory matrix by extending each channel $\{X_l(n)\}_{n=1,\ldots,N}$ $(1 \le l \le L)$ of $\mathbf{X}(t)$ with M lagged copies of itself:

$$\widetilde{X}_l = \begin{pmatrix} X_l(1) & X_l(2) & \cdots & X_l(N-M+1) \\ X_l(2) & X_l(3) & \cdots & X_l(N-M+2) \\ \vdots & \vdots & \ddots & \vdots \\ X_l(M) & X_l(M+1) & \cdots & X_l(N) \end{pmatrix} \qquad (l = 1, \ldots, L)$$

Step 2. Construct the grand trajectory $LM \times (N - M)$ matrix $\widetilde{X} = (\widetilde{X}_1, \ldots, \widetilde{X}_L)^T$, where the nth column is

$$\mathbf{X}^{(n)} = (X_1(n+1), \ldots, X_1(n+M), X_2(n+1), \ldots, X_2(n+M), \ldots, X_L(n+1), \ldots, X_L(n+M))^T, \tag{5.10.1}$$

where $n = 0, \ldots, N - M$. Then,

$$\widetilde{X} = \begin{pmatrix} X_1(1) & X_1(2) & \cdots & X_1(N-M+1) \\ \vdots & \vdots & \vdots & \vdots \\ X_1(M) & X_1(M+1) & \cdots & X_1(N) \\ X_2(1) & X_2(2) & \cdots & X_2(N-M+1) \\ \vdots & \vdots & \vdots & \vdots \\ X_2(M) & X_2(M+1) & \cdots & X_2(N) \\ \vdots & \vdots & \vdots & \vdots \\ X_L(1) & X_L(2) & \cdots & X_L(N-M+1) \\ \vdots & \vdots & \vdots & \vdots \\ X_L(M) & X_L(M+1) & \cdots & X_L(N) \end{pmatrix}$$

Step 3. Compute the estimate of the large covariance matrix $\widetilde{D}_{\mathbf{X}}$ as

$$\widetilde{D}_{\mathbf{X}} = \frac{1}{N-M+1} \widetilde{X}\widetilde{X}^T = (\widetilde{D}_{ll'})_{l,l'=1,\ldots,L} = \begin{pmatrix} \widetilde{D}_{11} & \cdots & \widetilde{D}_{1l'} & \cdots & \widetilde{D}_{1L} \\ \vdots & \vdots & \vdots & \vdots & \vdots \\ \widetilde{D}_{l1} & \cdots & \widetilde{D}_{ll'} & \cdots & \widetilde{D}_{lL} \\ \vdots & \vdots & \vdots & \vdots & \vdots \\ \widetilde{D}_{L1} & \cdots & \widetilde{D}_{Ll'} & \cdots & \widetilde{D}_{LL} \end{pmatrix},$$

where

$$\widetilde{D}_{ll'} = \left(\frac{1}{N-M+1} \sum_{n=1}^{N-M+1} X_l(n+j-1) X_{l'}(n+j'-1) \right)_{j,j'=1,\ldots,M}.$$

Since $\widetilde{D}_{ll'} = \widetilde{D}_{l'l}$ and each $\widetilde{D}_{ll'}$ is an $M \times M$ matrix, $\widetilde{D}_X$ is an $LM \times LM$ symmetric matrix.

Based on $\widetilde{D}_{\mathbf{X}}$, we will find the eigenvalues $\lambda_1 \geq \lambda_2 \geq \cdots \geq \lambda_{LM} \geq 0$ and the corresponding eigenvectors $\{\mathbf{E}^k\}_{k=1,\ldots,LM}$. Then, $\mathbf{E}^k$ describes space-time patterns of the vector stochastic processes $\mathbf{X}(t) = (X_1(t), \ldots, X_L(t))$. The importance of $\mathbf{E}^k$ decreases as k increases.

The space-time principal component $a_n^{(k)}$ of $\mathbf{X}^{(n)}$ is the projection of $\mathbf{X}^{(n)}$ onto $\mathbf{E}^k$. Denote

$$\mathbf{E}^k = (E_{11}^k, \ldots, E_{1M}^k, E_{21}^k, \ldots, E_{2M}^k, \ldots, E_{L1}^k, \ldots, E_{LM}^k)^T.$$

From this and (5.10.1), it follows that

$$a_n^{(k)} = (X^{(n)}, \mathbf{E}^k) = \sum_{m=1}^{M} \sum_{l=1}^{L} X_l(n+m) E_{lm}^k,$$

where $(\cdot, \cdot)$ is the inner product of the LM-dimensional vector space. It can be proved that the variance of $a_n^{(k)}$ is λ_k.

The *multichannel singular spectrum analysis* of the lagged copy $\mathbf{X}^{(n)}$ of the original $\mathbf{X}(t)$ with respect to eigenvectors $\{\mathbf{E}^k\}_{k=1,\ldots,LM}$ is $\mathbf{X}^{(n)} = \sum_{k=1}^{LM} a_n^{(k)} \mathbf{E}^k$. This implies that

$$X_l(n+m) = \sum_{k=1}^{LM} a_n^{(k)} E_{l,m}^k \qquad (l = 1, \ldots, L;\ m = 0, \ldots, M-1),$$

where $n = 1, \ldots, N-M$.

Further Reading

M.R. Allen, L.A. Smith, Monte Carlo SSA: detecting irregular oscillations in the presence of coloured noise. J. Clim. **9**, 3373–3404 (1996)

A.L. Berger, J.L. Melice, L. Hinnov, A strategy for frequency spectra of quaternary climate records. Clim. Dyn. **5**, 227–240 (1991)

P. Bloomfield, D. Nychka, Climate spectra and detecting climate change. Clim. Change **21**, 275–287 (1992)

M. Boiseau, M. Ghil, A. Juillet-Leclerc, Trends and interdecadal variability from south-central Pacific coral records. Geophys. Res. Lett. **26**, 2881–2884 (1999)

H.E. Brindley, R.J. Bantges, The spectral signature of recent climate change. Curr. Clim. Change Rep. **2**, 112–126 (2016)
J. Egger, POPs and MOPs. Clim. Dyn. **15**, 561–568 (1999)
N. Fagel, X. Boes, M.F. Loutre, Climate oscillations evidenced by spectral analysis of Southern Chilean lacustrine sediments: the assessment of ENSO over the last 600 years. J. Paleolimnol. **39**, 253–266 (2008)
M. Ghil, M.R. Allen, M.D. Dettinger, K. Ide, D. Kondrashov, M.E. Mann, A.W. Robertson, A. Saunders, Y. Tian, F. Varadi, P. Yiou, Advanced spectral methods for climatic time series. Rev. Geophys. **40**, 1003–1043 (2002)
M. Haddad, H. Hassani, H. Taibi, Sea level in the Mediterranean Sea: seasonal adjustment and trend extraction within the framework of SSA. Earth Sci. Inform. **6**, 99–111 (2013)
K. Hasselmann, PIPs and POPs: the reduction of complex dynamical systems using principal interaction and principal oscillation patterns. J. Geophys. Res. **93**, 11015–11021 (1988)
P. Hope, B.J. Henley, J. Gergis, J. Brown, H. Ye, Time-varying spectral characteristics of ENSO over the Last Millennium. Clim. Dyn. (2017, in press)
I.L. Hudson, M.R. Keatley, Singular spectrum analytic (SSA) decomposition and reconstruction of flowering: signatures of climatic impacts. Environ. Model. Assess. **22**, 37–52 (2017)
Y. Hui, S.D. Pagiatakis, Least squares spectral analysis and its application to superconducting gravimeter data analysis. Geo-spatial Inform. Sci. **7**, 279–283 (2004)
J.M. Lees, Reshaping spectrum estimates by removing periodic noise: application to seismic spectral rations. Geophys. Res. Lett. **21**, 199–236 (1995)
G.J. MacDonald, Spectral analysis of time series generated by nonlinear processes. Rev. Geophys. **27**, 449–469 (1989)
K. Mohammadi, H.R. Eslami, R. Kahawita, Parameter estimation of an ARMA model for river flow forecasting using goal programming. J. Hydrol. **331**, 293–299 (2006)
J. Park, C.R. Lindberg, F.L.I. Vernon, Multitaper spectral analysis of high-frequency seismograms. J. Geophys. Res. **92**, 12675–12684 (1987)
R. Poormirzaee, M.M. Oskouei, Use of spectral analysis for detection of alterations in ETM data Yazd, Iran. Appl. Geomatics **2**, 147–154 (2010)
A. Rossi, N. Massei, B. Laignel, A synthesis of the time-scale variability of commonly used climate indices using continuous wavelet transform. Glob. Planet. Change **78**, 1–13 (2011)
N. Scafetta, Discussion on the spectral coherence between planetary, solar and climate oscillations: a reply to some critiques. Astrophys. Space Sci. **354**, 275–299 (2014)
R. Vautard, P. Yiou, M. Ghil, Singular spectrum analysis: a toolkit for short noisy chaotic signals. Phys. D **58**, 95–126 (1992)
J.S. von Storch, P. Muller, E. Bauer, Climate variability in millennium integrations with coupled atmosphere-ocean GCMs: a spectral view. Clim. Dyn. **17**, 375–389 (2001)
W.D. Weeks, W.C. Boughton, Tests of ARMA model forms for rainfall-runoff modelling. J. Hydrol. **91**, 29–47 (1987)
P. Yiou, E. Baert, M.F. Loutre, Spectral analysis of climate data. Surv. Geophys. **17**, 619–663 (1996)
P. Yiou, M.F. Loutre, E. Baert, Spectral analysis of climate data. Surv. Geophys. **17**, 619–663 (1986)

Chapter 6
Climate Modeling

Global climate change is one of the greatest threats to human survival and social stability that has occurred in human history. The main factor causing climate change is the increase of global carbon emissions produced by human activities such as deforestation and burning of fossil fuels. The concentration of carbon dioxide in the atmosphere has increased from a pre-industrial value of about 280–391 ppm in 2011. In April 2007, the concentration reached more than 410 ppm. Understanding the different components of the climate system is critical for being able to simulate the system. In this chapter, we will first discuss basic facts on the climate system and how it might change, and then we will give a detailed qualitative analysis of how climate models are constructed and a introduction of framework of well-known Coupled Model Intercomparison Project Phase 6 (CMIP6) to simulate present and future changes of climate, environment, and ecosystem. Methods and tools in Chaps. 1–5 will be expected to be used widely to analyze CMIP6 outputs to reveal mechanisms of climate change and assess strategies for mitigating climate change in the following 5–10 years.

6.1 Greenhouse Gases

The *greenhouse effect* was discovered by Joseph Fourier in 1824. It is a process by which radiative energy leaving the Earth's surface is absorbed by some atmospheric gases, called *greenhouse gases*. Accurately, the term *greenhouse gas* refers to gases that are transparent to visible light emitted from the sun, but absorptive to the infrared radiation emitted by the earth. The major greenhouse gases are water vapor, carbon dioxide, methane, and ozone. Their percentage contribution to the greenhouse effect on the Earth is 36–70%, 9–26%, 4–9%, and 3–7%, respectively. Clouds are the major non-gas contributor to the Earth's greenhouse effect. Since the Earth reflects only part of the incoming sunlight, the Earth's blackbody temperature is actually below

Z. Zhang, *Multivariate Time Series Analysis in Climate and Environmental Research*, https://doi.org/10.1007/978-3-319-67340-0_6

its actual surface temperature. The mechanism that produces this difference between the surface temperature and the blackbody temperature is due to the atmosphere and is known as greenhouse effect. A recent climate warming of the Earth's surface and lower atmosphere is believed to be the result of a strengthening of the greenhouse effect mostly due to human-produced increases in atmospheric greenhouse gases.

The Earth receives energy from the Sun in the form of visible light whose wavelength lies between the violet at about 0.4–0.7 μm. About half of the sun's energy is absorbed at the Earth's surface, and the absorbed energy warms the Earth's surface. When the Earth's surface is warmed to a temperature around 255 K and above, it radiates photons corresponding to thermal infrared and far infrared wavelengths in the range 4–100 μm. The greenhouse gases are more absorbent for photons at these wavelengths, e.g., Water vapor absorbs strongly over a wide band of wavelengths near 6.3 μm and over a narrower band near 2.7 μm; Carbon dioxide is a strong absorber in a broad band near 15 μm and in a narrower band near 4.3 μm; Ozone absorbs strongly near 9.6 μm. In each layer of atmosphere, greenhouse gases absorb some of the energy being radiated upwards from lower layers, at the same time, to maintain its own equilibrium, it reradiates the absorbed energy both upwards and downwards. This results in more warmth below layers and land surface, while still radiating some energy back out into deep space from the upper layers of the atmosphere to maintain overall thermal equilibrium. Thus, the presence of the greenhouse gases results in the land surface receiving more radiation.

6.2 Impacts and Feedback of Climate Change

Over the period 1880–2012, the instrumental temperature record for observing climate change has shown an increased global mean temperature with 0.65–1.06°, especially the period from 1983 to 2012 was very likely the warmest 30-year period of the last 800 years in the Northern Hemisphere. Relative to 1850–1900, global surface temperature change for 2081–2100 is projected to likely exceed 1.5° for RCP4.5, RCP6.0, and RCP8.5 (IPCC AR5). This warming may bring various physical and ecological impacts on global ecosystem and environment. The probability of warming having unforeseen consequences will further increase with the rate, magnitude, and duration of climate change. With a global average temperature increase, the Greenland Ice Sheet and the West Antarctic Ice Sheet would have been losing mass and then contributed to sea level rising; the Arctic ocean would be free in the summer due to sea ice loss; the intense tropical cyclone activity would increases; the ocean would acidify due to the increase of dissolving carbon dioxide in seawater; diversity of ecosystems would likely be reduced and many species would likely extinct, and so on.

Regional effects of global climate warming are likely to be widely varying spatially over the planet. Some are the direct result of a generalized global change, while others are related to changes in certain ocean currents or weather systems. Low-latitude and less-developed areas are probably at the greatest risk from climate change. African continent faces are reductions in food security and agricultural productivity, increased water stress, and increased risks to human health. Asian megadeltas are facing large populations and high exposure to sea level rise, storm surge, and river flooding. Small islands are likely to experience large impacts due to higher exposure to sea level rise and storm surge and limited capacity to adapt to climate change.

Feedbacks represent the key process of the climate system and cannot be ignored. Feedback processes may amplify or diminish the effect of climate change, and so play an important part in determining the climate sensitivity and future climate state. The major climate feedback mechanisms include the *carbon cycle feedback*, the *water vapor feedback*, the *snow/ice feedback*, and the *cloud feedback*.

The feedback between climate change and the carbon cycle will amplify global warming. Warming will trigger methane and other greenhouse gases released from Arctic permafrost regions, so this will partially offset increases in land and ocean carbon sinks caused by rising atmospheric CO_2. As a result, the warming will be reinforced.

The water vapor feedback is due to increases in water vapor. If the atmosphere is warmed, the saturation vapor pressure increases, and the amount of water vapor in the atmosphere will increase. Since water vapor is one of greenhouse gases, the increase in water vapor content makes the atmosphere warm further, thus the atmosphere holds more water vapor, and so on until other processes stop the feedback loop.

The snow/ice feedback is due to decreases in snow/ice extent and then resulting in the decrease trend of global albedo. When snow/ice melts, land or open water takes place. Both land and open water are on average less reflective than snow/ice and, thus they will absorb more solar radiation. This results in more warming in a continuing cycle.

The cloud feedback lies in that changes in cloud cover can affect the cloud contribution to both the greenhouse effect and albedo. Fewer clouds mean that more sunlight reaches the Earth's surface, leading to further warming. At the same time, clouds emit infrared radiation back to the surface, resulting in a warming effect, while clouds reflect sunlight back to space, resulting in a cooling effect. As a consequence, the net effect of cloud feedback depends on the changes in cloud types, cloud temperatures, cloud height, and cloud's radiative properties.

6.3 Framework of Climate Models

The modeling of past, present, and future climates is of fundamental importance to predict climate change and mitigate future climate risks. Climate Models can provide a state-of-the-art tool to understanding processes and interactions in the climate system.

6.3.1 Basic Physical Laws Used in Climate Models

A climate model is a computer program that uses well-established physical, biological, and chemical principles to provide credible quantitative estimates of mass transfer, energy transfer, radiant exchange in the Earth system. It needs to divide the globe into a three-dimensional grid of cells representing specific geographic locations and elevations. Each of the components (atmosphere, land surface, ocean, and sea ice) has different equations calculated on the global grid for various climate variables. Climate models can simulate well the interactions of atmosphere, ocean, land, ice, and biosphere, and estimate climate change on seasonal, annual, decadal, and centennial timescales. It is also a powerful tool to investigate the degree to which observed and future climate changes may be due to natural variability, human activity, or a combination of both.

Fundamental equations governing atmospheric and oceanic dynamics are as follows.

(1) The *horizontal velocity equations* for atmosphere and ocean dynamics modeling is:

$$\frac{du}{dt} = fv - \frac{1}{\rho}\frac{\partial p}{\partial x} + F^x_{drag},$$
$$\frac{dv}{dt} = -fu - \frac{1}{\rho}\frac{\partial p}{\partial y} + F^y_{drag},$$

where p is the pressure, ρ is the density, f is the Coriolis parameter, $\mathbf{v} = (u, v, w)$ is the velocity, and F^x_{drag} and F^y_{drag} denote friction-like forces due to turbulent or surface drag on the flow in the x and y directions, respectively. These equations are also called *horizontal momentum equations*.

(2) The *vertical velocity equation* for atmosphere and ocean dynamics modeling is

$$\frac{\partial p}{\partial z} = -\rho g,$$

where p is the pressure, ρ is the density, and $g = 9.81\,\mathrm{m\,s^{-2}}$ be the gravitational acceleration. This equation is also called *hydrostatic balance equation*.

(3) The equation of state for the atmosphere is

$$\rho = \frac{p}{RT},$$

where ρ is the density, p is the pressure, $R = 287\,\mathrm{J\,kg^{-1}K^{-1}}$ is the ideal gas constant for air, and T is the temperature in kelvin. This equation is also called the *ideal gas law*.

(4) The *equation of state* for the ocean can be represented by a function $\rho = \mathcal{P}(T, S, p)$, i.e., the density ρ depends on temperature T, salinity S, and pressure p. When salinity does not change significant, the equation of state has an approximation as follows:

$$\rho = \rho_0(1 - \epsilon_T(T - T_0)),$$

where ϵ_T is called the *coefficient of thermal expansion for sea water*, especially, for the upper ocean,

$$T_0 = 22\,^\circ\mathrm{C}, \quad \epsilon_T = 2.7 \times 10^{-4}\,^\circ\mathrm{C}^{-1}, \quad \rho_0 = 1.03 \times 10^3\,\mathrm{kgm}^{-3}.$$

(5) *Atmospheric temperature equation* is

$$c_p \frac{dT}{dt} - \frac{1}{\rho}\frac{dp}{dt} = Q,$$

where T is temperature, ρ is density, p is pressure, c_p is the heat capacity of air at constant pressure, and Q is heating in $\mathrm{J\,kg^{-1}s^{-1}}$. Atmospheric temperature equation is also called a *thermodynamic energy equation*.

(6) *Oceanic temperature equation* is

$$c_w \frac{dT}{dt} = Q,$$

where T is temperature, $c_w \approx 4200\,\mathrm{J\,kg^{-1}K^{-1}}$ is the heat capacity of water, and Q is heating in $\mathrm{J\,kg^{-1}s^{-1}}$. It gives the time rate of change of temperature. Oceanic temperature equation is another thermodynamic energy equation.

(7) *Atmospheric continuity equation* is

$$\nabla \cdot \mathbf{v} = -\frac{\partial w}{\partial p},$$

where p is pressure, w is the vertical component of the velocity $\mathbf{v}$ in pressure coordinates, and $\nabla \cdot \mathbf{v}$ is the horizontal divergence defined along pressure surfaces.

(8) The vertical pressure velocity at the ocean surface is just the rate of change of the surface pressure p_s with time, i.e., $w = \frac{dp_s}{dt}$. Integrating vertically the atmospheric continuity equation, $w = -\int_{column} (\nabla \cdot \mathbf{v})\,\mathrm{d}p$. So

$$\frac{dp_s}{dt} = -\int_{column} (\nabla \cdot \mathbf{v})\,\mathrm{d}p.$$

This equation is called *atmospheric surface pressure equation*.

(9) *Oceanic continuity equation* is

$$\frac{\partial u}{\partial x} + \frac{\partial v}{\partial y} = -\frac{\partial w}{\partial z},$$

where the sum of two terms on the left-hand side is the horizontal divergence along the x and y directions and the velocity $\mathbf{v} = (u, v, w)$.

(10) The vertical velocity at the ocean surface is just the rate of change of the surface height η with time, i.e., $w = \frac{d\eta}{dt}$. Integrating vertically the oceanic continuity equation gives $w = -\int_{column} (\nabla \cdot \mathbf{v})\, dz$. So

$$\frac{d\eta}{dt} = -\int_{column} (\nabla \cdot \mathbf{v})\, dz,$$

This equation is called *oceanic surface height equation.*

(11) *Atmospheric moisture equation* is

$$\frac{dq}{dt} = P_{convection} + P_{mixing},$$

where q is the quantity of water vapor and is measured by the specific humidity, i.e., q is the ratio of mass of water vapor to total mass of air parcel; $P_{convection}$ includes the loss of moisture by condensation and precipitation as well as the part of vertical mixing associated with moist convection; and P_{mixing} includes mixing not associated with moist convection. This equation is an expression of the conservation of mass.

(12) *Oceanic salinity equation* is

$$\frac{dS}{dt} = P^S_{mixing},$$

where S is quantity of salt in the water and is measured by the salinity, i.e., S is the ratio of mass of salt to total mass of water parcel, and P^S_{mixing} is the change of salinity due to parameterized processes. This equation is another expression of the conservation of mass.

Fundamental equations used in land surface modeling are as follows.

(13) *Land soil temperature equation* is

$$C_T \frac{\partial T}{\partial t} = \frac{\partial}{\partial z}\left(K_T \frac{\partial T}{\partial z}\right),$$

where T is the soil temperature, C_T is thermal heat capacity and, K_T is soil thermal conductivity, both C_T and K_T are nonlinear functions of soil moisture and soil type.

(14) *Land soil moisture equation* is

$$\frac{\partial \Theta}{\partial t} = \frac{\partial K_\Theta}{\partial z} + \frac{\partial}{\partial z}\left(D_\Theta \frac{\partial \Theta}{\partial z}\right) + F_\Theta,$$

where D_Θ is soil water diffusivity, K_Θ is hydraulic conductivity, both D_Θ and K_Θ are nonlinear functions of soil moisture and soil type, and F_Θ is a source/sink term for precipitation/evapotranspiration. This equation is also called Richard's equation.

(15) *Canopy water equation* is

$$\frac{\partial C_w}{\partial t} = P - E_c,$$

where C_w is the canopy water, P is precipitation, E_c is canopy water evaporation. The precipitation P increases C_w, while the evaporation E_c decreases C_w.

Fundamental equation used in sea ice modeling is as follows:

(16) The *ice thickness distribution equation* is

$$\frac{\partial g}{\partial t} = -\nabla \cdot (g\mathbf{u}) - \frac{\partial}{\partial h}(fg) + \psi$$

where g is the thickness distribution function, $\mathbf{u}$ is the horizontal ice velocity, f is the rate of thermodynamic ice growth, and ψ is a ridging redistribution function.

6.3.2 Discretization and Parameterization

The dynamics and thermodynamics of fluids in the atmosphere and ocean can be governed by equations which comply with basic laws of physics. It is difficult in general to find out the accurate solutions of these equations, but their approximation solutions can be calculated numerically on computers.

Discretization is to replace approximately the continuous field of variables (e.g., temperature and velocity) in the atmosphere and ocean by a finite number of discrete values. One common approach to this discretization is to divide the fluid up into a number of *grid boxes* and then approximate the continuous field of variables by the average value across the grid box or the value at the center of the grid box. This approach can capture approximately the behavior of motions at the space scales much larger than the grid box but obviously omits the infinite number of values that the continuous field of variables has at different points within the grid box.

Parameterization is to represent all processes that occur at the space scales smaller than a climate model grid box. It is essential to model the physics of processes on the small scales. In climate models, the part of the computer code that deals with the physics of processes on the small scale is often called the *physics package*.

6.3.3 The Hierarchy of Climate Models

Among various types of climate models, some focus on few variables of the climate system that may be simple enough to run on a personal computer, while other models take into account deeply the interactions and feedbacks between different parts of the

Earth system and can only run on a supercomputer. The hierarchy of climate models in order of increasing complexity is as follows.

(1) Energy Balance Models

An energy balance model with a one-layer atmosphere is classified as a simple climate model. This model estimates the changes in the climate system from an analysis of the energy budget of the Earth by involving solar radiation flux, infrared fluxes, evaporation, and sensible heat.

(2) Intermediate Complexity Models

The Cane-Zebiak model is classified as an regional model of intermediate complexity. This model is used to simulate El Niño/Southern Oscillation (ENSO) and conduct experimental climate prediction. Its ocean component is essentially the one-layer model for the layer above the thermocline. More generally, the Earth system Models of Intermediate Complexity (EMIC) represents most aspects of the climate system but each in a simplified way. They extend the concept of energy balance models to include many latitude bands or a substantial number of regions of roughly continental width and parameterizations of many processes. For example, the UVic Earth System Climate Model is a standard EMIC. It consists of a three-dimensional ocean general circulation model coupled to a thermodynamic/dynamic sea-ice model, an energy-moisture balance atmospheric model with dynamical feedbacks, a thermomechanical land-ice model, and a reduced complexity atmosphere model.

(3) Coupled Ocean-Atmosphere General Circulation Models (GCM)

An atmospheric model coupled to a mixed-layer ocean is used in the early studies of global warming. The early studies include the effects of the ocean heat capacity in the surface layer of the ocean and a simple estimate of ocean heat transport that does not change in time. Atmospheric general circulation models coupled to mixed-layer oceans are very reasonable tools for a first approximation to global warming because many complex aspects occur in the atmosphere. These models do not have to be run for long periods of time to bring the deep ocean into equilibration.

(4) Regional Climate Models

Regional climate models can afford to have a smaller grid size than a global climate model because they cover only a region (e.g., the size of a continent). Regional climate models face the challenges in the boundary conditions at the edges of the region and in estimating regional climate model simulations.

(5) Earth System Models

Earth system models are most complex models in the model hierarchy (Table 6.1). These models include interactive carbon cycle in the atmospheric model, the land surface model, the ocean model, the sea-ice model, and the coupler. The atmospheric model carries carbon dioxide concentration as a predicted field. The land surface model includes the growth and decay of biomass. The ocean model carries a set of equations for dissolved carbon compounds both organic and inorganic.

Table 6.1 List of earth system models

Models	Units
BCC-CSM1.1	China Meteorological Administration
CanESM2	Canadian Centre for Climate Modelling and Analysis
CMCC-CESM	Centro Euro-Mediterraneo per I Cambiamenti Climatici
CNRM-CM5	Centre National de Recherches Meteorologiques
CSIRO-Mk3.6.0	CSIRO
EC-EARTH	EC-EARTH consortium
FIO-ESM	The First Institute of Oceanography, SOA
BNU-ESM	Beijing Normal University
FGOALS-g2	Chinese Academy of Sciences and Tsinghua University
MIROC-ESM	Japan Agency for Marine-Earth Science and Technology
HadGEM2-ES	Met Office Hadley Centre
MPI-ESM-LR	Max Planck Institute for Meteorology
MRI-ESM1	Meteorological Research Institute
GISS-E2-H	NASA Goddard Institute for Space Studies
CCSM4	National Center for Atmospheric Research
NorESM1-ME	Norwegian Climate Centre
GFDL-ESM2G	Geophysical Fluid Dynamics Laboratory

6.4 Coupled Model Intercomparison Project

Coupled Model Intercomparison Project (CMIP) began in 1995 under the auspices of the Working Group on Coupled Modeling (WGCM), which is in turn under auspices of CLIVAR and the Joint Scientific Committee for the World Climate Research Programme. Since then, phases 2–6 have subsequently been conducted and a number of CMIP experiments have been developed. The objective of CMIP is to assess the performance of various climate models and better understand past, present, and future climate changes arising from natural, unforced variability or in response to changes in radiative forcing in a multimodel context. The CMIP multimodel dataset has provided the key support for famous IPCC assessment reports of the United Nations: The results based on CMIP1 were used to inform the IPCC's Second Assessment Report (1995); the results based on CMIP2 were used to inform the IPCC's Third Assessment Report (2001); the results based on CMIP3 were used to inform the IPCC's Fourth Assessment Report (2007); and the results based on the ongoing CMIP5 are being used to inform the IPCCs Fifth Assessment Report (2014).

In 1995, the first phase of CMIP, called CMIP1, only collected output from coupled GCM control runs in which CO_2, solar brightness and other external climatic forcing are kept constant. A subsequent phase, CMIP2, collected output from both model control runs and matching runs in which CO_2 increases at the rate of 1% per year. The CMIP3 included realistic scenarios for both past and present climate forcing. The

CMIP4 is a transition between CMIP3 and CMIP4 and have relatively low impacts. In 2008, 20 climate modeling groups from around the world agreed to promote a new set of coordinated climate model experiments. These experiments comprise CMIP5. It promotes a standard set of model simulations in order to: evaluate how realistic the models are in simulating the recent past, provide projections of future climate change on near term (out to about 2035) and long term (out to 2100 and beyond), and understand some of the factors responsible for differences in model projections (e.g., clouds and the carbon cycle). All of the CMIP5 model output can be downloaded through four ESGF gateways at PCMDI (http://pcmdi9.llnl.gov/), BADC (http://esgf-index1.ceda.ac.uk), DKRZ (http://esgf-data.dkrz.de), NCI (http://esg2.nci.org.au).

In summer 2013, the sixth phase of CMIP was launched, denoted by CMIP6. Its scientific backdrop is associated with WCRP's seven Grand Science Challenges: clouds, circulation, and climate sensitivity; changes in cryosphere; climate extremes; regional sea-level rise; water availability; near-term climate prediction; biogeochemical cycles and climate change. In October 2014, CMIP6 experimental design was finalized by the WGCM and the CMIP Panel at the WGCM 18th Session, and 34 modeling groups were committed to participate CMIP6. The total amount of output from CMIP6 is estimated to be between 20 and 40 petabytes. CMIP6 will focuses on three following broad scientific questions:

- How does the Earth system respond to forcing?
- What are the origins and consequences of systematic model biases?
- How can we assess future climate changes given internal climate variability, predictability, and uncertainties in scenarios?

CMIP6 has a novel and more federated structure. It consists of the following several major elements:

(1) the Diagnostic Evaluation and Characterization of Klima (DECK) simulations include a historical Atmospheric Model Intercomparison Project (AMIP) simulation; a pre-industrial Control simulation; a simulation forced by an abrupt quadrupling of CO_2; a simulation forced by a 1% $year^{-1}$ CO_2 increase. The CMIP historical simulations span the period of extensive instrumental temperature measurements from 1850 to the present. The historical forcings, that will drive the DECK and CMIP6 historical simulations, are based as far as possible on observations and cover the period 1850–2014. These include: emissions of short-lived species and long-lived greenhouse gases; greenhouse gas concentrations; solar forcing; stratospheric aerosol dataset (volcanoes); AMIP Sea Surface Temperatures (SSTs) and Sea Ice Concentrations (SICs); aerosol forcing in terms of optical properties and fractional change in cloud droplet effective radius; the time-varying gridded ozone concentrations and nitrogen deposition;

(2) standardization, coordination, infrastructure, and documentation that will facilitate the distribution of model outputs and the characterization of the model ensemble;

Table 6.2 CMIP6-Endorsed MIPs

No.	Experiments
1	Aerosols and chemistry model intercomparison project (AerChemMIP)
2	Coupled climate carbon cycle model intercomparison project (C^4MIP)
3	Cloud feedback model intercomparison project (CFMIP)
4	Detection and attribution model intercomparison project (DAMIP)
5	Decadal climate prediction project (DCPP)
6	Flux-anomaly-forced model intercomparison project (FAFMIP)
7	Geoengineering model intercomparison project (GeoMIP)
8	Global monsoons model intercomparison project (GMMIP)
9	High-resolution model intercomparison project (HighResMIP)
10	Ice sheet model intercomparison project for CMIP6 (ISMIP6)
11	Land surface snow and soil moisture (LS3MIP)
12	Land-use model intercomparison project (LUMIP)
13	Ocean model intercomparison project (OMIP)
14	Paleoclimate modelling intercomparison project(PMIP)
15	Radiative forcing model intercomparison project (RFMIP)
16	Scenario model intercomparison project (ScenarioMIP)
17	Volcanic forcings model intercomparison project (VolMIP)
18	Coordinated regional climate downscaling experiment (CORDEX)
19	Dynamics and variability moder intercomparison projuct (DynVarMIP)
20	Sea-ice model intercomparison project (SIMIP)
21	Vulnerability, impacts, adaptation and climate services advisory board (VIACS AB)

(3) 21 CMIP-Endorsed MIPs (see Table 6.2) that build on the DECK and CMIP historical simulations to address a large range of specific questions with WCRP Grand Challenges as scientific backdrop. For each CMIP6-Endorsed MIPs, about eight modeling groups are required to commit to performing the Tier 1 experiments. Of the 21 CMIP6-Endorsed MIPs, four are diagnostic in nature, which means that they define and analyze additional output, but do not require additional experiments. In the remaining 17 MIPs, a total of around 190 experiments have been proposed resulting in 40,000 model simulation years with around half of these in Tier 1.

Further Reading

A. Amin, W. Nasim, M. Mubeen, S. Sarwar, Regional climate assessment of precipitation and temperature in Southern Punjab (Pakistan) using SimCLIM climate model for different temporal scales. Theor. Appl. Climatol. (2017, in press)

S. Bony, B. Stevens, D.M.W. Frierson, C. Jakov, M. Kageyama, R. Pincus, T.G. Shepherd, S.C. Sherwood, A.P. Siebesma, A.H. Sobel, M. Watanabe, M.J. Webb, Clouds, circulation, and climate sensitivity. Nat. Geosci. **8**, 261–268 (2015)

G. Brasseur, D. Carlson, Future directions for the world climate research programme. Eos **96**, 9–11 (2015)

C.M. Brierley, M. Collins, A.J. Thorpe, The impact of perturbations to ocean-model parameters on climate and climate change in a coupled model. Clim. Dyn. **34**, 325–343 (2010)

T.J. Crowley, Causes of climate change over the past 1000 years. Science **289**, 270–277 (2000)

A.V. Eliseev, I.I. Mokhov, K.E. Muryshev, Estimates of climate changes in the 20th–21st centuries based on the version of the IAP RAS climate model including the model of general ocean circulation. Russ. Meteorol. Hydrol. **36**, 73–81 (2011)

V. Eyring, S. Bony, G.A. Meehl, C.A. Senior, B. Stevens, R.J. Stouffer, K.E. Taylor, Overview of the coupled model intercomparison project phase 6 (CMIP6) experimental design and organization. Geosci. Model Dev. **9**, 1937–1958 (2016)

J.M. Gregory, Long-term effect of volcanic forcing on ocean heat content. Geophys. Res. Lett. **37**, L22701 (2010)

M.N. Islam, M. Almazroui, Direct effects and feedback of desert dust on the climate of the Arabian Peninsula during the wet season: a regional climate model study. Clim. Dyn. **39**, 2239–2350 (2012)

F. Kaspar, K. Prommel, U. Cubasch, Impacts of tectonic and orbital forcing on East African climate: A comparison based on global climate model simulations. Int. J. Earth Sci. **99**, 1677–1687 (2010)

V. Masson-Delmotte, M. Kageyama, P. Braconnot et al., Past and future polar amplification of climate change: climate model intercomparisons and ice-core constraints. Clim. Dyn. **27**, 437–440 (2006)

G.A. Meehl, G.J. Boer, C. Covey, M. Latif, R.J. Stouffer, The coupled model intercomparison project (CMIP). B. Am. Meteorol. Soc. **81**, 313–318 (2000)

G.A. Meehl, R. Moss, K.E. Taylor, V. Eyring, R.J. Stouffer, S. Bony, B. Stevens, Climate model intercomparisons: preparing for the next phase. Eos **59**, 77–78 (2014)

J.D. Neelin, *Climate Change and Climate Modeling* (University Press, Cambridge, 2011)

S. Noah, Diffenbaugh and Filippo Giorgi, climate change hotspots in the CMIP5 global climate model ensemble. Clim. Change **114**, 813–822 (2012)

A.S. Phillips, C. Deser, J. Fasullo, Evaluating model of variability in climate models. Eos. Trans. AGU. **95**, 453–455 (2014)

P.A. Stott, J.F.B. Mitchell, M.R. Allen, T.L. Delworth, J.M. Gregory, G.A. Meehl, B.D. Santer, Observational constraints on past attributable warming and predictions of future global warming. J. Clim. **19**, 3055–3069 (2006)

K.E. Taylor, R.J. Stouffer, G.A. Meehl, An overview of CMIP5 and the experiment design. B. Am. Meteorol. Soc. **93**, 485–498 (2012)

M. Verbitsky, B. Saltzman, Heinrich-type glacial surges in a low-order dynamical climate model. Clim. Dyn. **10**, 39–48 (1994)

K.J.E. Walsh, K.-C. Nguyen, J.L. McGregor, Fine-resolution regional climate model simulations of the impact of climate change on tropical cyclones near Australia. Clim. Dyn. **22**, 47–56 (2004)

Chapter 7
Regional Climate Change

Global warming of the climate system is unequivocal, as is now evident from observations of increases in global average air and ocean temperatures, widespread melting of snow and ice, and rising global average sea level. However, the impacts of global warming vary with regional differences in climate, geography, and land use. The most recent report from the IPCC has showed that a number of regional climate studies throughout the world have increased significantly during the last decade. In this chapter, we introduce some latest *case studies* on regional climate change to demonstrate how methods and tools in Chaps. 1–6 are applied in the research of climate change. Here, we choose Middle East and Mediterranean region, Asia-Pacific region, and Arctic region as three representative regions.

7.1 Middle East and Mediterranean Region

Middle East and Mediterranean region locates at a crossroad of global climatic patterns. There is a convergence of different maritime conditions over this region, extensive adjacent land masses transporting continental air and extreme differences in topographical features, which lead to a diverse climate in Middle East and Mediterranean region.

7.1.1 Precipitation

Precipitation is an important climate parameter for climate change and has been analyzed in different scales of time and space. Methods for precipitation forecast can be divided into parametric and non-parametric methods. Parametric methods include autoregressive, autoregressive moving average, and autoregressive integrated

Z. Zhang, *Multivariate Time Series Analysis in Climate and Environmental Research*, https://doi.org/10.1007/978-3-319-67340-0_7

moving average methods. Non-parametric methods include artificial neural network, self-organizing map, and wavelet-neural network methods.

Amiri et al. (2016) used a hybrid wavelet transform artificial neural network (WTANN) model and an artificial neural network (ANN) model to reveal temporal behavior and spatial distribution of mean monthly precipitation in Qara-Qum catchment which is located in northeastern of Iran. Four meteorological stations (Golmakan, Mashhad, Sarakhs, and Torbatejam stations) provided good quality, continuous data records for the period 1996–2010. The first 10 years data were used for calibration of the model, and the remaining 5 years data were used for validation. The weights of the neural network were adjusted by a feedforward neural network with backpropagation algorithm. The time series of dew point, temperature, and wind speed were also considered as ancillary variables in temporal prediction. Amiri et al. (2016) found that WTANN model was much better in modeling precipitation time series in Qara-Qum catchment because statistical indices showed better agreement between the measured and calculated values for WTANN model when compared with ANN model, moreover, using ancillary data could improve the accuracy of predicted precipitation data.

Khalili et al. (2016) used artificial neural network models to establish a prediction model for the monthly rainfall of Mashhad synoptic weather station, which is located in the Razavi-khorasan province of Iran. Since the weather in Mashhad synoptic weather station was interfered with different air masses caused by Polar continental, Maritime Tropical, and Sudanian air masses, it made the rainfall prediction extremely difficult. Khalili et al. (2016) used monthly rainfall data collected from 1953 to 2003 to train and validate the ANN models. The Hurst's rescaled range statistical analysis test indicated that rainfall in Mashhad had a long-memory effect, so ANNs was a suitable prediction model. Khalili et al. (2016) showed that among different structures of ANNs, M_{531} and M_{741} neural networks had relatively best performance, where M_{531} is a three-layer feedforward perceptron with an input layer of five source nodes, a single hidden layer of three neurons, and an output layer of one neuron, and M_{741} is another three-layer feedforward perceptron with an input layer of seven source nodes, a single hidden layer of four neurons, and an output layer of one neuron.

7.1.2 Air Temperature

Temperature extreme plays an important role in climate change due to socio-ecological sensitive responses. A lot of study results have revealed that extreme high temperatures and prolonged heat waves will damage greatly agricultural production, increase energy and water consumption, and exert a negative impact on human health, and that currently the cold extremes are decreasing and the warm extremes are increasing in the frequency aspect of cold and hot events. The CCl/CLIVAR Working Group on Climate Change Detection has been coordinating an international effort to develop a suite of indices for temperature extremes (Table 7.1).

Table 7.1 Indices for temperature extremes

Index	Definition
FD	Annual count when TN (daily minimum temperature) < 0 °C
SU	Annual count of days when TX (daily maximum temperature) > 25 °C
ID	Annual count of days when TX < 0 °C
TR	Annual count of days when TN > 20 °C
TXx	Monthly maximum value of daily maximum temperature
TNx	Monthly maximum value of daily minimum temperature
TXn	Monthly minimum value of daily maximum temperature
TNn	Monthly minimum value of daily minimum temperature
TN10p	Percentage of days when TN<10th percentile
TX10p	Percentage of days when TX<10th percentile
TN90p	Percentage of days when TN>90th percentile
TX90p	Percentage of days when TX>90th percentile
WSDI	Annual count of days with at least six consecutive days when TX>90th percentile
CSDI	Annual count of days with at least six consecutive days when TN<10th percentile
DTR	Monthly mean difference between TX and TN

Iran is in the temperate zone of the arid and semi-arid belt of the world and has four distinct seasons due to its wide range of latitude. The climate varies from cool temperature in the north and northwest to subtropical in the south and southeast, and the central Iran region has hot and dry summers and cool winters. The major precipitation in Iran region is produced by Mediterranean air masses with the westerly wind in the cold season. Based on the daily maximum and minimum temperatures provided by 663 meteorological stations in Iran during the period 1962–2004, Darand et al. (2015) analyzed the main extreme temperature indices used widely for global and regional climate extremes. Both the Mann–Kendall (MK) trend test and the simple linear regression were utilized to detect trends in these extreme temperature indices. Darand et al. (2015) found that 42.2 % of the surface area showed a significant increase in TX90p, as well as 70.9% in TN90p, 41.8% in TXx, 65.6% in TNx, 66.9% in TR, 35.5% in SU, and 26.5% in WSDI. There was a significant decrease of 47.3% in FD, 40.9% in TX10p, 68.5% in TN10p, 70% in CSDI, and 57.2% in DTR. These results indicated that the frequency of hot extreme temperature events increased while the frequency of cold extreme temperature events decreased in Iran within 1962–2004. About 66% of the surface area of Iran had a significant positive trend in frequency of hot days and nights, while about 40.9 and 68.5% of the surface area of Iran had a significant decrease trend in frequency of cold days and nights, respectively. The central and southern part of Iran region were more sensitive to climate warming. The strongest changes in temperature extremes were in southern regions. During 1962–2004, the cold days, cold nights, cold waves, and frost days became less frequent, while the hot days, hot nights, hot waves became more frequent.

7.1.3 Climate Modeling

The Middle East and North Africa (MENA) region mainly has an arid climate due to large desert areas. Therein, the southern part has a tropical climate and the northern part has a temperate climate. MENA is specifically susceptible to weather extremes such as dust storms, heavy rainfall, freak flash floods, rising sea levels, and multiannual droughts. These climate extremes are having a rising trend under climate change. Almazroui et al. (2016) assessed the performance of a regional climate model RegCM4 to simulate these climate characteristics of the MENA region. The RegCM4 model was forced with the boundary conditions obtained from the global reanalysis dataset ERA-Interim for the period 1979–2010. An east–west cold bias was found in the northern part of the MENA domain in RegCM4 that was absent in the ERA-Interim driving forcings, whereas a large warm bias was found over the southern Arabian Peninsula (Yemen/Oman) for both RegCM4 and ERA-Interim. RegCM4 model performed well in capturing the salient features of precipitation associated with Inter-Tropical Convergence Zone, Mediterranean cyclones as well as precipitation minima over the deserts. Moreover, the annual cycles of precipitation and mean temperature over the prominent river basins of the region could be captured by the RegCM4 model.

7.1.4 Desert Dust

The climate of the Arabian Peninsula is arid and semi-arid, and its temperature is also relatively high. The Rub Al-Khali desert, the largest sand desert in the world, is located in the southeastern parts of the Arabian Peninsula. The higher surface albedo in the Arabian Peninsula leads to less precipitation, especially its northern part is completely rainless during the dry season. Precipitation during the wet season becomes the main source of water for human consumption and agriculture. Therefore, the desert dust is the main impact on the Arabian Peninsula.

Islam and Almazroui (2012) investigated impacts of desert dust on the surface temperature and precipitation over the Arabian Peninsula during the wet season. The ICTP-RegCM4 climate model was used for investigating the dust radiative forcing and its feedback on the Arabian Peninsula. Based on the model simulations with and without dust load, Islam and Almazroui (2012) indicated that the desert dust effect could reduce incoming shortwave radiation by about 50–60 Wm^{-2} at the surface and reduce the outgoing longwave radiation by about 12–16 Wm^{-2} at the top of the atmosphere. This would lead to that the clear-sky radiative forcing at the top of the atmosphere is positive almost everywhere in the Arabian Peninsula, while it is negative over the surface of the entire Arabian Peninsula. The reduction of incoming and outgoing radiation could further result in a decrease in the surface air temperature and in an increase in the temperature gradient.

7.1.5 *Water Resources*

Climate change can affect water resources, while population growth, agricultural irrigation, economic development, household patterns, and the increasing production sectors are increasing the demand for water resources. Therefore, water resource is of vital important for human. The supply of water in both urban and rural area increases year by year, especially, the droughts strain the already scarce water resources. The Kingdom of Saudi Arabia (KSA) belongs to a tropical and subtropical desert region. It has a semi-arid to hyper-arid climate. Very low precipitation and extremely high evapotranspiration result in water scarcity and reduced vegetation cover. Therefore, KSA is highlighted as one of the most water scarce countries.

Amin et al. (2016) investigated the effects of climate change on water resources in the KSA using observational data and climate models. The analysis of more than 60 years (1950–2013) of climate data revealed that KSA's climate was characterized by very low precipitation (annual average of 112 mm) and extremely high evapotranspiration. In order to detect the drought frequency, intensity, and duration for KSA, Amin et al. (2016) computed the spatial and temporal variations in monthly precipitation over KSA using monthly precipitation data from 29 weather stations located across the KSA and gridded satellite precipitation data. The standardized precipitation index is defined by

$$\mathrm{SPI} = \frac{x_i - \overline{x}}{\sigma}, \tag{7.1}$$

where x_i is the monthly rainfall amount, and $\overline{x}$ and σ are the mean value and standard deviation of rainfall calculated from the whole precipitation time series. SPI represents standard deviations the precipitation deficit deviates from the normalized average and can be used to classify droughts. Amin et al. (2016) revealed that from 1950 to 1980, the monthly SPI values in KSA varied frequently from mild to moderate drought and from moderate to severe drought. Moreover, from 1980 to 2013, droughts became more severe due to the decrease in annual precipitation during this period.

Amin et al. (2016) further used seven climate models including NorESM1-M, IPSL-CM5A-LR, MRI-CGCM3, MIROC5, HadGEM2-ES, GISS-E2-R, and BCC-CSM1-1 to provide climate prediction for KSA up to 2050. Under the RCP 2.6 scenario, there would be a slight increase in the average annual precipitation. Under the RCP 4.5 scenario, the mean annual precipitation in 2050 would decrease by 1.5% over all of the country except for the northern borders. Under the RCP 6.0 scenario, the mean annual precipitation by 2050 would moderately decrease by 2% over the entire KSA except for the northern borders where the decrease would range from 5 to 11%.

7.1.6 Soil Temperature

Soil temperature is a meteorological parameter which determines the rates of physical, chemical, and biological reactions in the soil. Soil temperature is also an agricultural and environmental factor which influences strongly plant growth and soil formation. However, data measured for predicting soil temperature are very sparse in space and time, and often not available for a given site. The usual practice is to determine the soil temperature from other meteorological parameters using theoretical or empirical model.

Abyaneh et al. (2016) developed an artificial neural network (ANN) model and a co-active neuro-fuzzy inference system (CANFIS) model that are capable of estimating soil temperature fluctuations at six depths (5, 10, 20, 30, 50, and 100 cm) by using air temperature. Abyaneh et al. (2016) collected the daily climate data in 1992–2005 from Gorgan's meteorological station in North Iran and Zabol's meteorological station in South Iran. These data were separated into a training dataset and a testing dataset for both models. Mean air temperature was used as the model inputs, and soil temperature was used as the model outputs. The logistic function with slope parameter $a = 1$ was used as the activation function of ANN. Abyaneh et al. (2016) showed that the accuracy of the soil temperature predicted by both models decreases gradually from the surface down to the various depths, and the ANN model exhibited a better performance in predicting soil temperature compared to CANFIS model.

7.2 Asia-Pacific Region

The Asia-Pacific region is made up of thirty-two countries including China, Japan, Korea, US, and others. A broad range of climatological and geographic features exists within the Asia-Pacific region, from temperate areas with marked seasonal differences to tropical areas with relatively high temperatures throughout the year. Many countries within this region are highly sensitive to climate change.

7.2.1 Tibetan Plateau

Tibetan Plateau (TP) is one of Earth's extreme continental climate settings and is influenced by numerous climatic regimes such as the East Asian and Indian monsoons. It is located between 74–105° East and 25–40° North and covers an area of 2,545,000 km^2 approximately and is the world's highest elevated region averaging 4500 m above sea level. The Qilian Mountain, Kunlun Mountain, and Himalaya Mountain are, respectively, located in the northeastern, the northwestern, and the southern parts of the TP, and the Tangula Mountain is located in the middle of the

TP. Several large rivers such as Yellow river and Yangtze river originate there, and so the provision of water resources to most of the Asian continent depends directly on these large rives.

Permafrost is a key characteristic of the TP. *Permafrost* is referred to as the soil that has a temperature lower than 0 °C continuously for at least 24 consecutive months (Dobinski 2011). Permafrost in Tibetan Plateau mainly occurs in the plateau's interior between the Kunlun Mountain and the Tanggula Mountain, with Xidatan to the north of the Kunlun as its northern limit and Anduo to the south of the Tanggula as its southern limit. The present TP permafrost area is about 1,270,000 km^2, i.e., half of its total area approximately. The *active layer* is soils above the surface of permafrost that freezes during the winter and thaws during the summer. The trend in the soil freeze-thaw cycle in the active layer is positively correlated with climatic warming over the TP. Active layer thickness in the northern and western regions of the TP is larger than that in the eastern and southern regions (Pang et al. 2009). It depends on summer ground surface temperature, soil texture (thermal properties of the ground), vegetation cover, soil moisture content, snow cover, etc. (French 2007).

The permafrost on the TP is considered to be more sensitive to climatic warming than that in the Arctic region since Tibetan permafrost is relatively warm and thin. It is mostly warmer than −2.0 °C and less than 100 m in thickness (Cheng and Wu 2007). This sensitivity has significant influences on regional water balance, biological diversity, carbon cycle, and engineering construction in the TP. Significant *permafrost degradation* has occurred on the TP over recent decades. The process can be divided into five stages including starting stage, rising temperature stage, zero geothermal gradient stage, talik layer stage, and disappearing stage. Permafrost in the high and middle mountains is in the rising temperature stage; permafrost in the middle and lower mountains is in transition from late rising temperature stage to the zero geothermal gradient stage; permafrost in patchy permafrost area is in transition from the zero geothermal gradient stage to the talik layer stage, and permafrost is disappearing from the margins (Wu and Zhang 2010). Based on outputs from CLM4 climate model, Guo et al. (2012) estimated that under the A1B greenhouse gas emissions scenario, the permafrost area would decrease by approximately 39% by the mid-twenty-first century and by approximately 81% by the end of the twenty-first century. The *lower altitudinal permafrost limit* in the TP has risen from 4200 m above sea level in the north to 4800 m above sea level in the south. The empirical formula for the statistical relation between the lower limit of permafrost and latitude is given by expressing as a Gaussian function (Cheng 1983):

$$H = 3650\exp(-0.003(\varphi - 25.37)^2) + 1428, \tag{7.2}$$

where H is the lower limit of permafrost and φ is the latitude.

Permafrost temperatures are closely related to ground surface temperatures. The *mean annual ground surface temperatures* on the TP are highly variable and relate to the mean annual air temperature and are also influenced by land type, vegetation, snow cover, and summer monsoons. Wu et al. (2013) analyzed the ground surface temperature records from 16 meteorological stations located in or adjacent

Table 7.2 Trend of permafrost temperature in main permafrost observation stations

Location	Time	Soil depth (m)	Trend °C/year
Fenghuoshan pass	1996–2006	1	0.075
Wudaoliang	1996–2006	1	0.091
Kunlun Mountain pass	1995–2002	1	0.140
Tanggula Mountain pass	1998–2006	1	−0.016
Wudaoliang	1996–2006	6	0.061
Fenghuoshan pass	1996–2006	6	0.059
Kunlun Mountain pass	1995–2002	6	0.059
Tanggula Mountain pass	1998–2006	6	−0.017
Beiluhe	2002–2012	6	0.0013–0.0181
Beiluhe	2002–2012	10	0.085–0.0161

to permafrost regions on the central TP and then revealed a statistically significant warming. Mean annual ground surface temperature increased there at an average rate of 0.060 °C per year over the period of 1980–2007. The surface freezing index decreased at a rate of 111.2 °C days/decade, and the surface thawing index increased at a rate of 125.0 °C days/decade on the central TP. This means a significant weakening of seasonal frost penetration and a great thickening of active layer, especially prominent in the source regions of Yangtze River. Except for ground surface temperatures, permafrost temperatures at different depths had also been observed a significant increasing trend in the past forty years (Table 7.2) at main permafrost observation stations of the TP (Fig. 7.1) (Wu and Zhang 2008; Wu and Liu 2004; Zhang et al. 2016). Generally, an empirical formula of permafrost temperature is given through a linear regression (Wu et al. 2010):

$$T_{15} = 52.78 - 0.79\varphi - 0.57H_{100}, \tag{7.3}$$

where T_{15} is permafrost temperature at 15 m depth and φ is latitude, and H_{100} is altitude (in hectometers). In detail, permafrost temperature at 15 m depth decreases at a rate of 0.57 °C per 100 m altitude and increases at a rate of 0.79 °C per degree latitude moving north along the Qing-Tibet railway.

Permafrost degradation may result in formation of *thermokarst lakes* since the permafrost on the TP can be characterized by high ground temperature and high ice content. Niu et al. (2011) investigated thermokarst lakes between the Kunlun Mountain pass and the Fenghuo Mountain pass along the Qing-Tibet railway. Over 250 lakes are distributed within a 200 m wide band along both sides of the railway, covering 1,390,000 m^2 in area. The largest lake is about 60,000 m^2; the smallest is less than 200 m^2, and the mean area is 5580 m^2. The distribution of the thermokarst lakes is closely related to the ice content and the permafrost temperature; 83.8% of the lakes is in the ice-rich permafrost regions, and 54.9% of the lakes is in the high-temperature permafrost regions.

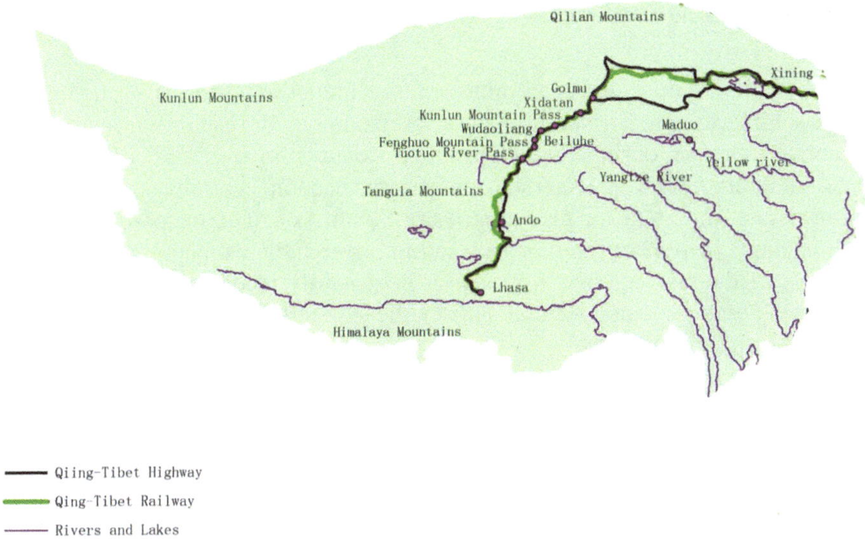

Fig. 7.1 Tibetan Plateau and the main permafrost observation stations

7.2.2 El Niño–Southern Oscillation

The El Niño–Southern Oscillation (ENSO), the most prominent natural climate variability on interannual timescales, has a significant impact on regional and global climate. Son et al. (2016) investigated the statistical relationship between the ENSO and precipitation variability in September over the Korean Peninsula during 1979–2012. The daily precipitation data at fourteen stations during that period were provided by Korea Meteorological Administration, and the ENSO index used is the monthly mean Niño3 sea surface temperature (SST) anomaly averaged over the region of 90°–150°W, 5°S–5°N, from the NOAA Extended Reconstructed SST for that period. Due to the strong seasonal dependency of the ENSO–precipitation relationship, 5-pentad running mean method was used. The lead-lag correlation between September Niño3 SST and 5-pentad moving-averaged precipitation data during 1979–2012 was calculated. Son et al. (2016) found that precipitation over the Korean Peninsula exhibited a negative correlation with September Niño3 SST during the period of 1979–2012. This negative correlation was partly attributed to the tropical cyclone (TC), particularly during La Niña phase. That is, TC tended to pass through Korean Peninsula more frequently during La Niña years, which led to more precipitation over the Korean Peninsula.

7.2.3 Indian Monsoon

As part of the larger scale Asian monsoon, the Indian monsoon is formed due to intense solar heating in late spring as the solar maximum moves north from the equator. Indian monsoon during June–September accounts for over 75% of the annual rainfall in South Asia. Due to agriculture-oriented economies, Indian monsoon rainfalls are quite important for the countries in South Asia. The meteorologists have made attempts to predict the monsoon rainfall over Indian subcontinent since the beginning of the past century. Although it is generally recognized that the ENSO influences Indian Monsoon Rainfall (IMR), the extent of ENSO influence seems to vary substantially over India.

The *Multivariate ENSO index* (MEI) is different from the ENSO index. An multivariate ENSO index is based on the six main observed variables over the tropical Pacific, namely, sea level pressure, zonal and meridional components of the surface wind, sea surface temperature, surface air temperature, and total cloudiness fraction of the sky. The MEI is calculated as the first principal component of above six observed fields. Since the MEI integrates the complete information on ENSO, it is more appropriate to study the links between the ENSO and Indian monsoon rainfall.

Singh (2001) examined the relationship between the ENSO and the IMR for entire monsoon season by using the MEI, and established the relationships between MEI and IMR for all thirty-five meteorological subdivisions of India for each monsoon mouth (i.e., June, July, August, and September). Monsoon rainfall data for the period 1960–1990 pertaining to each meteorological subdivision of India were provided by the India Meteorological Department. Bimonthly values of MEI for May–June, June–July, July–August, and August–September were used to compute the correlations with June, July, August, and September rainfall, respectively. MEI values of April–May, May–June, June–July, and July–August were used to compute the *lag correlations*. Monthly Southern Oscillation Index (SOI) for May, June, July, August, and September was used to compute *simultaneous and lag correlations* with the monsoon rainfall departures. Singh (2001) revealed that the premonsoon ENSO conditions had an adverse impact on the seasonal monsoon rainfall over the Northwestern and extreme South Indian Peninsular regions of India, while ENSO had no significant impact on the seasonal monsoon rainfall over major parts of Northeastern India.

7.2.4 Modeling Sea Surface Temperature

Sea surface temperature (SST) in the tropical Pacific Ocean is closely connected with global climate variations on seasonal and longer timescale. In order to understand the trend of regional climate under the influence of changing atmospheric composition, it is important that SST variability, particularly in the tropical Pacific associated with ENSO, is simulated well by climate models. The availability of the historical experiments from the Coupled Model Inter-comparison Project Phase 5 (CMIP5) provides such an opportunity to examine the ability of current generation of climate models in reproducing the observed climate variability.

Jha et al. (2014) showed that the majority of the coupled model inter-comparison project phase 5 (CMIP5) models could capture the relative large SST anomaly variance in the tropical central and eastern Pacific, as well as in North Pacific and North Atlantic. The models could reproduce the global averaged SST low-frequency variations, particularly since 1970s. However, the majority of models were unable to correctly simulate the spatial pattern of the observed SST trends. Therefore, there is still a challenge to reproduce the features of global historical SST variations with the state-of-the-art coupled general circulation model.

7.3 Arctic Region

The Arctic plays a fundamental role in the climate system and has shown significant climate change in recent decades; especially, the Arctic surface air temperature has risen twice as large as the global average in recent decades. Ice-albedo feedback, changes in clouds and water vapor, enhanced meridional energy transport in atmosphere and ocean and the vertical mixing in Arctic winter inversion are likely contributors to this Arctic warming amplification.

7.3.1 Sea Ice

Sea ice is one of the most visible components corresponding to climate changes. The first map of sea ice cover in Greenland and Barents Seas is made at the Danish Meteorological Institute on the basis of data from ship observations. With the development of navigation along the Siberian coast, the map of sea ice cover is spread to Siberian Arctic seas on the basis of data from aircraft observations (Alekseev et al. 2015). Now satellite-based remote sensing has become the main means for observation. The DMSP satellite systems of passive microwave sounding Special Sensor Microwave Imager (SSM/I) and Special Sensor Microwave Imager/Sounder (SSMIS) and EOS satellite system Advanced Microwave Scanning Radiometer (AMSR) have been used for monitoring sea ice cover. Two key sea ice data archiving and processing centers are the Arctic and Antarctic Research Institute (AARI, Russia) and the National Snow and Ice Data Center (NSIDC, United States), which are supported by AARI (Russia), NSIDC (United States), Canadian Ice Service (Canada), Swedish Meteorological and Hydrological Institute (Sweden), Finnish Meteorological Institute (Finland), and Japan Meteorological Agency (Japan). All observation data of sea ice show that a reduction of sea ice cover has accelerated for the last two decades in the twentieth century (Alekseev et al. 2015).

Sea ice is influenced strongly by the ocean. The decline of Arctic sea ice is one of the most striking changes to the Earth's climate in recent decades. It exhibits interannual timescale variability and has a strong impact on the surface atmosphere and large-scale atmospheric circulation. In return, sea ice predictability might be a source of predictability for the Arctic climate (Germe et al. 2014). In the Arctic, sea ice extent (SIE), sea ice area (SIA), and sea ice volume (SIV) are three major spatially integrated indices. The SIE and SIA interannual variability during winter arises from the marginal ice zones present in all peripheral seas. The *marginal ice zones* is the transition zones between the interior ice pack and the open ocean in the Arctic. In the *Central Arctic* covered completely by sea ice during winter, the reduction of SIV variability is only due to a reduction of sea ice thickness variability (Alekseev et al. 2015).

Climate warming in the Arctic region results from an intensification of the global atmospheric circulation which is accompanied by an increase in the west and south-west winds over the Northern Atlantic and Norwegian Sea and an increase in the migration of Atlantic waters in the Arctic with a simultaneous intensification of the ice and water backflow from the Arctic Basin to the Greenland Sea (Alekseev et al. 2015). The intensification of the atmospheric circulation is connected with an increase in the surface air temperature because the sensible and latent heat inflow with atmospheric circulation is accounted as the main part of the Arctic energy budget. Summer melting of Arctic sea ice depends strongly on the distribution of incoming solar radiation between the ice-reflected and ice and ocean absorbed parts. This division is subjected to seasonal fluctuations and strong interannual variability (Perovich et al. 2007). Koenigk et al. (2016) found significant correlations between sea ice areas in both September and November and winter sea level pressure, air temperature, and precipitation. However, these correlations do seldom exceed 0.6, indicating that Arctic sea ice variations can only explain a part of winter climate variations in northern mid and high latitudes.

7.3.2 Permafrost Carbon

Permafrost is perennially frozen soils stored as a result of processes unique to frozen soils. Recent scientific studies estimate that soils of the northern circumpolar permafrost zone contain almost 1700 petagrams of organic carbon (Schuur et al. 2008). This carbon accumulated over thousands of years under cold and sometimes water-logged conditions. Freeze-thaw mixing in combination with wind- and water-borne sediment deposition over hundreds and thousands of years has buried carbon many meters deep into permafrost soils (Ping et al. 2010).

Carbon emissions from the northern circumpolar permafrost zone will influence the future pace of climate change, especially Nzotungicimpaye and Zickfeld (2017) found that CH_4 could have a share of 20% in the warming caused by the total permafrost carbon released by 2100, and when CH_4 emissions from thermokarst lakes were considered, the contribution from permafrost CH_4 to the surface warming could increase to between 30 and 50%. The biotic and abiotic factors are important determinants of the overall release from permafrost carbon to climate. As soils above the surface of permafrost (active layer) thaw, ancient carbon is available for decomposition by soil organisms. Some carbon is metabolized easily and is consumed quickly by microorganisms, while some carbon is difficult to break down and remain within the soil for much longer. The bulk of the mixture of permafrost carbon is likely to be released slowly over decades after thaw with a small proportion of carbon persisting within the soil for much longer (Fan et al. 2008). Fire is another important abiotic mechanism for releasing permafrost carbon to the atmosphere. Fire frequency and severity are increasing in some parts of the boreal permafrost zone (Turetsky et al. 2011), and rare events such as the large Alaskan tundra wildfire may become more common in the future (Mack et al. 2011). Fires release carbon directly to the atmosphere via combustion and indirectly by warming surface soils and permafrost, thus increasing microbial activity (Chambers and Chapin 2002; Yoshikawa et al. 2002; Grosse et al. 2011).

Further Reading

H.Z. Abyaneh, M.B. Varkeshi, G. Golmohammadi, K. Mohammadi, Soil temperature estimation using an artificial neural network and co-active neuro-fuzzy inference system in two different climates. Arab. J. Geosci. **9**, 377 (2016)

G.V. Alekseev, E.I. Aleksandrov, N.I. Glok, N.E. Ivanov, V.M. Smolyanitsky, N.E. Kharlanenkova, A.V. Yulin, Arctic sea ice cover in connection with climate change, ISSN 0001–4338. Izv. Atmos. Oceanic Phys. **51**, 889–902 (2015)

L.V. Alexander, X. Zhang, T.C. Peterson, J. Caesar, B. Gleason, A.M.G. Klein Tank, M. Haylock, D. Collins, B. Trewin, F. Rahimzadeh, A. Tagipour, K.R. Kumar, J. Revadekar, G. Griffiths, L. Vincent, D.B. Stephenson, J. Burn, E. Aguilar, M. Brunet, M. Taylor, M. New, P. Zhai, M. Rusticucci, J.L. Vazquez-Aguirre, Global observed changes in daily climate extreme of temperature and precipitation. J. Geophys. Res. Atmos. **111**, 1–22 (2006)

M. Almazroui, M.D.N. Islam, A.K. Alkhalaf, F. Saeed, R. Dambul, M.A. Rahman, Simulation of temperature and precipitation climatology for the CORDEX-MENA/Arab domain using RegCM4. Arab. J. Geosci. **9**, 13 (2016)

M.T. Amin, S.H. Mahmoud, A.A. Alazba, Observations, projections, and impacts of climate change on water resources in Arabian peninsula: current and future scenarios. Environ. Earth Sci. **75**, 864 (2016)

M.A. Amiri, Y. Amerian, M.S. Mesgari, Spatial and temporal monthly precipitation forecasting using wavelet transform and neural networks, Qara-Qum catchment. Iran. Arab. J. Geosci. **9**, 421 (2016)

R.F. Carlson, A. MacCormick, D.G. Watts, Application of linear random model to four annual streamflow series. Water Resour. Res. **6**, 1070–1078 (1970)

S.D. Chambers, F.S. Chapin, Fire effects on surface-atmosphere energy exchange in Alaskan black spruce ecosystems: implications for feedbacks to regional climate. J. Geophys. Res. Atmos. **108**, D1 (2002)

G. Cheng, Vertical and horizontal zonation of highaltitude permafrost, in *Fourth International Conference on Permafrost* (Washington, USA, 1983), pp. 136–141

G. Cheng, T. Wu, Responses of permafrost to climate change and their environmental significance Qinghai-Tibet Plateau. J. Geophys. Res. **112**, F02S03 (2007)

M. Darand, A. Masoodian, H. Nazaripour, M.R.M. Daneshvar, Spatial and temporal trend analysis of temperature extremes based on Iranian climatic database (1962–2004). Arab. J. Geosci. **8**, 8469–8480 (2015)

W. Dobinski, Permafrost. Earth Sci. Rev. **108**, 158–169 (2011)

Z.S. Fan, J.C. Neff, J.W. Harden, K.P. Wickland, Boreal soil carbon dynamics under a changing climate: a model inversion approach. J. Geophys. Res. Biogeosci. **113**, G4 (2008)

H.M. French, *The Periglacial Environment*, 3rd edn. (Wiley, Sussex, 2007)

A. Germe, M. Chevallier, D.S. Mélia, E.S. Gomez, C. Cassou, Interannual predictability of Arctic sea ice in a global climate model: regional contrasts and temporal evolution. Clim. Dyn. **43**, 2519–2538 (2014)

P. Goovaerts, Geostatistics for Natural Resources Evaluation (Oxford, 1997)

G. Grosse, J. Harden, M. Turetsky, A.D. McGuire, P. Camill, C. Tarnocai, S. Frolking, E.A.G. Schuur, T. Jorgenson, S. Marchenko, V. Romanovsky, K.P. Wickland, N. French, M. Waldrop, L. Bourgeau-Chavez, R.G. Striegl, Vulnerability of high-latitude soil organic carbon in North America to disturbance. J. Geophys. Res. Biogeosci. **116**, G4 (2011)

D. Guo, H. Wang, D. Li, A projection of permafrost degradation on the Tibetan Plateau during the 21st century. J. Geophys. Res. **117**, D05106 (2012)

M.R. Haylock, N. Hofstra, A.M.G.K. Tank, E.J. Klok, P.D. Jones, M. New, A European daily high-resolution gridded data set of surface temperature and precipitation for 1950–2006. J. Geophys. Res. Atmos. **113**, 1–12 (2008)

N. Hofstra, M. Haylock, M. New, P. Jones, C. Frei, Comparison of six methods for the interpolation of daily, European climate data. J. Geophys. Res. Atmos. **113**, 1–19 (2008)

K.-C. Hsu, S.-Y. Li, Clustering spatial-temporal precipitation data using wavelet transform and self-organizing map neural network. Adv. Water Resour. **33**, 190–200 (2010)

M.N. Islam, M. Almazroui, Direct effects and feedback of desert dust on the climate of the Arabian Peninsula during the wet season: a regional climate model study. Clim. Dyn. **39**, 2239–2250 (2012)

B. Jha, Z.Z. Hu, A. Kumar, SST and ENSO variability and change simulated in historical experiments of CMIP5 models. Clim. Dyn. **42**, 2113–2124 (2014)

M.G. Kendall, *Rank Correlation Methods* (Hafner Publishing Company, New York, 1962)

N. Khalili, S.R. Khodashenas, K. Davary, M.M. Baygi, F. Karimaldini, Prediction of rainfall using artificial neural networks for synoptic station of Mashhad: a case study. Arab. J. Geosci. **9**, 624 (2016)

T. Koenigk, M. Caian, G. Nikulin, S. Schimanke, Regional Arctic sea ice variations as predictor for winter climate conditions. Clim. Dyn. **46**, 317–337 (2016)

M.Y. Kulakov, A.P. Makshtas, S.V. Shutilin, AARI-IOCM: combined model of water circulation and ices in the Arctic Ocean. Probl. Arkt. Antarkt. **2**, 6–18 (2012)

M.C. Mack, M.S. Bret-Harte, T.N. Hollingsworth, R.R. Jandt, E.A.G. Schuur, G.R. Shaver, D.L. Verbyla, Carbon loss from an unprecedented Arctic tundra wildfire. Nature **475**, 489–492 (2011)

H.B. Mann, Non-parametric test against trend. Econometrika **13**, 245–259 (1945)

N. Nakamura, A.H. Oort, Atmospheric heat budgets of the polar regions. J. Geophys. Res. **93**, 9510–9524 (1988)

L. Niu, B. Ye, J. Li, Y. Sheng, Effect of permafrost degradation on hydrological processes in typical basins with various permafrost coverage in Western China. Sci. China **54**, 615–624 (2011)

C. Nzotungicimpaye, K. Zickfeld, The contribution from methane to the permafrost carbon feedback. Curr. Clim. Change Rep. **3**, 58–68 (2017)

Q. Pang, G. Cheng, S. Li, W. Zhang, Active layer thickness calculation over the Qinghai-Tibet Plateau. Cold Reg. Sci. Technol. **57**, 23–28 (2009)

T. Partal, H.K. Cigizoglu, Prediction of daily precipitation using wavelet-neural netwarks. Hydrol. Sci. J. **54**, 234–246 (2009)

D.K. Perovich, S.V. Nghiem, T. Markus, A. Schweiger, Seasonal evolution and interannual variability of the local solar energy absorbed by the Arctic sea ice-ocean system. J. Geophys. Res. **112**, C03005 (2007)

P.E. Pfeifer, S.J. Deutrch, A three-stage iterative procedure for space-time modeling phillip. Technometrics **22**, 35–47 (1980)
C.L. Ping, G.J. Michaelson, E.S. Kane, E.C. Packee, C.A. Stiles, D.K. Swanson, N.D. Zaman, Carbon stores and biogeochemical properties of soils under black spruce forest. Alaska. Soil Sci. Soc. Am. J. **74**, 969–978 (2010)
J.D. Salas, Analysis and modelling of hydrologic time-series, in *Handbook of Hydrology*, ed. by D.R. Maidment (McGraw-Hill, New York, 1993)
E.A.G. Schuur, J. Bockheim, J.G. Canadell, E. Euskirchen, C.B. Field, S.V. Goryachkin, S. Hagemann, P. Kuhry, P.M. Lafleur, H. Lee, G. Mazhitova, F.E. Nelson, A. Rinke, V.E. Romanovsky, N. Shiklomanov, C. Tarnocai, S. Venevsky, J.G. Vogel, S.A. Zimov, Vulnerability of permafrost carbon to climate change: implications for the global carbon cycle. Bioscience **58**, 701–714 (2008)
E.A.G. Schuur, B.W. Abbott, W.B. Bowden, V. Brovkin, P. Camill, J.G. Canadell, J.P. Chanton, F.S. Chapin III, T.R. Christensen, P. Ciais, B.T. Crosby, C.I. Czimczik, G. Grosse, J. Harden, D.J. Hayes, G. Hugelius, J.D. Jastrow, J.B. Jones, T. Kleinen, C.D. Koven, G. Krinner, P. Kuhry, D.M. Lawrence, A.D. McGuire, S.M. Natali, J.A. O'Donnell, C.L. Ping, W.J. Riley, A. Rinke, V.E. Romanovsky, A.B.K. Sannel, C. Schädel, K. Schaefer, J. Sky, Z.M. Subin, C. Tarnocai, M.R. Turetsky, M.P. Waldrop, K.M. Walter Anthony, K.P. Wickland, C.J. Wilson, S.A. Zimov, Expert assessment of vulnerability of permafrost carbon to climate change. Climate Change **119**, 359–374 (2013)
D.T. Shindell, G. Faluvegi, D.M. Koch, G.A. Schmidt, N. Unger, S.E. Baner, Improved attribution of climate forcing to emissions. Science **326**, 716–718 (2009)
O.P. Singh, Multivariate ENSO index and Indian monsoon rainfall: relationships on monthly and subdivisional scales. Meteorol. Atmos. Phys. **78**, 1–9 (2001)
H.Y. Son, J.Y. Park, J.S. Kug, Precipitation variability in September over the Korean Peninsula during ENSO developing phase. Clim. Dyn. **46**, 3419–3430 (2016)
C. Tarnocai, J.G. Canadell, E.A.G. Schuur, P. Kuhry, G. Mazhitova, S. Zimov, Soil organic carbon pools in the northern circumpolar permafrost region. Global Biogeochem. Cycles **23**, 2 (2009)
M.R. Turetsky, W.F. Donahue, B.W. Benscoter, Experiment drying intensifies burning and carbon lossesin a northern peatland. Nat. Commun. **2**, 514 (2011)
V.Y. Vize, The source of Arctic warming. Sov. Akktika **1**, 1–7 (1937)
Q. Wu, Y. Liu, Ground temperature monitoring and its recent change in Qinghai-Tibet Plateau. Cold Reg. Sci. Technol. **38**, 85–92 (2004)
Q. Wu, T. Zhang, Recent permafrost warming on the Qinghai-Tibetan Plateau. J. Geophys. Res. **113**, D13 (2008)
Q. Wu, T. Zhang, Changes in active layer thickness over the Qinghai-Tibetan plateau from 1995 to 2007. J. Geophys. Res. **115**, D9 (2010)
Q. Wu, T. Zhang, Y. Liu, Permafrost temperatures and thickness on the Qinghai-Tibet plateau. Global Planet. Change **72**, 32–38 (2010)
T. Wu, L. Zhao, R. Li, Q. Wang, C. Xie, Q. Pang, Recent ground surface warming and its effects on permafrost on the central Qinghai-Tibet plateau. Int. J. Climatol. **33**, 920–930 (2013)
V. Yevjevich, *Stochastic Process in Hydrology*, 1st edn. (Water Resources Publications, Colorado, 1972)
K. Yoshikawa, W.R. Bolton, V.E. Romanovsky, M. Fukuda, L.D. Hinzman, Impacts of wildfire on the permafrost in the boreal forests of interior Alaska. J. Geophys. Res. Atmos. **108**, D1 (2002)
V.F. Zakharov, *The World Ocean and the Pleistocene Ice Periods* (Gidrometeoizdat, Leningrad, 1978)
V.F. Zakharov, *Arctic Ices and Current Natural Processes* (Gidrometeoizdat, Leningrad, 1981)
Z. Zhang, A. Rinke, J.C. Moore, Permafrost dynamic change on the Tibetan plateau under climatic warming since 1950s. Maejo Int. J. Sci. Technol. **10**, 242–255 (2016)
X. Zhang, G. Hegerl, F. Zwiers, J. Kenyon, Avoiding inhomogeneity in percentile-based indices of temperature extreme. J. Clim. **18**, 1641–1651 (2005)
S.A. Zimov, S.P. Davydov, G.M. Zimova, A.I. Davydova, E.A.G. Schuur, K. Dutta, F.S. Chapin, Permafrost carbon: stock and decomposability of a globally significant carbon pool. Geophys. Res. Lett. **33**, 20 (2006)

Chapter 8
Ecosystem and Carbon Cycle

The global carbon cycle is the fluxes of carbon among four main reservoirs: fossil carbon, the atmosphere, the oceans, and the land surface. Plants on land and in the ocean convert atmospheric carbon dioxide to biomass through photosynthesis, while the carbon returns to the atmosphere when the plants decay, are digested by animals, or burn in fires. Because plants and animals are an integral part of the carbon cycle, the carbon cycle is closely connected to ecosystems. Currently, ecosystem associated with the global carbon cycle has been greatly modified by human activities. Total anthropogenic carbon emissions have grown nearly at an average growth rate of 1.9% per year over the period 1850–2010, to reach about 10 Pg C per year in 2010; more than a half of total anthropogenic emissions are removed by land and ocean carbon sinks (Raupach et al. 2013). In this chapter, we will discuss key processes and related indicators of the interactions of global carbon cycle and terrestrial and oceanic ecosystems under a changing climate. These play key roles in improving the representation of terrestrial and oceanic ecosystem processes in Earth system models and supporting for time series analysis on examining the feedbacks between the ecosystem and climate system.

8.1 Terrestrial Ecosystems

An ecosystem is a community of living organisms and non-living objects that are interrelated. Most terrestrial ecosystems gain energy from the sun and materials from the air or rocks, transfer them among components within the ecosystem, and then release energy and materials to the environment.

Z. Zhang, *Multivariate Time Series Analysis in Climate and Environmental Research*, https://doi.org/10.1007/978-3-319-67340-0_8

8.1.1 Terrestrial Hydrologic Cycle

The hydrologic cycle controls all biogeochemical cycles on the Earth by influencing biotic processes, dissolving nutrients within terrestrial ecosystems, and transferring nutrients among terrestrial ecosystems. The terrestrial hydrologic cycle consists of three processes. (1) Water is transferred vertically from the atmosphere to the Earth via precipitation in form of rain and snow. (2) Water is infiltrated into the soil or intercepted by vegetation cover. When water inputs exceed the capacity of the soil to hold water, the excess water drains to groundwater or run off over the ground surface. Water within ecosystems moves along a gradient from high to low potential. The water potential is determined by pressure potential, osmotic potential, and matric potential, where the pressure potential is generated by gravitational forces and physiological precesses of organisms, the osmotic potential reflects presence of substances dissolved in water, and the matric potential is caused by absorption of water to surfaces. (3) Water is finally transferred from the Earth to the atmosphere via evaporation and evapotranspiration. *Evaporation* is a process that liquid water is converted to water vapor and removed from a lake, river, or soil into the air. *Evapotranspiration* is to remove liquid water from an area with vegetation and into the atmosphere by transpiration and evaporation, where transpiration is a process that water in plant tissues is lost to the atmosphere through stomata. The mechanism of evaporation and evapotranspiration is mainly driven by *net radiation*, which is defined as the balance of incoming and outgoing shortwave and longwave radiation. The shortwave radiation is the high-energy radiation emitted by the sun, while the longwave radiation is the thermal energy emitted by all bodies. The radiation balance equation can be expressed as

$$R_{net} = (S_{in} - S_{out}) + (L_{in} - L_{out})$$

with unit as watts per meter squared (Wm^{-2}), where R_{net} is the net radiation, S_{in}, L_{in} and S_{out}, L_{out} are the inputs and outputs of shortwave radiation and longwave radiation, respectively. The change in ecosystems could affect net radiation primarily through albedo.

Recent human activities significantly alter the hydrologic cycle: Human use of Earth's freshwater affects water management; human-driven land use changes terrestrial water budgets; and human alters the capacity of the atmosphere to hold water vapor. At the same time, as one of the major greenhouse gases, water vapor is transparent to solar radiation, but absorbs the Earth's longwave radiation, then it affects regional and global climate. Conversely, the trend for climate warming increases the quality of water vapor in the atmosphere, accelerates the hydrologic cycle, and increases evaporation and precipitation at the global scale.

8.1.2 Photosynthesis

Carbon enters terrestrial ecosystems via photosynthesis, transfers within terrestrial ecosystems, and losses from terrestrial ecosystem. Photosynthesis is to use solar energy, water, atmospheric carbon dioxide, and minerals to produce oxygen and energy-rich organic compounds. These organic compounds are transferred within ecosystems and are eventually released to the atmosphere by respiration or combustion. Photosynthesis consists of two major reactions: One is the light-harvesting reactions which transform light energy into chemical energy. The other is the carbon-fixation reactions which use the transformed chemical energy to convert carbon dioxide into sugars which can be stored, transported, or metabolized. Both groups of reactions occur simultaneously in the chloroplasts.

Carbon dioxide used in photosynthesis diffuses along a concentration gradient across the atmosphere outside the leaf and through the stomata in the leaf surface, and then into the chloroplast inside the leaf. The open stomata are a channel for plants to gain carbon. On the other hand, cell walls inside the leaf are covered with a thin film of water, the water evaporates readily, and water vapor diffuses out through the stomata to the atmosphere. The open stomata are also a channel for plants to lose water. Plants regulate CO_2 uptake and water loss by changing the size of stomatal openings. Plants with a high photosynthetic capacity have a high stomatal conductance. When plants reduce stomatal conductance to conserve water, photosynthesis declines. The *photosynthetic capacity* is referred to as the photosynthetic rate per unit leaf mass measured under favorable condition of light, moisture, and temperature. The *stomatal conductance* is referred to as the flux of water vapor or CO_2 per unit driving force between the leaf and the atmosphere.

Net photosynthesis by individual leaves is the net rate of carbon gain measured at the level of individual leaves. Net photosynthesis in the short term is affected by the following environment variables:

(a) CO_2 Concentration
When the CO_2 concentration inside the leaf is low, the net photosynthesis is limited by the rate of CO_2 diffusion into the chloroplast. When the CO_2 concentration inside the leaf is high, the net photosynthesis is limited by biochemical processes. Leaves can adjust the co-limitation of photosynthesis by diffusion and biochemical processes to be about equally limiting through altering stomatal conductance and changing the concentrations of light-harvesting pigments or photosynthetic enzymes.

(b) Light Intensity
Photosynthesis depends upon the absorption of light by pigments in the leaves of plants. Plants always spread out their leaves in such a way that each leaf maximizes the amount of light falling on them and the lower leaves are not shaded by the ones above. At low light intensities, the rate of the net photosynthesis increases linearly with increasing light. At the light intensities higher than certain threshold, the rate of the net photosynthesis stays constant. At extremely high

light intensities, the rate of the net photosynthesis declines as a result of photo-oxidation of photosynthetic enzymes and pigments. Leaves can adjust stomatal conductance and photosynthetic capacity in order to maximize carbon gain in different light environments. Here photosynthetic capacity is the photosynthetic rate per unit leaf mass measured under favorable conditions of light, moisture, and temperature.

(c) Nitrogen Concentration
Metabolic processes in plants depend on the adequate supply of nitrogen. Nitrogen deprivation reduces the leaf production, individual leaf area, and total leaf area, resulting in a reduced area for light interception for photosynthesis. Plants growing in high-nitrogen environment have higher tissue nitrogen concentration and net photosynthesis, while plants growing in low-nitrogen environment often produce long-lived leaves which have a low leaf nitrogen concentration and a low net photosynthesis.

(d) Temperature
When temperature is low, photosynthesis is limited directly by temperature. The higher the temperature, typically the greater the rate of photosynthesis. When temperature is higher than 40 °C, the rate of photosynthesis declines because the enzymes involved in the chemical reactions of photosynthesis are temperature sensitive and destroyed at higher temperatures.

(e) Water
Water is important for photosynthesis because it is the source of hydrogen for the organic compounds created through photosynthesis. When water is supplied abundantly, leaves open their stomata in response to high light intensities, regardless of the associated high rate of water loss. As leaf water stress drops, stomatal conductance trends to decline to avoid loss of too much water. This results in a decline in photosynthetic rate. In dry environment, due to the low stomatal conductance, the rate of photosynthesis is always low.

(f) Pollutants
Pollutants reduce the rate of photosynthesis through their effects on growth and the production of leaf area or by entering the stomata and damaging the photosynthetic capacity.

8.1.3 Gross and Net Primary Production

Photosynthesis is controlled by light intensity, atmospheric carbon dioxide, temperature, and nitrogen because light provides energy, temperature governs reaction rate, and nitrogen is used to produce photosynthetic enzymes. *Gross primary production*

(GPP) is the amount of carbon captured from the atmosphere through vegetation photosynthesis. Terrestrial ecosystem GPP is a key component in the global carbon cycle. It can be calculated as the sum of vegetation assimilated carbon flux, partitioned from net carbon exchange measured at eddy covariance tower sites (Reichstein et al. 2007).

The amount of GPP in terrestrial ecosystems is governed by various direct, interactive, and state factors, where the direct factors include leaf area, nitrogen, length of photosynthetic season, temperature, light intensity, and atmospheric carbon dioxide; the interactive factors include plant functional types and soil resource; and the state factors include biota, time, parent material, and climate. The key factors that account for most of the variations in GPP among terrestrial ecosystems are leaf area and length of the photosynthetic season which are ultimately determined by the interacting effects of soil resources, climate, vegetation, and disturbance regime.

Carbon enters the terrestrial ecosystem through photosynthesis by plants, while roots and aboveground portions of plants return about half of this carbon to the atmosphere as plant respiration. Respiration provides the energy for plants to acquire nutrients and to produce and maintain biomass. *Total plant respiration* (R_p) can be separated into three functional components including growth respiration (R_g), maintenance respiration (R_m), and the respiration cost of ion uptake (R_i), i.e.,

$$R_p = R_g + R_m + R_i.$$

Growth respiration biosynthesizes many classes of chemical compounds including cellulose, proteins, nucleic acids, and lipids for growth of new tissues. Maintenance respiration provides energy required all live cells to maintain ion gradients across cell membranes and to replace proteins, membranes, and other constituents. The respiration cost of ion uptake includes the cost of absorbed nutrients, the cost of absorbed nitrogen, and the cost of nitrate reduction. Maintenance respiration may account for about half of total plant respiration, and the other half is associated with growth respiration and respiration cost of ion uptake.

Net primary production (NPP) is the difference between carbon gain by photosynthesis and carbon loss through respiration, i.e.,

$$\text{NPP} = \text{GPP} - R_p. \tag{8.1.1}$$

NPP is one of the main causes for the observed seasonal changes in atmospheric CO_2 levels. Both the most common methods for estimating NPP include measuring the biomass produced during the growing season and measuring net gas exchange.

Variation in terrestrial NPP with climate originates from a direct influence of temperature and precipitation on plant metabolism. However, variation in NPP may also result from an indirect influence of climate by means of plant age, stand biomass, growing season length, and local adaptation. Over the past several decades, a significant warming of the global land surface has resulted in a remarkable increase in the NPP of terrestrial ecosystems. NPP is greatest in warm and moist environment and is lowest in cold and dry environment, but NPP will decline at extremely high precipitation due to indirect effects of excess moisture.

8.1.4 Net Ecosystem Production

Carbon enters terrestrial ecosystems through photosynthesis and then accumulates within the terrestrial ecosystems. *Net ecosystem production* (NEP) is referred to as the net accumulation of carbon by an ecosystem. NEP represents the total amount of organic carbon in an ecosystem available for storage, export as organic carbon, or non-biological oxidation to carbon dioxide through fire or ultraviolet oxidation.

Besides autotrophic respiration, the largest part of carbon loss from plants is carbon fluxes to the soil through litterfall, root exudation, and carbon fluxes to mycorrhizae and nitrogen-fixing bacteria. Plants lost also carbon from herbivores, disturbance of fire and human harvest of plants, and from emission of volatile organic compounds. Therefore, the net carbon accumulation of plants (NCA_p) is the balance between the net carbon gain by vegetation NPP and carbon losses of various pathways including carbon flux to the soil ($F_{p,s}$), heterotrophic respiration (R_h), emission (F_e), and disturbance of fire ($F_{p,f}$), and disturbance of harvest (F_h), i.e.,

$$\begin{aligned} NCA_p &= NPP - F_{p,s} - R_h - F_e - F_{p,f} - F_h \\ &= GPP - R_p - F_{p,s} - R_h - F_e - F_{p,f} - F_h. \end{aligned}$$

Similar to plants, the net carbon accumulation of animals (NCA_a) is the carbon gains by herbivores (R_h) and microbes to soil animals ($F_{m,a}$) minus the carbon losses by animal respiration (R_a) and fluxes from animal to soil ($F_{a,s}$), i.e.,

$$NCA_a = (R_h + F_{m,a}) - (R_a + F_{a,s}).$$

The net carbon accumulates of soil organic matter (SOM) is the balance between the annual carbon inputs to the soil from plants ($F_{p,s}$) and animals ($F_{a,s}$) and carbon loss through respiration of microbes (R_m), the consumption of microbes by soil animals ($F_{m,a}$), methane emissions to the atmosphere (F_{CH_4}), and leaching of organic and inorganic carbon to groundwater (F_l) and fire ($F_{s,f}$), i.e.,

$$NCA_{SOM} = (F_{p,s} + F_{a,s}) - (R_m + F_{m,a} + F_{CH_4} + F_l + F_{s,f}).$$

There are lateral carbon fluxes ($F_{lateral}$) into or out of the ecosystem due to deposition or erosion.

NEP is the sum of the net carbon accumulation in plants, animals, and the soil plus/minus lateral transfers of carbon among ecosystems, i.e.,

$$
\begin{aligned}
\mathrm{NEP} &= \mathrm{NCA}_p + \mathrm{NCA}_a + \mathrm{NCA}_{\mathrm{SOM}} \pm \mathrm{F}_{\mathrm{lateral}} \\
&= \mathrm{GPP} - \mathrm{R}_p - \mathrm{F}_{p,s} - \mathrm{R}_h - \mathrm{F}_e - \mathrm{F}_{p,f} - \mathrm{F}_h \\
&\quad + (\mathrm{R}_h + \mathrm{F}_{m,a}) - (\mathrm{R}_a + \mathrm{F}_{a,s}) \qquad (8.1.2)\\
&\quad + (\mathrm{F}_{p,s} + \mathrm{F}_{a,s}) - (\mathrm{R}_m + \mathrm{F}_{m,a} + \mathrm{F}_{\mathrm{CH}_4} + \mathrm{F}_l + \mathrm{F}_{s,f}) \pm \mathrm{F}_{\mathrm{lateral}} \\
&= \mathrm{GPP} - \mathrm{R}_p - \mathrm{F}_e - \mathrm{F}_{p,f} - \mathrm{F}_h - \mathrm{R}_a - (\mathrm{R}_m + \mathrm{F}_{\mathrm{CH}_4} + \mathrm{F}_l + \mathrm{F}_{s,f}) \pm \mathrm{F}_{\mathrm{lateral}}
\end{aligned}
$$

The net ecosystem autotrophic and heterotrophic respiration R_e consists of plant, animal, and microbe respirations, where the heterotrophic respiration R_h is from animals and microbes, i.e.,

$$
\begin{aligned}
\mathrm{R}_e &= \mathrm{R}_p + \mathrm{R}_a + \mathrm{R}_m, \\
\mathrm{R}_h &= \mathrm{R}_a + \mathrm{R}_m.
\end{aligned}
$$

The net disturbances are $\mathrm{F}_d = \mathrm{F}_{p,f} + \mathrm{F}_{s,f} + \mathrm{F}_h$. Finally, by (8.1.1) and (8.1.2), the net ecosystem production is

$$\mathrm{NEP} = \mathrm{GPP} - (\mathrm{R}_e + \mathrm{F}_e + \mathrm{F}_d + \mathrm{F}_{\mathrm{CH}_4} + \mathrm{F}_l) \pm \mathrm{F}_{\mathrm{lateral}}$$

or

$$\mathrm{NEP} = \mathrm{NPP} - (\mathrm{R}_h + \mathrm{F}_e + \mathrm{F}_d + \mathrm{F}_{\mathrm{CH}_4} + \mathrm{F}_l) \pm \mathrm{F}_{\mathrm{lateral}}.$$

NEP is a fundamental property of ecosystems and determines the impact of terrestrial biosphere on the quantity of carbon dioxide in the atmosphere. NEP equals zero at steady state.

The net carbon dioxide exchange between the ecosystem and the atmosphere is called *net ecosystem exchange* (NEE). NEE equals the difference between GPP and the ecosystem respiration, i.e.,

$$\mathrm{NEE} = \mathrm{GPP} - \mathrm{R}_e.$$

Therefore,

$$\mathrm{NEE} = \mathrm{GPP} - (\mathrm{R}_p + \mathrm{R}_h).$$

Since GPP equals zero at night,

$$\mathrm{NEE}_{night} = -(\mathrm{R}_p + \mathrm{R}_h) = -\mathrm{R}_e.$$

This shows that NEE is a direct measure of ecosystem respiration at night. During the daytime, NEE equals approximately the difference of GPP and R_e, i.e.,

$$\mathrm{NEE}_{daytime} = \mathrm{GPP} - \mathrm{R}_e.$$

8.1.5 Terrestrial Nutrient Cycle

Terrestrial nutrient cycle involves nutrient inputs to terrestrial ecosystems and nutrient outputs from terrestrial ecosystems as well as the internal transfers of nutrients within ecosystems. Nutrients enter terrestrial ecosystem through the weathering of rocks, the biological fixation of atmosphere nitrogen, and the deposition of nutrients. The internal transfers include the conversion of nutrients, chemical reactions of elements, biological uptake, and exchange of nutrients. Nutrients are lost from terrestrial ecosystems by leaching, trace-gas emission, wind and water erosion, and disturbance. Nutrient cycle through vegetation is essential, where the key steps are uptake, use, and loss of nutrient by vegetation.

Nutrient uptake by vegetation is governed by nutrient supply rate from the soil, root length, and root activity. The plant absorbs nutrient through three approaches: diffusion of nutrient to the root surface, mass flow of nutrient to the root surface, and root interception for available nutrients. *Diffusion* delivers most nutrient to plant roots. It is based on the movement of ions along a concentration gradient. In general, anion exchange capacity of soils is much lower than that of cation exchange capacity, so most anions diffuse more rapidly in soils than do cations. The driving force for diffusion is produced by plant reducing nutrient concentration at the root surface. *Mass flow* augments the supply of ions. It is the movement of dissolved nutrients to the root surface in flowing soil water. Mass flow is an important mechanism for supplying nutrients that the plant needs in small quantities. *Root interception* occurs in new soil where the plant creates new root surface as its roots elongate.

Nutrient use supports the production of new tissues. Plants store nutrients only when nutrient uptake exceeds the requirement for production. The hormonal balance of the plant governs the allocation of nutrient within the plant: Nitrogen is incorporated into protein with lesser amounts in nucleic acids and lipoproteins; Phosphorus is incorporated into sugar phosphates involved in energy transformations, nucleic acids, and phospholipids; Calcium is an important component of cell walls; Potassium is important in osmotic regulation. Magnesium, Manganese, and other cations serve as cofactors for enzymes. *Nutrient use efficiency* (NUE) is the ratio of nutrient to biomass lost in litterfall. This ratio is highest in unproductive sites. NUE is equal to the product of nutrient productivity (a_n) and residence time (t_r), i.e.,

$$\text{NUE} = a_n \times t_r,$$

where nutrient productivity a_n is a high instantaneous rate of carbon uptake per unit nutrient, and residence time t_r is the average time that the nutrient remains in the plant. Nutrient use efficiency is maximized by prologing tissue longevity.

Nutrient loss from plants to soil is an internal transfer within ecosystem. Nutrient loss primarily includes tissue senescence loss, leaching loss from plants and consumption of tissues by herbivores. Disturbance, such as fire, wind, and diseases that kill plants, can cause occasional large nutrient losses. Nutrient loss from plants to soil has very different consequences from the ecosystem to the atmosphere or groundwater.

Terrestrial decomposition is the primary mechanism by which nutrients are returned to the soil. It breaks down dead organic matter and releases carbon to the atmosphere and nutrients used again by plant and microbe. This process is done through the action of leaching, fragmentation, and chemical alteration. Decomposition rate is regulated by physical environment, substrate quality, and the composition of soil fauna including microfauna, mesofauna, and macrofauna. *Leaching* of litter by water is a physical process. It removes soluble materials from decomposing organic matter. These soluble materials are absorbed by organisms, react with the mineral phase of the soil, or lost from the system in solution. Leaching begins when litter first falls to the ground. *Fragmentation* by soil animals is to break large pieces of organic matter into small pieces. This process provides a food source for soil animals and creates fresh surfaces for microbial colonization. Fragmentation also mixes the decomposing organic matter into the soil. Fragmentation of litter can greatly enhance microbial decomposition by destructing the cuticle on plants or the skin on animals and/or by increasing the ratio of litter surface area to mass. *Chemical alteration* of dead organic matter is primarily a consequence of the activity of fungi and bacteria. Fungi have hyphal networks which enable fungi to acquire carbon from the litter and nitrogen from the mineral soil. Fungi have a competitive advantage over bacteria in decomposing tissues with low nutrient concentrations. Bacteria are another main decomposers of terrestrial dead plant material. Bacteria have a small size and large surface-to-volume ratio which enables bacteria to absorb substrates rapidly and to grow and divide quickly in substrate-rich zones.

Nitrogen cycle is the most important component in nutrient cycle. Although nitrogen has an abundant well-distributed quantity in Earth's atmosphere, it most frequently limits plant production. It is a necessary component of proteins, DNA, and chlorophyll. The nitrogen cycle consists of nitrogen input to ecosystems, internal transfer, and nitrogen loss from ecosystems. *Nitrogen input* to terrestrial ecosystems is mainly through deposition and fixation. Nitrogen is deposited in ecosystems in dissolved and gaseous forms. Nitrogen is fixed by nitrogen-fixed bacteria through transferring N_2 into ammonium (NH_4^+). *Internal transfer* of nitrogen is a cycling process. As microbes break down dead organic matter derived from plants, animals, and soil microbes, the nitrogen is released as dissolved organic nitrogen (DON), and then plants absorb these DON to support their growth. The soil microbes always absorb more DON than they require for growth and release ammonium into the soil solution. Nitrifiers further convert ammonium through nitrite to nitrate. The conversion from ammonium to nitrate is called *nitrification*. Finally, nitrogen is consumed by plants, animals, and soil microbes and is returned to the soil as DON due to decomposition of dead organic matter. *Nitrogen loss* from ecosystems is a *denitrification* process. Nitrogen losses include gaseous loss, leaching loss, and erosional loss. *Gaseous loss* means that nitrogen is released as ammonia gas, nitrous oxide, nitric oxide, and dinitrogen. *Leaching loss* means that nitrogen leached from terrestrial ecosystems moves to lakes and rivers. It includes nitrate loss and dissolved organic nitrogen loss. *Erosional loss* occurs in areas exposed to high winds.

8.2 Ocean Ecosystems

The oceans and adjacent seas cover 70.8% of the surface of Earth, which amounts to 361,254,000 km^2. It is about two and a half times the land area. Comparing the depth of the ocean with the altitude of the land, much of the oceans are more than 3000 m deep, while little of the land surface is above 3000 m in altitude. Rising atmospheric CO_2 and climate change are increasing ocean temperatures and lead to serious negative impacts on ocean ecosystem. Monitoring these important changes and impacts using ships and other platforms has generated large amounts of data for further time series analysis.

The ocean and the atmosphere share a common boundary which is called the *air–sea interface*. This direct physical contact enables them to exchange energy and matter. Saltwater can be sprayed into the atmosphere through whitecaps of waves and freshwater can be added to the ocean through precipitation. Carbon dioxide from the atmosphere can dissolve in the surface waters of the ocean. Some of the carbon dioxide stays as dissolved gas, but much of it is converted into organic matter through photosynthesis of phytoplankton near the sunlit ocean surface. Although oceans have a large capacity to absorb CO_2 and bringing carbon into the ocean ecosystem, the level of carbon dioxide in the atmosphere is still rising. The ocean would continue to soak up more and more carbon dioxide until global warming heated the ocean enough to slow down ocean circulation.

8.2.1 Solubility and Air–Sea Gas Exchange

Seawater are exposed to the atmosphere at the ocean surface, so various atmosphere gases have opportunities to enter the ocean across the air–sea interface. The efficiency of the transfer is controlled by complex interaction of a variety of processes in the air and in the water near the interface. When the accumulation of an atmosphere gas in the seawater reaches the equilibrium, the sea water is said to be *saturated* with respect to this gas. In general, atmospheric gases are close to a state of saturation in surface waters of the ocean, but their concentrations are often slightly above saturation.

Gases have a temperature-dependence solubility. Higher temperatures mean the molecules have a greater energy. This implies a higher rate of exchange of molecules between the air and water so that the equilibrium situation occurs with fewer molecules of gas within the liquid. An increase in temperature results in a decrease in gas solubility in water, while a decrease in temperature results in an increase of gas solubility in water. Gas solubility is also weakly dependent on salinity and pressure. Solubility decreases with higher salinity. The higher pressure often results in a higher concentration of gas can be sustained.

Greenhouse gases, including carbon dioxide, methane, and nitrous oxide, have demonstrated anomalously high solubilities due to the fact that greenhouse gases are involved in chemical reactions with water. Greenhouse gases also show an enhanced temperature dependence of their solubility since related chemical reactions are temperature-dependent. If the chemical reaction soaks up gas faster than the gas enters the ocean, then saturation is not reached.

The basic driving force for exchange of gas across the air–sea interface is the difference in gas concentration (or partial pressure) between the air and seawater. The flux of a gas across the air–sea interface into the ocean is modeled as

$$F = k_\tau (p_a - p_s), \tag{8.2.1}$$

where F is the flux of some gas, p_a and p_s are the atmosphere partial pressure and the ocean surface partial pressure, respectively, and k_τ is the transfer velocity (or piston velocity). The transfer velocity represents the variability of the rate of exchange related to the sea state and the atmosphere stability. This flux formula shows that the flux depends on transfer velocity and partial pressures.

Wind speed is a main factor affecting the air–sea flux rate. A calm sea and stable air allow only slow exchange because the surface air is renewed infrequently and the seawater has little bubble environment. When the wind is blowing over the ocean, energy is transferred from the wind to the surface layers, and then it results in that the sea surface becomes rough and the surface air is renewed frequently. When the wind becomes strong enough for breaking waves or driving currents, there is an abrupt change in transfer velocity, which affect the air–sea flux rate.

As the ocean has a higher heat capacity than that of the atmosphere, the ocean trends to change the temperature slowly, and this results in that the surface of the sea is usually a different temperature to the air immediately above it. Heat transfer between the atmosphere and ocean can also affect the air–sea flux rate. If the latent heat flux is directed toward the ocean producing condensation on the water surface, then gas flux into the ocean is inhibited. Basically, the solubility increase with decreasing temperature establishes a chemical potential gradient toward the ocean. If the sea surface temperature is less than the air temperature, then the resulting flux of a gas in atmosphere across the air–sea interface into the ocean is given by

$$F = -k_\tau \frac{P_m Q_{sol}}{RT_m} \left(\ln \frac{T_s}{T_a} - \ln \frac{P_s}{P_a} \right), \tag{8.2.2}$$

where P_a, P_s are stated as above, T_a, T_s are the gas's temperature and the ocean surface temperature, respectively, $P_m = 0.5(P_s + P_a)$, $T_m = 0.5(T_s + T_a)$, and Q_{sol} is enthalpy of solution (i.e., the energy released when a mole of the gas is dissolved in seawater). If the sea and air temperatures are identical, i.e., $T_s = T_a$, then the flux of the gas becomes:

$$F = k_\tau \frac{P_m Q_{sol}}{RT_s} (\ln P_s - \ln P_a).$$

Let $k_\tau^* = k_\tau \frac{P_m Q_{sol}}{RT_s}$, $P_s^* = \ln P_s$, and $P_a^* = \ln P_a$ in this formula. Then

$$F = k_\tau^* (P_s^* - P_a^*).$$

Namely, when the air and sea temperatures are identical, formula (8.2.2) is reduced to formula (8.2.1).

8.2.2 Oceanic Carbon Sink

Carbon dioxide is a major product of the combustion of carbon-containing substances such as wood, coal, and oil. It is very reactive in water and is absorbed easily by the ocean. Approximately one-third of the carbon dioxide released into the atmosphere as a result of human activity has been absorbed by the oceans, so the ocean is a sink for carbon dioxide. When carbon dioxide dissolves in water, a small part is transformed to carbonic acid H_2CO_3, which in turn dissociates to bicarbonate HCO_3^-, carbonate CO_3^{2-}, and hydrogen H^+ ions, i.e., the following full equilibrium:

$$CO_2 + H_2O(\longleftrightarrow H_2CO_3) \longleftrightarrow H^+ + HCO_3^- \longleftrightarrow 2H^+ + CO_3^{2-}. \quad (8.2.3)$$

This chemical reaction results in the acidity of seawater which has increased by 30% over the past 150 years, and some regions have already become corrosive enough to inhibit the growth of corals.

The chemical reaction (8.2.3) is to distribute carbon atoms among the gas phase, the carbonate and bicarbonate ions in solution. The *rapidity of reaction* can be measured by the following two *dissociation constants* K_1 and K_2:

$$K_1 = \frac{[\,H^+\,][\,HCO_3^-\,]}{[\,H_2O\,][\,CO_2\,]}$$

is the gas to bicarbonate equilibrium in the left-hand side of (8.2.3) and

$$K_2 = \frac{[\,H^+\,][\,CO_3^{2-}\,]}{[\,HCO_3^-\,]}$$

is the bicarbonate to carbonate equilibrium in the right-hand side of (8.2.3), where the square brackets represent the concentration of the respective chemical ions. These dissociation constants K_1 and K_2 depend on temperature, pressure, and salinity. If additional CO_3^{2-} can be available, the net reaction will become

$$CO_2\,(\text{gas}) + H_2O + CO_3^{2-} \longleftrightarrow 2HCO_3^{2-} \quad (8.2.4)$$

and the corresponding dissociation constant K is

$$K = \frac{[\,HCO_3^-\,]}{[\,CO_2\,][\,H_2O\,][\,CO_3^{2-}\,]} = \frac{K_1}{K_2}.$$

The bicarbonate and carbonate ions are not produced solely from the equilibrium (8.2.4) with atmospheric carbon dioxide. They can also derived from deposition of the weathering products of rocks containing calcium carbonate, such as limestone. The preexisting oceanic carbonate source from calcium carbonate weathering permits to absorb much more of carbon dioxide. The reaction of carbon dioxide with calcium carbonate, i.e.,

$$CaCO_3 + CO_2 + H_2O \longleftrightarrow Ca^{2+} + 2HCO_3^-$$

might be significant. The surface waters are generally super-saturated with respect to calcium carbonate, so this kind of reaction is often inhibited. However, the bottom waters in coastal zones are under-saturated with respect to calcium carbonate, so anthropogenic carbon dioxide can penetrate and be absorbed by the oceans by this approach. In a word, much of carbon dioxide that enters the ocean from the atmosphere does not remain in the surface waters but is transferred into the intermediate and deep ocean. Conversely, if atmospheric carbon dioxide level was suddenly decreased, then the ocean would leak carbon dioxide back to the atmosphere.

The input of carbon dioxide to the ocean encourages biological activity. Phytoplankton in pelagic ecosystems use primarily CO_2 as their carbon source through photosynthesis, while some marine algae in the littoral zone, such as the macroalga, use bicarbonate as their carbon source. Some of the carbon is stored within marine life-forms which result in that its return to the atmosphere is delayed. At the same time, some of the carbon is finally added to the geosphere's reservoir, and it may not be returned to the atmosphere for millions of years.

8.2.3 Compounds in Seawater

Marine organisms play an important role in redistributing the inorganic compounds which are present in seawater, including carbon, nitrogenous, and sulfureous compounds, methyl iodide and methyl chloride. These compounds are active and closely related to global biogeochemical cycles.

(i) Carbon Compounds
Carbon dioxide is a major carbon compound for marine biological activity. It moves between the various geosphere reservoirs through carbon cycle. Marine organisms (e.g., phytoplankton) can fix carbon from dissolved carbon dioxide through photosynthesis. When these marine organisms die, their remains sink toward the sediment and form detritus. The suspended detritus can transfer carbon from the surface waters to the geosphere reservoir.
Methane (CH_4) is a principal carbon compound other than carbon dioxide. Oceanic methane production may be associated with regions of high phytoplankton biomass. It is produced by marine bacterial decay in the absence of oxygen, i.e.,

$$2CH_2O \longrightarrow CH_4 + CO_2.$$

At the same time, huge amounts of methane are stored around the world in the seafloor in the form of solid methane hydrates. Climate warming, however, could cause these hydrates to destabilize

Carbon monoxide (CO) is another principal carbon compound. Like methane, carbon monoxide is also produced by biological activity in the oceans. There is a large flux of carbon monoxide to atmosphere from the ocean. Some of it will have been formed by photochemical oxidation of methane, but much of it is produced by microbiological activity. Carbon monoxide is a product of incomplete respiration in the situation of an inadequate oxygen supply, i.e.,

$$HCO^- + O_2 \longrightarrow CO + HOO^-.$$

(ii) Nitrogenous Compounds

Nitrous oxide (N_2O) is a principal nitrogen-based gas for marine biological activity. During the oxidation of organic material by phytoplankton, some of the nitrogen is converted to nitrous oxide, i.e.,

$$2O_2 + 2NH_4^+ \longrightarrow N_2O + 3H_2O + 2H^+.$$

The ocean is a source of nitrous oxide. It contributes about a quarter of the net input to the atmosphere.

Ammonia (NH_4) is produced by cell protein decomposition in both aerobic and anaerobic conditions. Ammonia can be utilized as a nitrogen source by some phytoplankton and as the ammonium ion in its dissolved form, i.e.,

$$NH_3 + H_2O \longleftrightarrow NH_4^+ + OH^-$$

Ammonia is a weak greenhouse gas and is a supply of nutrient for marine biological activity.

(iii) Sulfureous Compounds

Hydrogen sulfide (H_2S) is a principal sulfureous gas in seawater. It is a major oxidation product of anaerobic decay. This gas is produced by marine biological processes and can be photochemically oxidized very rapidly to sulfuric acid (H_2SO_4) in air, i.e.,

$$H_2S + 2O_2 \longleftrightarrow H_2SO_4.$$

The sulfate particles from this acid can contribute to cloud condensation nuclei.

Dimethyl sulfide (DMS, $(CH_3)_2S$) is another principal sulfureous gas. DMS has been observed in considerable concentrations during plankton blooms. DMS is a product of algal cell destruction and is excreted by plankton during oxidation of dimethylsulfoniopropionate (DMSP, $CH_3C_2H_4CO_2SCH_3$). Most of DMS is destroyed within the ocean through biological consumption, biologically and photochemically aided oxidation, and adsorption onto particles, while only a few tens of micromoles of DMS per square meter per day are transferred to the atmosphere. Within the atmosphere, DMS can be oxidized via reaction with hydroxyl ions to form sulfur dioxide:

$$(CH_3)_2S + 2OH^- \longleftrightarrow SO_2 + 2CH_4$$

or form methane sulfonic acid (MSA, CH_3SO_3H):

$$2(CH_3)_2S + O_2 + 2HO_2 \longleftrightarrow 2CH_3SO_3H + 2CH_3.$$

(iv) Methyl Compounds
Methyl (CH_3I) is also a product of algal cell destruction. Methyl reacts with dissolved chloride ions to form methyl chloride (CH_3Cl), i.e.,

$$CH_3I + Cl^- \longleftrightarrow CH_3Cl + I^-.$$

Methyl chloride is a natural source of chloride, and its most prolific source is the ocean.

8.2.4 Biogeochemical Cycles

Biogeochemical cycling involves the various reservoirs that store elements, the fluxes between these reservoirs, and related physical, chemical, and biological processes. The biogeochemical cycles of carbon and other important elements such as nitrogen, phosphorus, and sulfur are now being massively perturbed by recent climate change, and therefore, understanding the biogeochemistry of the surface ocean and its interactions with both the atmosphere and the deep ocean is important for impact assessments of climate change.

(i) The Carbon Cycle
The ocean influences the climate by absorbing and storing carbon dioxide. Phytoplankton are a small but important role in the carbon cycle through influencing the exchange of gases between the air and sea. Atmospheric carbon dioxide can dissolve easily in water, and then, phytoplankton fix carbon from dissolved carbon dioxide through photosynthesis in order to build block of phytoplankton cells. When phytoplankton die, their remains sink and form the known detritus. These detritus transfer carbon from the surface waters to the geosphere reservoirs. A large proportion of detritus is recycled within the ocean through direct consumption by other organisms. The remainder of detritus sinks to the seafloor and adds to the marine sediment. Over decades to centuries, some of the carbon may returns to the surface by upwelling and mixing from depth. If atmospheric carbon dioxide level once is suddenly decreased, the ocean leaks immediately carbon dioxide back to the atmosphere.

(ii) The Nitrogen Cycle
Nitrogen is a predominant gaseous constituent of the atmosphere. The proportion of nitrogen to the atmosphere is 78% approximately. Like carbon, as a key component to form the cell structure, nitrogen can control oceanic species' growth and affect the fixation of carbon. Therefore, the nitrogen cycle is closely linked to the fixation of atmospheric carbon dioxide and export of carbon from the ocean's surface.
The nitrogen cycle is a complex biogeochemical cycle in which nitrogen is converted from gas form into a form that is useful in biological processes: Nitrogen gas is not useful to most living things, but marine microbes can absorb the nitrogen and make it into ammonium (NH_4^+) and nitrate(NO_3^-), which is most easily consumed by various marine organisms. When marine organisms die, nitrogen will be disposed of in decay and wastes.

(iii) The Phosphorus Cycle
Phosphorus availability can impact primary production rates in the ocean as well as species distribution and ecosystem structure. River runoff is the main phosphorus source in the ocean. It delivers annually about 1.5 Mt of dissolved phosphorus into the ocean (Baturin 2003). These phosphates can stimulate the growth of plankton, but excess growth will consume large amounts of dissolved oxygen in the seawater, potentially suffocating fish and other marine animals, while also blocking available sunlight to bottom dwelling species. When phosphorous-containing compounds from dead bodies or wastes of marine organisms sink to the ocean floor, they form new sedimentary layers.

(iv) The Oxygen Cycle
Oxygen is another predominant gaseous constituent of the atmosphere. The proportion of oxygen to the atmosphere is 20.9% approximately. The surface ocean water has an oxygen concentration 30-fold lower than in air, i.e., the ratio of water to air is 1:30. Oxygen concentration in deep waters is much lower. The ocean is not only a sink for oxygen due to respiration of marine organisms, but also a source of oxygen due to photosynthesis of marine plants and chemical changes of marine sediments. Much of the oxygen in the ocean is recycled, only a small amount of oxygen is leaked to the atmosphere, and excess oxygen is used in the atmosphere for weathering of surface materials to maintain the balance in oxygen levels.

(v) The Sulfur Cycle
Sulfur is one of the constituents of many proteins, vitamins, and hormones. It is provided to marine organisms through recycling of dead organic material, dissolution of atmospheric sulfur dioxide, input of sulfates from rivers and precipitation, as well as anaerobic decay of organic material. Much of the sulfur in the ocean is deposited to marine sediments, and the rest is lost from the ocean as sulfureous gas (e.g., dimethyl sulfide).

Further Reading

G.N. Baturin, Phosphorus cycle in the ocean. Lithology and Mineral Resources **38**, 101–119 (2003)
G.R. Bigg, *The Oceans and Climate* (Cambridge Univ. Press, Cambridge, 1996)
W.S. Broecker, T.H. Peng, Interhemispheric transport of carbon dioxide by ocean circulation. Nature **356**, 587–589 (1992)
G.S. Campbell, J.M. Norman, *An Introduction to Environmental Biophysics* (Springer-Verlag, New York, 1998)
F.S. Chapin, The mineral nutrition of wild plants. Ann. Rev. Ecol. Syst. **11**, 233–260 (1980)
F.S. Chapin, P.A. Matson, H.A. Mooney, *Principles of Terrestrial Ecosystem Ecology* (Springer-Verlag, New York Berlin Heidelberg, 2002)
F.S. Chapin, E.D. Schulze, H.A. Mooney, The ecology and economics of storage in plants. Ann. Rev. Ecol. Syst. **21**, 423–448 (1990)
P.G. Falkowski, R.T. Barber, V. Smetacek, Biogeochemical controls and feedbacks on ocean primary production. Science **281**, 200–206 (1999)
R.W. Howarth (ed.), *Nitrogen Cycling in the North Atlantic Ocean and its Watershed* (Kluwer, Dordrecht, 1996)
T. Ingestad, G.I. Ägren, Nutrient uptake and allocation at steady-state nutrition. Physiol. Plant. **72**, 450–459 (1988)
B. Mary, S. Recous, D. Darwis, D. Robin, Interactions between decomposition of plant residues and nitrogen cycling in soil. Plant Soil **181**, 71–82 (1996)
H.A. Mooney, The carbon balance of plants. Annual Review of Ecology and Systematics **3**, 315–346 (1972)
E.A. Paul, F.E. Clark, *Soil Microbiology and Biochemistry*, 2nd edn. (Academic Press, San Diego, CA, 1996)
M.R. Raupach, Ecosystem Services and the Global Carbon Cycle, in *Ecosystem Services and Carbon Sequestration in the Biosphere*, ed. by R. Lal, et al. (Springer, Dordrecht, 2013)
M. Reichstein et al., Determinants of terrestrial ecosystem carbon balance inferred from European eddy covariance flux sites. Geophys. Res. Lett. **34**, G4 (2007)
W.H. Schlesinger, Carbon balance in terrestrial detritus. Annual Review of Ecology and Syatematics **8**, 51–81 (1977)
E.D. Schulze, F.M. Kelliher, C. Körner, J. Lloyd, R. Leuning, Relationship among maximum stomatal conductance, ecosystem surface conductance, carbon assimilation rate, and plant nitrogen nutrition: A global ecology scaling exercise. Ann. Rev. Ecol. Syst. **25**, 629–660 (1994)
I. Valiela, *Marine Ecological Processes* (Springer-Verlag, New York, 1995)
P.M. Vitousek, Nutrient cycling and nutrient use efficiency. American Naturalist **119**, 553–572 (1982)
N. Wells, *The Atmosphere and Ocean: A Physical Introduction* (Taylor Francis, London and Philadelphia, 1986)

Chapter 9
Paleoclimate Proxies

The ability to reconstruct past climates has expanded in recent years with a better understanding of present and future climate variability and climate change. Historical and instrumental records are short, only spanning the last few decades in many places. To extend the record beyond this period, scientists turn to proxies or proxy records that are indirect measures of past climates or environments preserved in natural archives, such as tree-rings, ice cores, and speleothems. The character of deposition or rate of growth of these proxies' materials was influenced significantly by the climatic conditions of the time in which they were laid down or grew. In this chapter, we will introduce the main reconstruction methods and tools of past climate and environmental conditions derived from various proxies.

9.1 Tree-Rings

As a fundamental indicator for climate, ecosystem, and environment, tree-rings have been widely applied in studies of environment and climate change. In certain cases, trees can live for many hundreds of years and in an extraordinary case, like the bristlecone pine, thousands of years. Since tree growth is influenced by climatic conditions such as temperature and precipitation, patterns in tree-ring widths, density, and isotopic composition reflecting annual variations in climate. The study of the relationship between climate and tree growth in an effort to reconstruct past climates is known as dendroclimatology. Not only living trees but also dead or sub-fossil trees can be used to develop tree-ring chronologies. Tree ring records can be combined to create climate records that span a time frame longer than the life of a single tree. The world's longest continuous tree-ring chronology extends over more than 7000 years (Frank et al. 2010).

Z. Zhang, *Multivariate Time Series Analysis in Climate and Environmental Research*, https://doi.org/10.1007/978-3-319-67340-0_9

9.1.1 Field Works

Tree-ring core samples are acquired through fieldworks. First of all, the location of sample sites needs to be chosen carefully, for example, the sample sites should be chosen in upper-elevation tree-line locations and cold mountain valley environments in the case of temperature reconstruction, or in a steep, rocky, south facing slope in the case of precipitation reconstruction. Secondly, enough old trees should be sampled in each sample site. The key of this fieldwork is to collect tree-ring core samples. Since most reliable meteorological records are only about several decades long, trees selected for sampling must be older than meteorological records. After removing a cylinder of wood roughly 4–5 mm in diameter along the radius of a tree, the tree-ring core samples are collected at breast height (about 1.3 m above the ground) from trees by using an increment borer. Finally, these tree-ring core samples are brought back to the laboratory and then are treated according the standard process, including *tree-ring width measurement*, *cross-dating*, and *tree-ring width chronologies*. In order to measure tree-ring width, tree-ring core samples are first treated to be air dried and polished with successively finer grades of sandpaper until annual rings could be distinguished easily although some core samples may be discarded due to irregular rings or missing rings. After this work is done, the tree-ring widths are measured with a precision of 0.01 or 0.001 mm by a LINTAB system or similar system. The tree-ring cross-dating is to match patterns of wide and narrow rings over time to ensure the exact dating of each annual ring to its calendar year of growth. If a similar pattern of wide and narrow rings in living trees could be found on the samples taken from old dead trees, then these samples would be considered cross-dated, so cross-dating can extend tree-ring chronologies based only on living trees much further back in time. On the other hand, it is well known that some growth rings are often missing in trees because of critical hydrological conditions. Cross-dating permits the reliable identification of "false rings" and missing or partially missing rings. Cores with any ambiguities of cross-dating are often excluded from further analysis. Finally, the cross-dated tree-ring series can be quality checked by the COFECHA software. Tree-ring width chronologies are a final step of the standard process. In order to combine samples with differences in growth rates and eliminate non-climatic, age-related growth trends, with the help of the negative exponential function/linear regression function and the spline function, the untreated tree-ring width data are detrended and/or standardized by program ARSTAN. Finally, three categories of tree-ring width chronologies, including the standard chronology, the residual chronology, and the autoregression standard chronology, are obtained.

Tree-ring isotopic data are also a powerful tool for reconstructing climatic conditions with perfect annual resolution. Tree-ring isotopic data have an important advantage that they can be used for climate reconstructions without detrending. To acquire isotopic information, researchers always cut each ring of tree core samples using a scalpel blade under a binocular microscope and extract cellulose, and then measure the $\delta^{18}O$ and $\delta^{13}C$ values of tree-ring cellulose using a stable isotope ratio mass spectrometer (e.g., Thermal Chemical Elemental Analyzer). The $\delta^{18}O$ values can record past changes in temperature, precipitation, relative humidity, or cloud cover since the principal source of water for tree growth is precipitation and the oxygen isotopic ratio ($\delta^{18}O$) is related to temperature and precipitation amount. The $\delta^{13}C$ values show generally a prominent downward trend of 1–2% starting around 1850 AD due to its depletion in atmospheric CO_2 by fossil fuel emissions. It must be removed prior to any climatic reconstruction.

9.1.2 Statistical Analysis

After tree-ring chronologies are obtained in tree-ring laboratories, various advanced statistical methods (von Storch and Zwiers 1999; Marengo et al. 2013; Pascual et al. 2013) are used to analyze these chonologies, develop tree-ring-based climate reconstruction and examine the link between local/regional climate and global climate.

Correlation Analysis is used to examine the relationship between tree-ring parameters and climate factors. Main tree-ring parameters include *tree-ring width, density*, and *isotopic data*. Main climate factors include *temperature, precipitation, runoff, drought, cloud cover*. In the tree-ring study, researchers always first compute the correlation coefficient between each tree-ring parameter and each climate factor, then researchers pick up a tree-ring parameter and a climatic factor such that their correlation coefficient is largest, and finally, researchers take this tree-ring parameter to reconstruct this climate factor that extends back several centuries to millennia. In addition, correlation analysis is also used to examine the link between this reconstructed climate series and global climate.

Linear Regression and *Curvilinear Regression* are main tools to develop a tree-ring-based climate reconstruction. Since tree-ring data have shown to have strong linear correlations with climate factors, with the help of tree-ring width, density, or isotope data, researchers use tree-ring chronologies to reconstruct temperature, precipitation, runoff, drought, cloud cover, and so on that extend back several centuries to millennia. In detail, if researchers find a significant relationship between a tree-ring parameter and a climatic factor (e.g., temperature) in correlation analysis, with the help of linear regression, a climate reconstruction is estimated using this climatic factor as the dependent variable and the tree-ring parameter as the independent variable. Recently, in order to obtain better climate reconstruction, many researchers begin to use curvilinear regression rather than linear regression.

Principal Component Analysis can detect spatial pattern of tree-ring-based climate reconstructions from different regions of China and discover the dominant model of variability.

Power Spectral Analysis, Multi-Taper Method, and *Wavelet Analysis* can be used to analyze interannual to multi-decadal variability contained in the reconstructed climatic series. Researchers often choose red noise as climatic background noise and combine power spectral analysis/multi-taper method/wavelet analysis with *statistical significance test*. By this way, researchers can extract significant cycles in the tree-ring-based climate reconstructions and large-scale climate patterns such as the El

Niño-Southern Oscillation (ENSO), the Asian monsoon variability, and the North Atlantic Oscillation (NAO). If, there are some common significant cycles between the reconstructed climatic series and some large-scale climate pattern, it may suggest a link between local climate and this large-scale climate pattern.

9.2 Ice Cores

Located high in mountains in glaciers and deep in polar ice caps, such as on Antarctica or Greenland, ice has accumulated from snowfall over many centuries. Glacial ice contains dust, air bubbles, or isotopes that can be used to interpret the past climate at the time the snow fell and formed the ice. An *ice core* is a cylinder-shaped sample of ice drilled from a glacier. Analysis of the physical and chemical properties of an ice core can reveal past variations in climate ranging from seasons to hundreds of thousands of years. Ice core records provide the most direct and detailed way to investigate past temperatures and atmospheric composition including greenhouse gases, volcanism, solar activity, dustiness, and biomass burning.

9.2.1 Ice and Isotopes

Snow falls in Antarctica, Greenland, and smaller glaciers and accumulates there. Somewhat compressed old snow is called *firn*. When firn is buried beneath subsequent snowfall, it becomes ice. The transition from snow to firn to ice occurs as the weight of overlying material causes the snow crystals to compress, deform, and recrystallize in more compact form. The firn-ice transition usually occurs at a depth of around 70–100 m. High rates of snow accumulation provide excellent time resolution, and bubbles in the ice core preserve actual samples of the world's ancient atmosphere.

There are different isotopic forms of water, in which hydrogen is either H or D and oxygen is either ^{16}O or ^{18}O. The 99.99% of water is $H_2\,^{16}O$, other forms are particularly $H_2\,^{18}O$ and $HD\,^{16}O$. When water evaporates, a higher concentration of the lighter form of water appears in water vapor. When water vapor condenses, there is an increase in the concentration of the heavier form in the product liquid compared with the original vapor. The isotopic composition of water is introduced as deviations of the concentrations of its heavy components from the standard mean ocean water (SMOW) as follows:

$$\delta^{18}O = \left(\frac{\left(\frac{^{18}O}{^{16}O}\right)_{\text{sample}}}{\left(\frac{^{18}O}{^{16}O}\right)_{\text{standard}}} - 1 \right) \times 1{,}000$$

$$\delta D = \left(\frac{\left(\frac{D}{H}\right)_{\text{sample}}}{\left(\frac{D}{H}\right)_{\text{standard}}} - 1 \right) \times 1{,}000$$

where the notation δ^{18}O stands for $\delta(H_2\,^{18}O)$ for oxygen and the notation δD stands for $\delta(HD\,^{16}O)$ for hydrogen. The units of δ are given in parts per thousand. The SMOW reference contains 997,680 ppm of $H_2^{16}O$, 320 ppm of $HD^{16}O$, and 2000 ppm of $H_2^{18}O$. The isotopic ratios δ^{18}O and δD of the most abundant stable water isotopes in glacier ice are related primarily to the local temperature at the time of precipitation. They vary over the seasons and preserve a signal of annual cycle. At high-accumulation sites, the annual signal of these isotopes may be traced several thousands of years back in time. At lower accumulation sites, diffusion of water molecules in the firn reduces or completely erases the seasonal variation.

The δ^{18}O and δD are used to reconstruct past temperatures at the core site through a simple relation:

$$\delta = AT + B,$$

where A and B are to be determined, δ stands for either δ^{18}O or δD, and T is environmental temperature at the core site. This relation converting isotope measurements to temperature is called the *isotopic paleothermometer*.

Except that δ^{18}O and its counterpart δD has been used as markers, Methane (CH_4), Carbon dioxide (CO_2), and Nitrous oxide (N_2O) are the most prominent trace gases measured on extracted air from ice cores.

9.2.2 Ice Core Samples

Ice cores provide a basis for studying paleoclimate. Paleoclimate information derived from ice cores comes from four principal analyzes including analysis of the stable isotopes of hydrogen and oxygen; analysis of other gases in the air bubbles in the ice; analysis of dissolved and particulate matter in the firn and ice; and analysis of other physical properties such as thickness of the firn and ice.

Many ice cores have been drilled in Greenland and have provided a wealth of information on historical temperatures and cover a maximum time range of up to about 150,000 years before the present. The first ice core with 1390 m in depth was drilled at Camp Century in Greenland from 1963 to 1966. This core was dated back to about 100,000 years before the present, permeated by many fluctuations. The Greenland Ice Core Project (GRIP), a multinational European research project involving eight nations: Belgium, Denmark, France, Germany, Iceland, Italy, Switzerland, and the UK, successfully drilled a 3029 m ice core down to bedrock in central Greenland from 1989 to 1992. The ice core contained records of past climate-temperature, precipitation, gas content, and chemical composition. The Greenland Ice Sheet Project-2 (GISP2) successfully drilled a 3053 m ice core from 1989 to 1993. The GISP2 was located at the highest point on the Greenland ice sheet which is 3208 m above sea level on the ice divide of central Greenland. Since here the ice extends almost vertically down to the bottom, this is a geologically ideal location for drilling. The GISP2 ice core data provide a good evidence for the Medieval Warm Period around 1000 AD and the Little Ice Age from about 1400 to 1850 AD. The North GRIP drilled a 3080 m ice core at near the center of Greenland at 75°N from 1999 to 2003. This was the deepest ice core recovered in the world. The North GRIP ice core data extended back further in time, reaching the Eemian interglacial period prior to the most recent ice age.

Many ice cores have also been drilled at Antarctica. These ice cores have provided longer term records than Greenland and have recently been extended to about 800,000 years before the present. The Vostok ice core at Antarctica provided data for 420,000 years and revealed four past glacial cycles. Drilling stopped just above Lake Vostok. The EPICA Dome C core in Antarctica was drilled at 75° S near Dome C. The core went back 800,000 years and revealed eight previous glacial cycles. Near the Dome F summit in Antarctica at 77°S, the first core was drilled from 1995 to 1996, and it covered a period back to 320,000 years. The second core was drilled from 2003 to 2007, and it extended the climate record of the first core to about 720,000 years. The West Antarctica Ice Sheet Divide (WAIS Divide) drilled a 3405 m ice core in 2011. The WAIS Divide ice core would make up the ice core contain uniquely detailed information on past environmental conditions during the last 68,000 years, such as the atmospheric concentration of greenhouse gases, surface air temperature, wind patterns, the extent of sea ice around Antarctica, and the average temperature of the ocean.

The website

http://www.ncdc.noaa.gov/paleo/icecore/current.html

provides various ice core data at Greenland, Antarctica, and elsewhere.

9.3 Speleothems

The increased markedly interest in speleothems during past several decades reflects the need to provide reliable paleoclimate records from a range of continental settings. Speleothems are mostly calcite, but occasionally aragonite. They are deposited slowly by degassing of cave drip water (i.e., meteoric water-fed drips in caves). Most studies utilize stalagmites rather than stalactites because their simple geometry, rapid growth rates and tendency to precipitate close to isotope equilibrium with the cave drip water, facilitates paleoclimate reconstruction. Variations in δ^{18}O and δ^{13}C of speleothem calcite and δD of water extracted from stalagmite fluid inclusions have been widely recognized as powerful tools to reconstruct the paleoclimate/environment of terrestrial areas.

9.3.1 Oxygen Isotope Ratio

Oxygen isotope data are often used to recover paleotemperature. A relative enrichment in ^{18}O caused by a rapid loss of CO_2 from solution has been observed in fast drip stalactites (Johnson et al. 2006). The oxygen isotope ratio $\delta^{18}O$ of speleothem calcite can be used as an indicator reconstructing past temperature. The $\delta^{18}O$ of speleothem calcite depends not only on the temperature at the time of deposition but also on the $\delta^{18}O$ of the cave seepage waters. In order to evaluate more critically the extent to which drip water $\delta^{18}O$ might reflect cave air temperature, McDermott et al. (2005) analyzed the oxygen isotope data and recent carbonate precipitates for modern cave waters with mean annual temperatures ranging from 2.8 to 26.6 °C from 22 caves. Their study reveals that there is a strong correlation between cave drip water $\delta^{18}O$ and cave air temperature.

The known paleotemperature equations representing a relationship between oxygen isotope and temperature in speleothem are as follows.

(1) Craig equation:

$$T_p = 16.9 - 4.2(\delta^{18}O_{ct} - \delta^{18}O_w) + 0.13(\delta^{18}O_{ct} - \delta^{18}O_w)^2,$$

where T_p is the predicted value of temperature, $\delta^{18}O_{ct}$ is the $\delta^{18}O$ in calcite deposited in speleothems, and $\delta^{18}O_w$ is the $\delta^{18}O$ in cave drip water. This paleotemperature equation is an empirically-derived O isotope-temperature relationship equation.

(2) Grossman-Ku equation:

$$1000\ln\alpha = 2.559(10^6 T^{-2}) + 0.715,$$

where T is in Kelvins and

$$\alpha = (1000 + \delta^{18}O_{CaCO_3}/1000 + \delta^{18}O_{H_2O})\ \text{SMOW},$$

here SMOW is the standard mean ocean water.

(3) Patterson-Smith-Lohmann equation:

$$1000\ln\alpha = 18.56(10^3 T^{-1}) - 33.49,$$

where T is in Kelvins and α is the fractionation factor between calcite ($CaCO_3$) and water.

(4) Kim-Neil equation:

$$\begin{aligned} 1000\ln\alpha &= 18.03(10^3 T^{-2}) - 32.42, \\ 1000\ln\alpha &= 2.76(10^6 T^{-2}) - 3.96, \\ 1000\ln\alpha &= 2.63(10^6 T^{-2}) - 4.04, \end{aligned}$$

where T is in Kelvins and α's are the fractionation factors between calcite ($CaCO_3$) and water, octavite ($CdCO_3$) and water, and witherite ($BaCO_3$) and water, respectively.

9.3.2 *Carbon Isotope Ratio*

Likewise, a relative enrichment in ^{13}C caused by a rapid loss of CO_2 from solution has been observed in fast drip stalactites. The carbon isotope ratio $\delta^{13}C$ of speleothem calcite can be also used as an indicator reconstructing past temperature (Johnson et al. 2006). Carbon isotope variations in speleothems are usually taken to reflect climate-driven paleovegetation signals.

Carbon isotope ratios in speleothems reflect the balance between isotopically light biogenic carbon derived from the soil CO_2 and the heavier carbon dissolved from the limestone bedrock. The downward percolating drip waters acquire calcium carbonate in the soil and host-rocks above a cave. In an open system, continuous equilibration is maintained between the downward percolating drip water and an infinite reservoir of soil CO_2, and the $\delta^{13}C$ of the dissolved species reflects the carbon isotopic ratio of the soil CO_2. In a closed system, the downward percolating drip water loses contact with the soil CO_2 reservoir as soon as carbonate dissolution commences, the CO_2 consumption in the reaction $H_2O + CO_2 = H_2CO_3$ limits the extent of limestone dissolution, and the $\delta^{13}C$ of the host-rock above a cave exerts a strong influence on the carbon isotope ratio of the dissolved inorganic carbon. In the arid region, changes of the $\delta^{13}C$ values of speleothem calcite have been interpreted to reflect climate-driven changes in vegetation type. In many temperate-zone speleothems, $\delta^{13}C$ values is higher than the predicted values to be in equilibrium with expected C3 vegetation (Baker et al. 1997; McDermott et al. 2005). The geochemical criteria to infer reliable paleoclimate information from the $\delta^{13}C$ record of temperate-zone speleothems are essential but they have not been established so far. One promising line of research is to combine trace element and carbon isotope data, several possible mechanisms for changes of the $\delta^{13}C$ can be ruled out. As an example, a study of a 31,000 year old speleothem from New Zealand ruled out a reduction in cave seepage water flow rates as an explanation for elevated $\delta^{13}C$ (Hellstrom and McCulloch 2000).

9.3.3 *Hydrogen Isotope Ratio*

The hydrogen isotope ratio δD of water extracted from stalagmite fluid inclusions is also an indicator of past temperature. A study of a speleothem from Coldwater Cave showed that although the oxygen isotopy ratio $\delta^{18}O$ of calcite was almost constant throughout the growth history of the 83 cm long stalagmite, the hydrogen isotope ratio δD varied by 1.2% (Harmon et al. 1979). Today, drip waters from many caves have been shown to resemble the global meteoric water line of Craig for which $\delta D = 8\delta^{18}O + 10$ (McDermott et al. 2005).

Thermal decrepitation and crushing techniques are two primary techniques of the early studies in use for extracting inclusion water for isotope measurement that are still used today. Both decrepitation and crushing techniques require that the water is recovered crygenically to a liquid nitrogen trap at $-196\,^{\circ}C$ and to raise the temperature to remove all trapped CO_2. The recovered water is stored in a pyrex tube for the measurements of both oxygen and hydrogen isotopes. During this process, some CO_2 with known isotope composition is put into the sample pyrex tube with the water to simulate exchange between H_2O and CO_2 at room temperature until equilibration is reached (Kishima and Sakai 1980; Socki et al. 1999). The CO_2 and water are cryogenically separated at $-75\,^{\circ}C$, and then the water is raised to high temperature to reduce hydrogen using zinc, chromium, manganese, or uranium (Coleman et al. 1982; Tanweer et al. 1988; Donnelly et al. 2001; Stash et al. 2000; Hartley 1980). The isotope composition of the CO_2 and the hydrogen are then measured directly by mass spectrometry.

Further Reading

A. Baker, E. Ito, P.L. Smart, R.F. McEwan, Elevated and variable value of $\delta^{13}C$ in speleothems in a British cave system. Chem. Geol. **136**, 263–270 (1997)

M.L. Coleman, T.J. Shepherd, J.J. Durham, J.E. Rouse, G.R. Moore, Reduction of water with zinc for hydrogen isotope analysis. Anal. Chem. **54**, 993–995 (1982)

F.W. Cruz, X.F. Wang, A. Auler, M. Vuille, S.J. Burns, L.R. Edwards, I. Karmann, H. Cheng, Orbital and millennial-scale precipitation changes in Brazol from speleothem records, in *Past Climate Variability in South America and Surrounding Regions, Developments in Paleoenvironmental Research*, ed. by F. Vimeux et al. (Springer, Berlin, 2009)

C.J. de Cisneros, E. Caballero, B. Andreo, J.J. Durán, Climate variability during the middle-late pleistocene based on stalagmite from Órganos cave (Sierra de Camorra, Southern Spain), in *Hydrogeological and Environmental Investigations*, ed. by B. Andreo et al. (Springer, Berlin, 2015)

M. Deininger, F. McDermott, M. Mudelsee, M. Werner, N. Frank, A. Mangini, Coherency of late holocene European speleothem $\delta^{18}O$ records linked to north Atlantic ocean circulation. Clim. Dyn. (2016, in press)

T. Donnelly, S. Waldron, A. Tait, J. Dougans, S. Bearhop, Hydrogen isotope analysis of natural abundance and deuterium-enriched water by reduction over chromium on-line to a dynamic dual inlet isotope-ratio mass spectrometer, rapid commun. Mass Sp. **15**, 1297–1303 (2001)

I.J. Fairchild, A. Borsato, A.F. Tooth, F. McDermott, S. Frisia, C.J. Hawkesworth, Y. Huang, B. Spiro, Controls on trace element (Sr-Mg) compositions of carbonate cave waters: implications for speleothem climate records. Chem. Geol. **166**, 255–269 (2000)

D. Frank, J. Esper, E. Zorita, R. Wilson, A noodle, hockey stick, and spaghetti plate: a perspective on high-resolution paleoclimatology. WIREs Clim. Change **1**, 507–C516 (2010)

D. Genty, A. Baker, B. Vokal, Intra- and inter-annual growth rate of modern stalagmites. Chem. Geol. **176**, 191–212 (2001)

E.L. Grossman, T.L. Ku, Oxygen and carbon isotope fractionation in biogenic aragonite: temperature effects. Chem. Geol. **59**, 59–74 (1986)

R. Harmon, H.P. Schwarcz, J.R. O'Neil, D/H ratios in speleothem fluid inclusions: a guide to variations in the isotopic composition of meteoric precipitation? Earth Planet. Sci. Lett. **42**, 254–266 (1979)

P.E. Hartley, Mass spectrometric determination of Deuterium in water at natural levels. Anal. Chem. **52**, 2232–2234 (1980)

J.C. Hellstrom, M.T. McCulloch, Multi-proxy constraints on the climatic significance of trace element records from a New Zealand speleothem. Earth Planet. Sci. Lett. **179**, 287–297 (2000)

C.H. Hendy, The isotopic geochemistry of speleothems I. The calculation of the effects of the different modes of formation on the isotopic composition of speleothems and their applicability as palaeoclimatic indicators. Geochim. Cosmochim. Ac. **35**, 801–824 (1971)

K. Holmgren, J.A. Lee-Thorp, G.R.J. Cooper, K. Lundblad, T.C. Partridge, L. Scott, R. Sithaldeen, A.S. Talma, P.D. Tyson, Persistent millennial-scale climatic variability over the past 25,000 years in Southern Africa. Quat. Sci. Rev. **22**, 2311–2326 (2003)

K.R. Johnson, C. Hu, N.S. Belshaw, G.M. Henderson, Seasonal trace-element and stable-isotope variations in a Chinese speleothem: the potential for high-resolution paleomonsoon reconstruction. Earth Planet. Sci. lett. **244**, 394–407 (2006)

J. Kang, R. Larsson, What is the link netween temperature and carbon dioxide levels? A granger causality analysis based on ice core data. Theor. Appl. Climatol. **116**, 537–548 (2014)

S.T. Kim, J.R. O'Neil, Equilibrium and nonequilibrium oxygen isotope effects in synthetic carbonates. Geochim. cosmochim. Ac. **61**, 3461–3475 (1997)

N. Kishima, H. Sakai, Oxygen-18 and deuterium determination on a single water sample of a few milligrams. Anal. Chem. **52**, 356–358 (1980)

M. Küttel, E.J. Steig, Q. Ding, A.J. Monaghan, D.S. Battisti, seasonal climate information preserved in west Antarctic ice core water isotopes: relationships to temperature, large-scale circulation, and sea ice. Clim. Dyn. **39**, 1841–1857 (2012)

J. Marengo, L. Borma, D. Rodriguez, P. Pinho, W. Soares, L. Alves, Recent extremes of drought and flooding in Amazonia: vulnerabilities and human adaptation. Am. J. Clim. Change **2**, 87–96 (2013)

F. McDermott, H. Schwarcz, P.J. Rowe, Isotopes in speleothems, in *Isotopes in Palaeoenvironmental Research*, ed. by M.J. Leng (Springer, Berlin, 2005)

A. Pascual, F. Valero, M. Martin, C. Garcia-Legaz, Spanish extreme winds and their relationships with Atlantic large-scale atmospheric patterns. Am. J. Clim. Change **2**, 23–35 (2013)

W.P. Patterson, G.R. Smith, K.C. Lohmann, Continental paleothermometry and seasonality using the isotopic composition of aragonitic otoliths of freshwater fishes. in *Continental Climate Change from Isotopic Records*, vol. 78. ed. by P. Swart, K.C. Lohmann, J. McKenzie, S. Savin (Amer. Geophys. Union Monogr., 1993) pp. 191–202

D. Rapp, *Measurements, Interpretation, and Models* (Springer, Berlin, 2012)

C. Scarchilli, M. Frezzotti, P.M. Ruti, Snow precipitation at four ice core sites in East Antarctica: provenance, seasonality and blocking factors. Clim. Dyn. **37**, 2107–2125 (2011)

Z. Sharp, *Principles of Stable Isotope Geochemistry* (Pearson Prentice Hall, Upper Saddle River, 2007)

R.A. Socki, C.S. Romanck, E.K. Gibson, On-line technique for measuring stable oxygen and hydrogen isotopes from microlitre quantities of water. Anal. Chem. **71**, 2250–2253 (1999)

O.S. Stash, R. Drimmie, J. Morrison, S.K. Frape, A.R. Heemskerk, W.A. Mark, On-line D/H analysis for water, natural gas, and organic solvents by manganese reduction. Anal. Chem. **72**, 2664–2666 (2000)

A. Svensson, Ice cores, in *Encyclopedia of Scientific Dating Methods*, ed. by J.W. Rink, J.W. Thompson (Springer, Dordrecht, 2015)

A. Tanweer, G. Hut, J.O. Burgman, Optimal conditions for the reduction of water to hydrogen by zinc for mass spectrometric analysis of deuterium content. Chem. Geol. **73**, 199–203 (1988)

M. Thamban, S.S. Naik, C.M. Laiuraj, A. Chaturvedi, R. Ravindra, in *Antarctic Climate Variability During the Past Few Centuries Based on Ice Core Records from Coastal Dronning Maud Land*

and Its Implications on the Recent Warming, ed. by R. Sinha, R. Ravindra. Earth system processes and disaster management, Society of Earth scientists series 1 (Springer, Berlin, 2013)
E.R. Thomas, T.J. Bracegirdle, Precipitation pathways for five new ice core sites in Ellsworth Land, West Antarctica. Clim. Dyn. **44**, 2067–2078 (2015)
P.C. Treble, W.F. Budd, P.K. Hope, P.K. Rustomji, Synoptic-scale climate patterns associated with rainfall $\delta^{18}O$ in Southern Australia. J. Hydrol. **302**, 270–282 (2005)
P. Vallelonga, A. Svensson, Ice core archives of mineral dust, in *Mineral Dust: A Key Player in the Earth System*, ed. by P. Knippertz, J.B.W. Stuut (Springer, Berlin, 2014)
H. von Storch, F.W. Zwiers, *Statistical Analysis in Climate Research* (Cambridge University Press, Cambridge, 1999)
Z. Zhang, J.C. Moore, Comment on "significance tests for the wavelet power and the wavelet power spectrum" by Ge. Ann. Geophys. **30**, 1743–1750 (2012)
Z. Zhang, J.C. Moore, A. Grinsted, Haar wavelet analysis of climatic time series. Int. J. Wavelets, Multiresol. Inf. Process. **12**, 1450020 (2014)
Z. Zhang, J.C. Moore, New significance test methods for Fourier analysis of geophysical time series. Nonlin. Process. Geophys. **18**, 643–652 (2011)
Z. Zhang, Tree-rings, a key ecological indicator of environment and climate change. Ecol. Ind. **51**, 107–116 (2015)

Chapter 10
Strategies for Climate Change Mitigation

Climate change and its social, environmental, and ethical consequences are widely recognized as the major set of interconnected problems facing human societies. Its impacts and costs will be large, serious, and unevenly spread, globally for decades. Future climate change will largely be determined by strategies society makes about emissions. Lower emissions of greenhouse gases mean less future warming and less severe impacts. In this chapter, we start from three popular quantitative assessment methods, and then give some important results on carbon emissions reduction potentials which are derived directly from these assessment methods. In the second half of this chapter, we will focus on some novel approaches to mitigating climate change, including carbon capture and storage and geoengineering.

10.1 Assessment Methods and Tools

In this section, we will introduce basic quantitative assessment methods and tools which are widely used in economics, policy and governance of climate change.

10.1.1 Data Envelopment Analysis

Data envelopment analysis (DEA) is a very powerful service management and bench-making technique to evaluate carbon emissions reduction efficiency of different organizations (i.e., so-called decision-making units (DMUs)). DEA measures the relative performance of DMUs using a linear optimization technique. The most important models in DEA are Charnes-Cooper-Rhodes (CCR) DEA models. It can be divided into the input-oriented CCR DEA models and the output-oriented CCR DEA models.

Z. Zhang, *Multivariate Time Series Analysis in Climate and Environmental Research*, https://doi.org/10.1007/978-3-319-67340-0_10

Assume that there are n DMUs, and $\mathbf{y}_j = (y_{1j}, \ldots, y_{qj})$ and $\mathbf{x}_j = (x_{1j}, \ldots, x_{mj})$ are the output and the input of the jth DMU. For the particular DMU_0 with the input $\mathbf{x}_0$ and the output $\mathbf{y}_0$ being evaluated, the input-oriented CCR DEA model is

$$(L_{CCR}^{ip}): \quad \max\ (\boldsymbol{\mu}, \mathbf{y}_0) \quad \text{subject to}$$
$$\begin{cases} (\boldsymbol{\mu}, \mathbf{y}_j) - (\boldsymbol{\nu}, \mathbf{x}_j) \le 0 \quad (j = 1, \ldots, n), \\ (\boldsymbol{\nu}, \mathbf{x}_0) = 1, \\ \boldsymbol{\mu} \ge 0, \quad \boldsymbol{\nu} \ge 0, \end{cases}$$

and the dual model of the input-oriented CCR DEA model is

$$(L_{CCR}^{id}): \quad \max\ \theta \quad \text{subject to}$$
$$\begin{cases} (\boldsymbol{\lambda}, \mathbf{y}^r) \ge y_{r0} \quad (r = 1, \ldots, q), \\ (\boldsymbol{\lambda}, \mathbf{x}^i) \le \theta x_{i0} \quad (i = 1, \ldots, m), \\ \boldsymbol{\lambda} \ge 0, \quad \theta > 0, \end{cases}$$

where $\mathbf{y}^r = (y_{r1}, \ldots, y_{rn})\ (r = 1, \ldots, q)$, $\mathbf{x}^i = (x_{i1}, \ldots, x_{in})\ (i = 1, \ldots, m)$, and $\boldsymbol{\lambda} = (\lambda_1, \ldots, \lambda_n)$.

The output-oriented CCR DEA model is

$$(L_{CCR}^{op}): \quad \max\ (\boldsymbol{\nu}, \mathbf{x}_0) \quad \text{subject to}$$
$$\begin{cases} (\boldsymbol{\nu}, \mathbf{x}_j) - (\boldsymbol{\mu}, \mathbf{y}_j) \ge 0 \quad (j = 1, \ldots, n), \\ (\boldsymbol{\mu}, \mathbf{y}_0) = 1, \\ \boldsymbol{\nu} \ge 0, \quad \boldsymbol{\mu} \ge 0 \end{cases}$$

and the dual model of the output-oriented CCR DEA model is

$$(L_{CCR}^{od}): \quad \max\ \varphi \quad \text{subject to}$$
$$\begin{cases} (\boldsymbol{\lambda}, \mathbf{x}^i) \le x_{i0} \quad (i = 1, \ldots, m), \\ (\boldsymbol{\lambda}, \mathbf{y}^r) \ge \varphi y_{r0} \ (r = 1, \ldots, q), \\ \boldsymbol{\lambda} \ge \mathbf{0}. \end{cases}$$

10.1.2 Risk Assessment

The risks that a changing climate presents are now widely acknowledged. The risk assessment provides a systematic procedure for estimating the probability of harm to, or from, the environment, the severity of harm, and uncertainty. The key point of risk assessment is to solve the multiobjective optimization problems by Haimes–Lasdon–Wismer ϵ−constraint method.

Let $f_i : \mathbb{R}^n \to \mathbb{R}\ (i = 1, \ldots, N)$ and $c_i : \mathbb{R}^n \to \mathbb{R}\ (i = 1, \ldots, m)$ all be continuously differentiable functions. The optimization

$$\min_{\mathbf{x}} \{f_1(\mathbf{x}), \ldots, f_N(\mathbf{x})\} \quad \text{subject to} \quad c_i(\mathbf{x}) \leq 0 \quad (i = 1, \ldots, m) \qquad (10.1.1)$$

is an multiobjective optimization model with N objective functions $f_i(\mathbf{x})$ and m constraint conditions $c_i(\mathbf{x}) \leq 0$. Haimes–Lasdon–Wismer $\epsilon-$constraint method is to rewrite this multiobjective optimization into an equivalent single-objective optimization model with $N - 1\epsilon-$constraint conditions as follows:

$$\min_{\mathbf{x} \in \mathbf{X}} f_i(\mathbf{x}) \quad \text{subject to} \quad f_j(\mathbf{x}) \leq \varepsilon_j \quad (j \neq i;\; j = 1, \ldots, N). \qquad (10.1.2)$$

where ϵ_j $(j \neq i,\; j = 1, \ldots, m)$ are variables and

$$X = \{\mathbf{x} \in \mathbb{R}^n \mid c_i(\mathbf{x}) \leq 0 \;\; (i = 1, \ldots, m)\},$$
$$\varepsilon_j = \min_{\mathbf{x} \in X} f_j(\mathbf{x}) + \epsilon_j, \quad \epsilon_j > 0 \;\; (j \neq i;\; j = 1, \ldots, N).$$

From (10.1.2), it is seen that $N - 1$ objective functions in (10.1.1) are replaced by $N - 1\epsilon-$constraint conditions, i.e., one objective $f_i(\mathbf{x})$ is the principal objective, and the others $f_j(\mathbf{x})$ $(j \neq i;\; j = 1, \ldots, N)$ are the constraining objectives.

10.1.3 Life Cycle Assessment

Life cycle assessment (LCA) is widely used to systematically record and analyze carbon emissions and their impacts throughout the entire life cycle of a product. This involves an end-to-end analysis of the product. The LCA considers all raw materials, transports, production processes, usage, and disposal of the product. In general, the LCA to estimate carbon essimions consists of four steps:

(1) Goal definition and scoping: the LCA's purpose is to assess the total carbon emissions of certain product (i.e., product carbon footprints). The scope is the life cycle of certain product;
(2) Life cycle inventory: quantifying the energy and raw material inputs and related carbon emissions associated with each stage of production;
(3) Impact analysis: assessing the impacts of carbon emissions associated with energy and raw material inputs;
(4) Improvement analysis: evaluating opportunities to slow down energy consumption and reduce carbon emissions at each stage of the product life cycle.

10.2 Carbon Emissions Reduction

Reduction of fossil carbon emissions from diverse sectors is central to efforts to reduce fossil carbon emissions due to the large material's flows they process and to the large quantities of energy they consume. If the energy is used inefficiently, this will lead to higher carbon emission levels. Data envelopment analysis, risk assessment and life cycle assessment in Sect. 10.1 have widely used to calculate product carbon footprint and analyze carbon emissions reduction potentials. In this section, we will list some important results derived from these assessment methods.

10.2.1 Industrial Sector

Carbon emissions are generated in almost all activities of industrial sectors, including extraction of materials from the Earth's crust, production, procurement, inventory management, order processing, transportation, usage, and end-of-life management of used products.

The *iron and steel industry* is a major source of anthropogenic CO_2 emissions. Among different pathways to achieve CO_2 emissions reduction, one pays more attention to industrial symbiosis, a system's approach which is designed to build upon win-win synergies between environment and economic performance through physical sharing of waste energy, exchanging of waste materials, by-products and infrastructure sharing among co-located entities.

Cement is the basic and most widely used building material in civil engineering. Ishak and Hashim (2015) reviewed carbon emissions of all stages of cement manufacturing including raw materials preparation, clinker production, combustion of fuels in the kiln, and the production of the final cement products. Results showed that 90% of CO_2 emissions from cement plants were generated from clinker production while the remaining 10% was from raw materials preparation and the finishing stage of producing cement. Liu et al. (2015), Crossin (2015), and Yang et al. (2015) showed that the use of supplementary cementations materials, such as fly ash, silica fume, copper slag, sewage sludge, and ground-granulated blast furnace slag, are often promoted as ways to reduce carbon emissions.

The *rubber industry* is also a source of carbon emissions. Carbon emissions in the rubber industry are closely connected to energy consumption. Among rubber product manufacturing processes, the rubber material milling process, the extruding process, and the rolling process all have a relatively high electricity consumption rate. Dayaratne and Gunawardana (2015) investigated three rubber-band manufacturing factories. Results showed that rubber manufacturing should adapt cleaner manufacturing model and implement energy-efficient measures to achieve sustainable production, and the corresponding financial barriers can be solved through clean development mechanism.

The global primary *aluminum industry* is responsible for 1% of global carbon emissions. China is the biggest aluminum producer in the world. China's aluminum industry will be confronted with restrictions on the high-quality bauxite import and degradation of domestic bauxite quality. For futher CO_2 emission reduction, China will be expected to modernize existing smelters and eliminating smaller and backward smelters, and accelerate technology evolution of the aluminum industry.

The *paper industry* is one of the most energy-intensive sectors and one of the largest carbon emissions among manufacturing industries with a direct emission of about 40 Mt of CO_2 per year in Europe. Conventional manufacturing of paper consists of processing wood fiber streams into planar structures. With the development of future manufacturing concepts (FMCs), the final paper product will have a tailor-made layered structure. Leon et al. (2015) quantified carbon emissions reduction potentials in super-calendered-(SC) paper production and lightweight-coated (LWC) paper production. Results showed that carbon emissions in the FMC strategies applied to the SC paper were reduced by 23% with a total of 10.7 gCO_2-eq emissions saved per square meter of SC paper while carbon emissions in the FMC strategies applied to the LWC paper were reduced by 20% with a total of 19.7 gCO_2-eq emissions saved per square meter of LWC paper. This means that the environmental benefits gained through the application of the FMC manufacturing in the paper industry are significant.

In the *chemical fiber industry*, natural gas, oil, and other low-molecular weight raw materials are used to synthesize polymers through chemical addition or condensation reactions. The polymers may then be spun into synthetic fibers. So energy saving and carbon emissions reduction are important and can provide immense benefits. Lin and Zhao (2015) investigated the chemical fiber production in China. Results showed that GDP, R&D expenditure, and energy price were the main factors which exert a great impact on energy consumption in the chemical fiber industry, and predicted the energy-saving potential for China's chemical fiber industry.

In the *hydraulic industry*, hydraulic presses are machine tools which use a hydraulic cylinder to generate compressive forces which are commonly used for forging, molding, blanking, punching, deep drawing, and metal forming operations in many manufacturing fields. Energy losses within hydraulic systems with high pressure and large flows are serious. Zhao et al. (2015) divided hydraulic press systems into electric-mechanical energy conversion units, mechanical-hydraulic energy conversion units, hydraulic-hydraulic energy conversion units, hydraulic-mechanical energy conversion units, mechanical to deformation energy conversion units, and thermal to thermal energy conversion units. They proposed an analytical approach for calculating energy flows in large- and medium-sized hydraulic press systems. Results showed that energy storage and recycling units should be considered to be included in hydraulic presses in order to reduce carbon emissions.

In the *methanol production industry*, Taghdisian et al. (2015) proposed a so-called green-integrated methanol case (GIMC), which is an eco-design method for sustainable production of methanol by implementing a multiobjective optimization CO_2-efficiency model that was formulated to maximize methanol production and minimize carbon dioxide emissions. The source of CO_2 in GIMC is the methanol plant itself where the injected CO_2 is supplied from reformer flue gas. Results showed that using the multiobjective approach in the GIMC would lead to the carbon emissions reduction.

The carbon linkage caused by the intermediate trade among industrial sectors has typically been ignored. Zhao et al. (2015) integrated the environmental input–output model with the modified hypothetical extraction method to investigate the carbon linkage among industrial sectors in South Africa. They gave the total carbon linkage of industrial systems, total backward carbon linkage, and total forward carbon linkage in South Africa in 2005. They suggested that adjusting industrial structure, improving energy efficiency, developing new energy, and establishing clean energy mechanisms are conducive to reduce the carbon emission in South Africa.

10.2.2 Agriculture Sector

The agriculture sector contributes substantial quantities of CO_2 and methane emission. Improved energy use efficiency in agriculture, as one of the principal requirements of sustainable development, can reduce carbon emission, help to minimize climate change risks, and prevent the destruction of natural resources.

In the *mushroom production sector*, Ebrahimi and Salehi (2015) studied Iran's button mushroom production. They gave the average total energy input and the total carbon emissions of mushroom production for efficient and inefficient units. Results showed that the carbon emissions of mushroom production was reduced in efficient units compared with inefficient units, and management of diesel fuel and electricity consumption in all mushroom production facilities helped the more efficient systems to achieve such reductions.

In the *lucerne production sector*, Mushtaq et al. (2015) studied the irrigated lucerne crops in Australia. They presented a novel-integrated assessment framework to analyze the potential trade-offs among water savings, energy consumption, carbon emission, and economic costs/benefits. Results showed that efficient sprinkler technology not only saved water but also reduced energy use and carbon emissions.

In the *cotton production sector*, Visser et al. (2015) investigated Australia's "farm to ship" cotton production. They estimated the total carbon emission of producing a bale of cotton from the farm to the ship's side. Results showed that if the waste is broadcasted and incorporated into the soil at the farm level, then it could generate a 27% reduction in the farm emissions footprint and a 15% reduction in the whole farm to ship carbon footprint.

In the *livestock production sector*, Riano and Gonzalez (2015) estimated carbon emissions reduction of a swine manure treatment plant in Spain. Results showed that compared with conventional storage in anaerobic tanks, implementing the manure treatment plant could lead to a total annual carbon emission reduction of 62%, including CO_2 emission reduction by 72%, CH_4 emission reduction by 69%, and no change of N_2O emission.

In the *fisheries sector*, fuel use intensity of fisheries varies with regard to target species, equipment employed, region of fishing, technologies used, skipper behavior, and other factors. Park et al. (2015) measured fuel inputs to purse seining vessels targeting primarily skipjack and yellow fin tuna. Results showed that the use of fish aggregating devices (FADs) in purse seine fisheries for tuna was found to be inversely correlated with efficiency, going against conventional logic that FAD use improves efficiency.

Scientific regulation of carbon flows under conservation *agriculture tillage* is of great significant for mitigating carbon emissions and for increasing carbon sequestration potential in soil. Chen et al. (2015) investigated conventional tillage without residue retention, conventional tillage with residue retention, rotary tillage with residue retention, and no-till with residue retention. They gave the annual increase in rate of soil organic carbon stocks and the annual carbon emissions under these four types of agricultural production. Results showed that widespread adoption of conservation tillage would be beneficial in the reduction of carbon emissions.

10.2.3 The Building Sector

The *building sector* contributes a quarter of the global total carbon emissions. Although the construction phase in a building's life cycle is relatively short, the density of the carbon emissions in the construction phase is higher than that in the operations and maintenance phases. In the building sector, carbon emissions embodied in the manufacturing of material and the transformation of energy into products for the construction. Improved energy efficiency standards and strict control of the increase in urban civil building floor areas will be the most effective ways to reduce carbon emissions in this sector. Ma et al. (2015) showed that although the building life span is also an important factor for carbon emissions, its influence is less sensitive than improved technology and energy efficiency standards. To monitor, evaluate, and forecast carbon emissions for building construction projects better, Kim et al. (2015) developed an integrated carbon dioxide cost and schedule management system. This system can support faster and more accurate evaluation and forecasting of the project performance and CO_2 emissions based on the construction schedule.

10.2.4 The Transportation Sector

In the transportation sector, Wang et al. (2015) proposed an empirical method to estimate the total carbon emissions from the raw materials production process, material transportation process, and onsite construction process for different project types in highway construction (e.g., subgrade, pavement, bridge, and tunnels). Results showed that over 80% of the carbon emissions are generated from the raw materials production process. In order to reduce these emissions, low fossil carbon systems in material's production are preferred.

10.2.5 The Household Sector

The need of the household sector to reduce the energy use and carbon emissions has been emphasized recently. A large proportion of energy consumption and associated carbon emissions is from the household sector. Zhang et al. (2015) and Han et al. (2015) studied main factors influencing household carbon emissions. Results showed that household income is the most important contributor to the difference of household carbon emissions, and its positive effect increases as household carbon emissions rise; household house ownership and deposits contribute little to household carbon emissions, while household car ownerships contribute more; young people and children will emit more household carbon emissions than adults, and the employed persons emit more than the unemployed or retired persons; and education increases household carbon emissions overall but mainly at the low quintiles.

10.2.6 Low-Carbon Energy

Mainstream low-carbon renewable energy sources include biomass, wind power, hydropower, solar power, ocean thermal, wave, tidal, and geothermal energy sources. For biomass, Muench (2015) thoroughly analyzed the greenhouse gas mitigation potential of biomass systems for electricity generation. He showed that electricity from biomass can be an appropriate strategy for greenhouse gas mitigation in the European Union and recommended to promote the employment of dedicated and non-dedicated lignocellulosic biomass with thermochemical conversion because these biomass systems yield the highest carbon mitigation. For wind power, Hacatoglu et al. (2015) introduced a new approach to assess the environmental sustainability of wind-battery systems. Comparing a wind-battery system with a gas-fired power plant, they showed that a wind-battery system could produce fewer potential global warming, stratospheric ozone depletion, air pollution, and water pollution impacts.

Brazil is undoubtedly a country with considerable renewable energy generation capacity. The structure of the Brazilian energy matrix defines Brazil as a global leader in power generation from renewable sources. Guerra et al. (2015) showed that the current composition of Brazilian energy matrix has outstanding participation of hydropower, even through Brazil has great potential for the exploitation of other renewable energy sources such as wind, solar, and biomass.

10.3 Carbon Capture, Transport, Utilization, and Storage

Coal is the most abundant energy resource and more than 38% of the world's electricity supply is derived from coal-fired power plants. Since coal offers many advantages, coal will remain a significant source of energy for several decades. Due to the carbon-rich character of coal, flue gas emission from coal-fired power plants is a major source for the release of anthropogenic carbon dioxide into the atmosphere, which leads to global warming and climate change. Carbon capture and storage (CCS) is a process consisting of carbon dioxide separation from coal-fired power plants and other industrial sources, transport to a storage location, and long-term isolation from the atmosphere.

10.3.1 Carbon Capture

Carbon capture is the first stage of CCS. There are three carbon capture technologies including pre-combustion capture, post-combustion capture, and oxy-combustion capture.

(i) Pre-Combustion Capture Process

Pre-combustion capture process is based on the coal gasification process and to use membranes to separate hydrogen and carbon dioxide in the coal-derived synthesis gas. Finally, the CO_2-free gas, which is now composed almost entirely of hydrogen, is used in a gas turbine generator to produce electricity. This capture process requires a significant quantity of additional energy to generate the heat required for coal gasification.

Gasification of coal is a process that converts coal into carbon monoxide and hydrogen by reaction of coal at high temperatures with a controlled amount of oxygen and/or steam, then a reaction between the carbon monoxide and the high-temperature steam will product more hydrogen and carbon dioxide. One advantage of gasification technology is the possibility to process lower grade coals which are more widely available than the high grade coals.

Integrated gasification combined cycle (IGCC) is a super clean and high efficiency power generation technology. This technology integrates with coal gasification and gas turbine-combined cycle. IGCC technology has the highest potential to capture carbon dioxide with lowest penalties in term of energy efficiency and capital and operational costs and is advancing toward the target of approaching zero emission of carbon dioxide. Introduction in the IGCC-based schemes of pre-combustion carbon capture step can significantly reduce carbon dioxide emission. Its carbon capture rate is higher than 90%.

Co-gasification of coal with renewable energy source coupled with carbon capture paves the way toward zero or negative emission power plants. This technology requires low ash fusion temperature, so it can increase the cold gas efficiency and reduce the oxygen consumption.

Underground coal gasification (UCG) technology was proposed recently. UCG can provide an economic approach to utilize coal reserves which due to great seam depths and complex geological boundary conditions are not exploitable by conventional coal mining. The simplest route to the exploitation of UCG for power generation is to adopt a UCG-based stem cycle, termed as *Integrated Underground Gasification Steam Cycle*. The overall thermal efficiency of the steam cycle is only 25.2% with CCS, primarily due to the significant amount of energy spent on the air separation unit to produce the oxygen required for gasification of coal.

Pre-combustion capture process is also combined with the coal liquefaction. *Liquefaction* of coal is a process that produces liquid fuels from coal. In detail, coal is gasified to produce synthesis gas and then is catalytically treated in a Fischer–Tropsch process to produce liquid fuels, such as gasoline and diesel. During this process, the excess carbon in coal is emitted in the form of CO_2 which will be captured by pre-combustion capture technology.

(ii) Post-Combustion Capture Process

In a modern coal-fired power plant, pulverized coal is mixed with air and burned in a furnace or boiler. The heat released by combustion generates steam which drives a turbine generator. *Post-combustion carbon capture process* is currently the most developed and popular technique in pulverized coal plants. This technology can be retrofitted at relatively low cost to existing coal-fired power stations and allows the combustion process to kept relatively unchanged. However, when a power plant is retrofitted with a post-combustion capture process, the net efficiency decreases by a significant 19.4%. The most mature post-combustion capture technology involves chemical absorption of carbon dioxide from the exhaust gases. Amine and chilled ammonia are likely to be the most cost-effective solvents. The amine-based system consists of an absorber where carbon dioxide is removed and a regenerator where carbon dioxide is released and the original solvent is recovered. Providing energy for solvent regeneration significantly reduces the plant's net efficiency. In a chilled ammonia-based system, carbon dioxide is absorbed at $(0 - 20\,°C)$ and released at above $100\,°C$. In addition, as a physical approach, membranes are often used to selectively separate carbon dioxide from other components of a flue gas. Their selectivity to CO_2 over N_2 determines the purity of the captured CO_2 stream.

(iii) Oxy-Combustion Capture Process

Oxy-combustion capture process is one of the three promising technologies designed to support CCS from coal-fired power plants. Oxy-combustion is to replace the air (O_2, N_2) by a mixture of pure O_2 and recycled flue gas. The pure O_2 is conventionally supplied by an air separation unit which cryogenically separates O_2 from N_2 from the air. The principal attraction of oxy-combustion capture is to avoid the need for a costly post-combustion CO_2 capture system, however, it requires an air separation unit to generate the relatively pure (95–99%) oxygen needed for combustion. Currently, a commercial oxy-fired operation is not expected to become economic until a carbon price (or equivalent) is legislated.

Using the above-carbon capture technologies, the captured CO_2 mixture contains many impurities such as N_2, H_2, water, SO_x, and fly ash. These impurities will change the thermodynamic properties of the CO_2 mixtures and then affect the transport of captured CO_2. For example, the N_2 can effect the CO_2 transport process by its low boiling point. A small amount of N_2 can change flow conditions from single phase flow to two-phase flow. A 2% concentration of H_2 in CO_2 can reduce the molar density up to 25% compared to pure CO_2 (Sanchez-Vicente et al. 2013). The presence of water in the CO_2 stream can form carbonic acid or hydrate when CO_2 dissolves in water with dispersed water droplet in CO_2 fluid being saturated; the water also reacts with other acidic compounds to form acids (e.g., H_2SO_3, H_2SO_4) which may produce a durability risk due to internal corrosion damage of steel pipelines (Sim et al. 2014).

These impurities can be controlled by the *air pollution control devices* during the stage of CO_2 capture. In general, approximately $80-95\%$ of the SO_2 and 50% of the SO_3 is removed by *wet flue gas desulfurization* scrubbers. Mercury concentrations is also controlled by a *similar wet flue gas desulfurization* system. The NO_x is controlled by *physical membrane*. Fly ash is collected and removed by *electrostatic precipitations*, and the water content can be lowered by *gas conditioning* (Lee et al. 2009; Rubin et al. 2012).

10.3.2 Transport of CO_2

The transport of CO_2 is the second stage of CCS. When the captured CO_2 is transported to geological storage sites, trucks, pipelines, and ships are three options for CO_2 transport.

Truck transport has relatively large leakage risk, high transport costs, and only a relatively small amount is transported per load, so it is not suitable for large-scale CCS projects (Ming et al. 2014).

Pipeline transport is considered to be the most reliable transportation method. Industry has more than 40 years of experience with pipeline transportation of CO_2. Most of that CO_2 was transported for usage in enhanced oil recovery field. The main technical problems involve pipeline integrity, flow assurance, safety, and operational considerations.

Ship transport may be economically attractive for long distance CO_2 transport for sea storage sites. The ship-based transport of CO_2 would require that CO_2 is compressed or liquefied. Furthermore, a liquefied-CO_2 transport ship would have to be capable of processing boil-off gas while at sea. For a low storage temperature, more of liquid is transported and the cost of re-liquefaction is reduced. However, extra cost is incurred for liquefaction. For a high storage temperature, while the energy costs are reduced, the cost of tank manufacture will be higher and less CO_2 will be transported. For ocean storage sites, one option is the CO_2 injection at great depth where it dissolves or forms hydrates or heavier than water plumes that sinks at the bottom of the ocean. It may cause ocean acidification and threat ocean ecosystem, so this option is not considered again (Amouroux et al. 2014). Another option is the CO_2 injection at geologic structures beneath the continental shelf. Similar to geological storage sites in land, it is considered to be of low risk.

10.3.3 Geological Storage of CO_2

Geological Storage of CO_2 is the most important stage of CCS. It entails injecting CO_2 emitted from fossil fuel burning power stations and factories into underground geological structures.

(i) Storage in Deep Saline Aquifers

Deep saline aquifers possess much larger storage capacities with widespread distributions than oil and gas reservoirs and coal seams. Moreover, deep saline aquifers have greater regional coverage, so they may be possibly near many CO_2 emission sites. The safe and reliable geological storage sites depend on (Cooper 2009; Warwick et al. 2013):

(a) adequate porosity and thickness for storage capacity, and permeability for injection;
(b) a satisfactory sealing caprock which ensures the containment of appropriate fluids;
(c) a stable geological environment which avoids compromising the integrity of the storage site;
(d) the depth of the storage reservoir which must be larger than 914 m which ensures that the CO_2 would be in a supercritical state with high density, low viscosity, good fluidity, and then minimize the storage volume and easily flow within pores or fractures in rock masses.

Deep aquifers are geologic layers of porous rock that are saturated with brine and are located at 700–3000 m below ground level. It was estimated that the global deep aquifers can store about 10,000 billion tons of CO_2 (Silva et al. 2015). The capability of deep aquifers to store CO_2 is controlled by the depositional environment, structure, stratigraphy, and pressure/temperature conditions. At the same time, injection of large volumes of CO_2 into the deep saline aquifers can perturb the subsurface environment, leading to physical geochemical, and biogeochemical changes of geological reservoirs.

The main CO_2 trapping mechanisms in deep aquifers include *hydrodynamic trapping*, *solubility trapping*, and *geochemical trapping*. In hydrodynamic trapping, the CO_2 is held within porous formations below a caprock of low permeability. The CO_2-isolated blobs are the size of the pores of the rocks, tens to hundreds of micrometers. In solubility trapping, the CO_2 is dissolved into the groundwater, where its solubility decreases as temperature and salinity increase. In geochemical trapping, the CO_2 reacts with natural fluids and minerals in the subsurface, which leads to the safest and most effective approach to permanently trap CO_2 for a long time. The geochemical trapping process is significantly affected by temperature, pressure, salinity, aquifer thickness, tilt angle, anisotropy, aquifer layers, as well as the mineral composition of the formation rock (Zhao et al. 2014; Streimikiene 2012; Silva et al. 2015).

(ii) Storage of CO_2 with Enhanced Industrial Production

Since the storage of CO_2 with enhanced industrial production can help to reduce CO_2 emissions and enhance industrial production at the same time, it has a great potential to enable large-scale CO_2 storage at reasonable cost. Mainly, enhanced industrial production includes oil, natural gas, coalbed methane, shale gas, geothermal energy, and uranium.

(a) Enhanced Oil Recovery

Depleted oil reservoirs are a leading target for storage of CO_2 and offer one of the most readily available and suitable storage solutions. When CO_2 is turned into a supercritical fluid at about 73.8 bar pressure and 31.1 °C, it is soluble in oil. The resulting solution has lower viscosity and density than the parent oil, thus enabling production of some of the oil in place from depleted reservoirs. After that, the produced fluids are separated on a platform with CO_2 recycled in situ (Amouroux et al. 2014). The technology of CO_2 flooding to achieve enhanced oil recovery (EOR) can increase oil production significantly and reduce the life cycle carbon emissions of conventional oil production by 25–60%. A relatively high percentage, about 75%, of the injected CO_2 is safely stored after production is stopped (Hasanvand et al. 2013). The reasons for CO_2 retention are (Olea 2015):

- dissolution into oil not flowing to the producing wells;
- dissolution into the formation waters;
- chemical reaction with minerals in the formation matrix;
- accumulation in the pore space vacated by the produced oil;
- leakage and dissolution into the subjacent aquifer;
- loss into nearby geological structure.

Currently, CO_2-EOR is mature and has been practiced for many years. Globally, CO_2-EOR has the potential to produce 470 billion barrels of additional oil and to store 140 billion metric tons of CO_2 which is equivalent to the greenhouse gas emissions from 750 large one GW size coal-fired power plants over 30 years (Carpenter and Koperna 2014).

(b) Enhanced Natural Gas Recovery (EGR)

Similar to EOR, by injecting CO_2 into depleted gas wells, the pressure of the well would be increased to a level that would make the gas being easily forced out of the well by the CO_2 (Jeon and Kim 2015). Since high density and viscosity of CO_2 relative to methane create a high displacing efficiency, the EGR process is technically feasible (Hussen et al. 2012). The incremental natural gas recovery enhanced by CO_2 can give additional revenue, so the overall cost of CO_2-EGR may be reduced compared to pure CO_2 geological storage in depleted gas field. However, the technology of CO_2 injection into natural gas reservoirs is still at the very early stage of development (Khan et al. 2013).

(c) Enhanced Coalbed Methane Technology

Injection of CO_2 in coalbeds with adequate permeability and high gas saturation is considered to be an attractive option for CO_2 storage. Methane is predominantly physically absorbed to the large internal surface area of the micro-pores in the coal. Because CO_2 is adsorbed more strongly than methane, the injection of CO_2 will result in expelling methane. CO_2-enhanced coalbed methane technology envisages the injection and storage of CO_2 with the concomitant production of methane. Another advantage of this technology is that coalbeds are often located in the vicinity of many current or future coal-fired power plants, so CO_2 transportation costs can be reduced.

(d) Enhanced Shale Gas Recovery

The potential storage of CO_2 in organic-rich gas shales is attracting increasing interest. The process of CO_2-enhanced shale gas recovery (ESGR) is to inject CO_2 into a shale stratum to increase the recovery efficiency of shale gas. In shale gas reservoirs, natural gas exists as free gas in the pores and open or partially open natural fractures and also as absorbed phase on clay and kerogen surfaces. Similar to CO_2-ECBM, gas shale reservoirs appear to adsorb methane while preferentially absorbing CO_2.

(e) Enhanced Geothermal System

Instead of water or brine, the use of supercritical CO_2 as the heat exchange fluid in the enhanced geothermal system (EGS) has significant potential to increase their productivity, contribute further to reducing carbon emissions and increase the economic viability of geothermal power generation (Brown 2000). The higher pressure within the reservoir compared with its surroundings will force the supercritical CO_2 fluid to diffuse into the surrounding rock masses through faults, fractures, and pores.

(f) Enhanced in Situ Uranium Leaching

CO_2-enhanced in situ uranium leaching (IUL) is a novel technology for sandstone-type uranium mining. The key process is to inject CO_2 and leach uranium ore out of geological formation through reaction with ore and minerals in ore deposits (Wei et al. 2015). The main risk linked to CO_2-IUL is radiation exposure.

10.3.4 Utilization of CO_2

Industrial utilization of CO_2 represents a promising approach for reducing carbon emissions. Some industrial utilization schemes can only store CO_2 temporarily and emit CO_2 usually to the atmosphere at the end of the product's life, which can range from days or weeks to years, while other industrial utilization schemes can store CO_2 permanently (Bruhn et al. 2016).

Various innovative construction products can be manufactured while storing CO_2 at the same time. Instead of traditional high-temperature clinkering method, a novel approach is provided for the production of a CO_2−stored cementitious material, where the hydrothermal synthesis of a cementitious material is performed utilizing carbonated lime infused with silica fume and hydrated alumina (Jo et al. 2015). In iron and steel industry, due to high calcium-silicate content, all types of steel slag (EAF, BOF, and ladelalag) show potential to react with CO_2 for production of cementitious material. The key carbonation reactions of dicalcium silicate and tricalcium silicate are

$$2(2CaO \cdot SiO_2) + CO_2 + 3H_2O = 3CaO \cdot 2SiO_2 \cdot 3H_2O + CaCO_3$$
$$2(3CaO \cdot SiO_2) + 3CO_2 + 3H_2O = 3CaO \cdot 2SiO_2 \cdot 3H_2O + 3CaCO_3$$

In general, these carbonation processes can be carried out at steel mill by using the locally produced ladle slag and flue gas CO_2 to make building products with a much reduced embodied energy in comparison to Portland cement products (Mahoutian et al. 2014). In addition, Higuchi et al. (2014) suggested using the additive (dicalcium silicate γ phase: $\gamma - 2CaO \cdot SiO_2$) and coal fly ash to produce concrete with CO_2 storage, where $\gamma - 2CaO \cdot SiO_2$ can be manufactured using a by-product containing $CaOH_2$ and SiO_2 power.

CO_2 can be used as a feed stock for chemical engineering. Currently CO_2 chemical feedstock accounts for only about $0.5-2\%$ of emissions, but in the future, it could be expected to mitigate 700 megatons of CO_2 per year (Leung et al. 2014; Morrison et al. 2016). Using highly purified CO_2, many high-added value chemicals can be synthesized for the benefit of a wide variety of sectors of the chemical industry. At high pressure and high temperature, methane can be synthesized by reaction with CO_2 and H_2 using a metallic catalyst (copper and zinc oxides on an alumina-based ceramic, $Cu/ZnO/Al_2O_3$). Here H_2 is often generated by electrolysis of seawater using a renewable energy such as wind or solar (Amouroux et al. 2014). CO_2 can also be utilized to make organic carbonates like dimethyl carbonate, propylene carbonate, etc., or inorganic carbonates like sodium carbonate or calcium carbonate. In addition, CO_2 can also be used to prepare salicylic acid, an important intermediate for pharmaceuticals (Yang and Wang 2015).

CO_2 can be used as thermochemical energy storage. Methane reforming with carbon dioxide is a good approach for solar thermochemical storage and other high-temperature energy storage. The product syngas, including hydrogen and carbon monoxide, can efficiently store the absorbed solar energy. As the operating temperature is 800 °C, the total energy efficiency is about 70% (Lu et al. 2016).

10.4 Geoengineering

The current world is facing a series of unprecedented major global environmental problems caused by global warming. Most of the observed warming over the last 50 years is likely to have been due to the increasing concentrations of greenhouse gases produced by human activities such as deforestation and burning fossil fuel.

Scientists have proposed to use geoengineering (or climate engineering) to artificially cool the planet. The main attraction of geoengineering lies in its low-energy costs and short lead time for technical implementation.

According to the location where geoengineering carry out, geoengineering is divided into four categories:

- Space-Based Geoengineering;
- Atmosphere-Based Geoenginnering;
- Land-Based Geoengineering;
- Ocean-Based Geoengineering.

Among all geoengineering schemes, two fundamental difference methodologies, including the carbon dioxide removal (CDR) and the solar radiation management (SRM), are employed. The CDR is to use physical, chemical, or biological approaches to remove atmospheric carbon dioxide, while the SRM is to increase albedo.

10.4.1 Space-Based Geoengineering

The most common space-based geoengineering is to position sun-shields in space to reflect the solar radiation. The ideal place for sun-shields is the Lagrangian point (1.5×10^6 km from the Earth), where the gravitational fields of the Earth and the Sun are in balance and allow a small mass to remain stationary relative to the Earth. Except for sun reflector, dust ring and dust cloud placed in Earth orbit are also space-based geoengineering schemes (Bewick et al. 2012). Currenlty, there are two experiments in the Geoengineering Model Intercomparison Project (GeoMIP) to simulate these geoengieering schemes:

(G1) The experiment is started from a control run. The instantaneous quadrupling of carbon dioxide concentration from pre-industrial levels is balanced by a reduction in the solar constant (is equivalent to increasing of albedo in the real world) until year 50.

(G2) The experiment is started from a control run. The positive radiative forcing of an increase in carbon dioxide concentration of 1% per year is balanced by a decrease in the solar constant (is equivalent to increasing of albedo in the real world) until year 50.

Until now, twenty mainstream Earth system modeling groups, such as CESM, HadCM3, CanESM2, CSIRO Mk3L, GISS-E2-R, NorESM1-M, BNU-ESM, and MIROC-ESM, have participated in GeoMIP and at least twelve groups have submited the corresponding experiment results on G1/G2. The outputs of GeoMIP are used for analyzing the impacts of space-based geoengineering on global climate system.

G1 is a completely artificial experiment and cannot be interpreted as a realistic geoengineering scheme, so the results from G1 are just extreme responses that may help to interpret the results of more realistic geoengineering experiments. Under G1 scenario, Kravitz et al. (2013) showed that the global temperatures are well constrained to pre-industrial levels, the polar regions are relatively warmer by approximately 0.8 °C, while the tropics are relatively cooler by approximately 0.3 °C. Tilmes et al. (2013) indicated that global decrease in precipitation of 0.12 mm day^{-1} (4.9%) over land and 0.14 mm day^{-1} (4.5%) over the ocean. For the Arctic region, Moore et al. (2014) showed that G1 returns Arctic sea ice concentrations and extent to pre-industrial conditions with intermodel spread of seasonal ice extent being much greater than the difference in ensemble means of pre-industrial and G1.

Compared with G1, G2 is relatively realistic geoengineering experiment. Jones (2013) focus on the impact of the sudden termination of geoengineering after 50 years of offsetting a 1% per annum increase in CO_2 and find that significant climate change would ensue rapidly upon the termination of geoengineering, with temperature, precipitation, and sea ice cover very likely changing considerably faster than would be experienced under the influence of rising greenhouse gas concentrations in the absence of geoengineering.

Space-based geoengineering can increase global albedo. However, many proposals in the following sections only deal with increasing regional albedo.

10.4.2 Atmosphere-Based Geoenginnering

Stratospheric geoengineering with sulfate aerosols is one of the major geoengineering schemes. It is inspired by Volcanic Mount Pinatubo which in 1991 erupted between 15 and 30 million tons of sulfur dioxide gas, and then reflected more sunlight back into space and reduced global temperatures greatly. In 1992–1993, the average temperature of the entire planet was cooled 0.4 to 0.5 °C. Stratospheric geoengineering with sulfate aerosols is the simulation of the effect of large volcanic eruptions on global climate. Moore et al. (2010) showed that if large quantities of sulfur dioxide (equivalent to almost a Pinatubo per year) are injected, sea-level drops for several decades until the mid twenty-first century before starting to rise again. Robock et al. (2009) pointed out that stratospheric geoengineering with sulfate aerosols could have unintended and possibly harmful consequences (e.g., ocean acidification).

The fundamental experiments-related stratospheric geoengineering in GeoMIP are as follows:

(G3) Assume a RCP4.5 scenario (representative concentration pathway, with a radiative forcing of 4.5 Wm^{-2} in the year 2100). Injects sulfate aerosols beginning in 2020 to balance the anthropogenic forcing and attempt to keep the net forcing constant (at 2020 levels) at the top of the atmosphere.

(G4) Assume a RCP4.5 scenario, starting in 2020, injects stratospheric aerosols at a rate of 5 Tg SO_2 per year (equivalent to a 1991 Pinatubo eruption every four years) to reduce global average temperature to about 1980 values.

Five Earth system models (BNU-ESM, GISS-E2-R, HadGEM2-ES, MIROC-ESM, MIROC-ESM-CHEM) have been used to run G3 and G4 experiments. Several terabytes of data have been produced by the GeoMIP consortium with output from these modeling experiments. Based on analyzing these GeoMIP outputs, Berdahl et al. (2014) indicated that stratospheric geoengineering is successful at producing some global annual average temperature cooling. During the geoengineering period from 2020 to 2070, the global mean rate of warming in RCP4.5 from 2020 to 2070 is 0.03 K/a, while it is 0.02 K/a for G4 and 0.01 K/a for G3. In Arctic region, summer temperature warming for RCP4.5 is 0.04 K/a, while it is 0.03 K/a and 0.01 K/a for G4 and G3, respectively. But neither G3 nor G4 experiment is capable of retaining 2020 September sea ice extents throughout the entire geoengineering period.

Goes et al. (2011) used economic model, carbon cycle model, and climate model to analyze potential economic impacts of aerosol geoenginerring strategies, they indicated that substituting aerosol geoengineering for carbon dioxide abatement can fail an economic cost-benefit test. Moreover, aerosol geoenginerring threatens some person's access to adequate food and drinking water resources, and poses serious risks to future generations. Research on ethical and scientific analysis of stratospheric geoengineering is just at the beginning, more and more comprehensive researches will be carried out in very near future (Tuana et al. 2012).

10.4.3 Land-Based Geoengineering

Large-scale afforestation and reforestation can affect and alter global carbon cycle. With increasing atmospheric carbon dioxide concentrations caused by human activity, afforestation and reforestation can increase the plant and soil sink of atmospheric CO_2 through photosynthesis and so mitigate climate change. Soil carbon sink is significantly different in different region since soil carbon cycling is affected by local temperature, local precipitation, soil types, previous land use, and forest types. Soil carbon sink increase in tropical and moist regions while decrease in temperate regions. Zomer et al. (2008) indicated that globally more than 760 Mha of land were found to be suitable to develop Clean Development Mechanism (CDM) projects on afforestation and reforestation, including 138 Mha for avoided tropical deforestation, 217 Mha for regeneration of tropical forests, and 345 Mha for plantations and agroforestry. Therefore, the development potential for CDM projects on afforestation and reforestation is huge and should will play a larger, increasingly important role in the future.

Large-scale afforestation and reforestation also affect global and regional climate directly and alter global water cycle. In the global scale, if one converted global potentially suitable land to forest, annual evapotranspiration will increase directly. Afforestation will affect runoff more in large river basins than that in small river basins, and runoff in South American will be affected most compared with other regions (Iroumé and Palacios 2013). Afforestation of upland catchments with fast-growing plantations can have significant impact on in situ water use, with consequent impacts on water availability downstream.

Biochar productions can be used to increase land carbon sink through creating biochar and mixing it with soil. However, this process will involve additional energy cost which will bring additional carbon emissions. Lehmann et al. (Lehmann et al. (2006)) estimate that current global potential production of biochar is about 0.6 gigatons (Gt) per year and by 2100, production of biochar could reach between 5.5 and 9.5 gigatons (Gt) per year.

Chemical weathering on land can reduce atmospheric carbon dioxide concentration and governs atmospheric/soil carbon dioxide uptake. Many human activities, such as acid rain, can accelerate the weathering process (Pierson-Wickmann et al. 2009). Chemical weathering on land depends on lithology, runoff or drainage intensity, hydrological flow path and seasonality, temperature, land cover/use, plant composition and ecosystem processes, and so on (Hartmann 2009). The main chemical reactions involved in weathering-based geoengineering schemes are

$$\begin{aligned} CaSiO_3 + 2CO_2 + H_2O &\longrightarrow Ca^{2+} + 2HCO_3^- + SiO_2, \\ CaCO_3 + CO_2 + H_2O &\longrightarrow Ca^{2+} + 2HCO_3^-. \end{aligned}$$

Two methods evaluating effects of weathering-based geoengineering on the removal of atmospheric CO_2 are the *reverse methodology* decomposing river chemistry into rock-weathering products (Gaillardet et al. 1999; Velbel and Price 2007) and the *forward-modeling approach* based on relations between rock-weathering rates for lithological classes and dominant control (Amiotte-Suchet et al. 2003).

Bio-energy with carbon sequestration are extensively used to inhibit the increase of the concentration of atmospheric CO_2 and mitigate global warming. It provides a powerful tool for reducing CO_2 levels fast and is free of the risks. The implementation of a global bio-energy program will provide numerous side-benefits (Read and Lermit 2005). Liquid or solid fuels derived from biomass, such as corn-based ethanol, is a carbon-neutral energy source. Recently, scientists further suggest decreasing the amount of CO_2 emitted from a corn-based ethanol biorefinery through the co-cultivation of microalgae (Rosenberg et al. 2011).

Sea-level rise will be a big disaster for human being. To prevent the melting of glaciers, Zhang et al. (2015) further presented a glacier-related geoengineering in the Greenland fjords, i.e., by building a dam in the fjord which would both block incoming warmer Atlantic waters from melting the ice shelves, and serve as a pinning point for the ice shelf to attach to as it advances. The generally cooler local climate induced by reduced melting and a more extensive ice cover compared with open water in the fjords would then serve to act as a larger scale climate feedback as the ice sheet grows and sea-level rise is slowed.

The *bio-geoengineering approach* is to engineering climate with the help of the albedo differences between plants (Ridgwell et al. 2009). Crop plants in agriculture often have a higher albedo than natural vegetation. Even if for the same crop, different varieties also have different albedo. So the bio-geoengineering only require a change in the variety of crop grown, and would not be threaten to food production. In order to assess the impact of crop albedo bio-geoengineering on regional climate, Singarayer et al. (2009) indicated that if one increases crop canopy albedo by 0.04 (which represents a potential 20% increase in canopy albedo), the largest cooling of about 1 °C will occur in the summer of Europe, while the greatest cooling in winter is expected in South East Asia.

The *White roof method* is one of cheap and easy geoengineering schemes. Using light-colored roofing materials or simply painting roofs white can increase urban surface albedo. But white roof method does not work very well in mitigating global warming (Jacobson and Hoeve 2012).

Two main *desert geoengineering* schemes are designed to carry out in desert regions. One is through afforestation, the other is through desert reflectors. Ornstein et al. (2009) suggested to plant fast-growing trees in desert regions to sequester the heat of the desert. But the side effect is that desert afforestation may be at greater risk of avian-borne disease (Manfready 2011). Gaskill et al. (2004) suggested covering the deserts by a reflective polyethylene-alumibium surface to increase mean albedo from 0.36 to 0.8 and produce a big decrease in global radiative force.

10.4.4 Ocean-Based Geoengineering

Covering 70% of the Earth's surface, the ocean contains approximately 50 times the carbon present in the atmosphere. The annual carbon flux between the atmosphere and the ocean is approximately 100 Pg (Raven and Falkowski 1999). Hence ocean-based geoengineering has big potential of development.

Ocean iron fertilization is the most important ocean-based geoengineering scheme. Photosynthesis by marine phytoplankton not only consumes carbon dioxide but also consumes nitrogen, phosphorus, and iron. Since nitrogen and phosphorus level remain high compared with the concentration of iron in ocean, adding iron into ocean can stimulate phytoplankton growth, which can potentially enhance carbon sequestration and reduce atmospheric carbon dioxide concentrations (Williamson 2012). If ocean iron fertilization is implemented for 10 years, 0.4–2.2 Gt/a carbon will be stored in the Southern Pacific Ocean. However, increased phytoplankton growth by iron fertilization could cause positive effects on overfished fish stocks and negative effects on the development of toxic algal blooms (Bertram 2010). The first ocean-fertilization field experiment done by Russ George is to spread 100 tons of iron sulfate into the Pacific Ocean from a fishing boat in the west of the islands of Haida Gwaii.

Ocean alkalinity is to put more lime into ocean to increase ocean carbon storage. The basic principle for this geoengineering project is

$$Ca(OH)_2 + 2CO_2 \longrightarrow Ca^{2+} + 2HCO_3^-.$$

Increasing ocean albedo can reflect more sunlight into the space. Since ice albedo can be much higher than seawater, a feasible geoengineering scheme is to break sea ice and increase sea ice cover in the winter of Arctic region.

10.4.5 Conclusions

Scientific discussion and research on geoengineering are today far more acceptable than that in just a few years ago. IPCC AR4(2007) does not consider geoengineering worth more than a passing mention while IPCC AR5(2013) has several sections on geoengineering (Sect. 6.5 for CDR, and Sect. 7.7 for SRM). Most of the proposed CDR geoengineering schemes are to be carried out on land or in the ocean, while most of the SRM geoengineering schemes are to be carried out in the atmosphere or space. CDR schemes are able to only sequester an amount of atmospheric CO_2 that is small compared with cumulative anthropogenic emissions. SRM geoengineering schemes can act rapidly to mitigate climate change with significant global mean temperature decreases, however, unwanted side-effects, such as diminished rainfall in some regions, would certainly also occur alongside the intended effect. The costs and benefits of SRM geoengineering schemes are likely to be widely varying spatially over the planet with some countries and regions gaining considerably while others may be faced with a worse set of circumstances than would be the case without geoengineering. Moreover, once SRM goengineering started, it must be maintained for a very long period. Otherwise, when it is terminated, climate reverts rapidly to maintain a global energy balance.

Current geoengineering research has mostly focused on physical science aspects while research on law, governance, economics, ethics, and social policy of geoengineering is very limited, so geoengineering idea is still far from deployment ready.

Further Reading

P. Amiotte-Suchet, J.L. Probst, W. Ludwig, Worldwide distribution of continental rock lithology: implications for the atmospheric/soil CO_2 uptake by continental weathering and alkalinity river transport to the oceans. Global Biogeochem. Cycles **17**, 1038 (2003)

J. Amouroux, P. Siffert, J.P. Massué, S. Cavadias, B. Trujillo, K. Hashimoto, P. Rutberg, S. Dresvin, X. Wang, Carbon dioxide: a new material for energy storage. Prog. Nat. Sci.: Materials Int. **24**, 295–304 (2014)

S. Bachu, Carbon dioxide storage capacity in uneconomic coalbeds in Alberta, Canada: Methodology, potential and site identification. Int. J. Greenhouse Gas Control **1**, 374–385 (2007)

V. Badescu, R.B. Cathcart, *Macro-Engineering Seawater in Unique Environments* (Environmental Science and Engineering, Springer, Heidelberg, 2011)

M. Berdahl, A. Robock, B. Ji, J. Moore, A. Jones, B. Kravitz, S. Watanabe, Arctic cryosphere response in the geoengineering model intercomparison project (GeoMIP) G3 and G4 scenarios. J. Geophys. Res. **119**, 3 (2014)

C. Bertram, Ocean iron fertilization in the context of the Kyoto protocol and the post-Kyoto process. Energy Policy **38**, 1130–1139 (2010)

R. Bewick, J.P. Sanchez, C.R. McInnes, The feasibility of using an L1 positioned dust cloud as a method of space-based geoengineering. Adv. Space Res. **49**, 1212–1228 (2012)

D.W. Brown, A hot dry rock geothermal energy concept utilising supercritical CO_2 instead of water, in *Proceedings of the 25th workshop on Geothermal Reservoir Engineering*, Stanford, California, 2000

T. Bruhn, H. Naims, B. Olfe-Kräutlein, Separating the debate on CO_2 utilization from carbon capture and storage. Environ. Sci. Policy **60**, 38–43 (2016)

S.M. Carpenter, G. Koperna, Development of the first internationally accepted standard for geologic storage of carbon dioxide utilizing enhanced oil recovery (EOR) under the International Standards Organization (ISO) Technical Committee TC-265. Energy Procedia **63**, 6717–6729 (2014)

A. Charnes, W.W. Cooper, E. Rhodes, Measuring the efficiency of decision making units. Eur. J. Operational Res. **2**, 429–444 (1978)

Z. Chen, S.B. Dikgwatlhe, J.F. Xue, H.L. Zhang, F. Chen, X. Xiao, Tillage impacts on net carbon flux in paddy soil of the Southern China. J. Clean. Prod. **103**, 70–76 (2015)

C. Cooper, A technical basis for carbon dioxide storage. Energy Procedia **1**, 1727–1733 (2009)

E.C. Crossin, The greenhouse gas implications of using ground granulated blast furnace slag as a cement substitute. J. Clean. Prod. **95**, 101–108 (2015)

S. Dayaratne, K.D. Gunawardena, Carbon footprint reduction: a critical study of runner production in small and medium scale enterprises in Sri Lanka. J. Clean. Prod. **103**, 87–103 (2015)

R. Ebrahimi, M. Salehi, Investigation of CO_2 emission reduction and improving energy use efficiency of button mushroom production using data envelopment analysis. J. Clean. Prod. **103**, 112–119 (2015)

L. Favier, O. Gagliardini, G. Durand, T. Zwinger, A three-dimensional full stokes model of the grounding line dynamics: effect of a pinning point beneath the ice shelf. The Cryosphere **6**, 101–112 (2012)

S. Fischer, A. Liebscher, M. Wandrey, CO_2SINK group, CO_2–brine-rock interaction-first results of long-term exposure experiments at in situ P-T conditions of the Ketzin CO_2 reservoir. Chemical Erde. **70**, 155–164 (2010)

J. Gaillardet, B. Dupre, P. Louvat, C.J. Allegre, Global silicate weathering and CO_2 consumption rates deduced from the chemistry of large rives. Chem. Geol. **159**, 3–30 (1999)

A. Gaskill, Desert area coverage, Global albedo enhancement project, 2004

M. Goes, K. Keller, N. Tuana, The economic (or lack thereof) of aerosol geoengineering. Clim. Change **109**, 79–744 (2011)

J. Guerra, L. Dutra, N. Schwinden, S. Andrade, Future scenarios and trends in energy generation in Brazil: supply and demand and mitigation forecasts. J. Clean. Prod. **103**, 197–210 (2015)

K. Hacatoglu, I. Dincer, M.A. Rosen, A new model to assess the environmental impact and sustainability of energy systems. J. Clean. Prod. **103**, 211–218 (2015)

Y.Y. Haimes, L.S. Lasdon, D.A. Wismer, On the bicriterion formulation of the integrated system identification and system optimization. IEEE Trans. Syst. Man Cyber. **1**, 296–297 (1971)

L. Han, X. Xu, L. Han, Applying quantile regression and shapley decomposition to analyzing the determinants of household embedded carbon emissions: evidence from urban China. J. Clean. Prod. **103**, 219–230 (2015)

J. Hartmann, Bicarbonate-fluxes and CO_2-consumption by chemical weathering on the Japanese Archipelago-Application of a multi-lithological model framework. Chem. Geol. **265**, 237–271 (2009)

M.Z. Hasanvand, M.A. Ahmadi, S.R. Shadizadeh, R. Behbahani, F. Feyzi, Geological storage of carbon dioxide by injection of carbonated water in an Iranian oil reservoir: a case study. J. Pet. Sci. Eng. **111**, 170–177 (2013)

T. Higuchi, M. Morioka, I. Yoshioka, K. Yokozeki, Development of a new ecological concrete with CO_2 emissions below zero. Constr. Building Mater. **67**, 338–343 (2014)

D. Huisingh, Z. Zhang, J.C. Moore, Q. Qiao, Q. Li, Resent advances in carbon emissions reduction: policies, technologies, monitoring, assessment, and modeling. J. Clean. Prod. **103**, 1–13 (2015)

C. Hussen, R. Amin, G. Madden, B. Evans, Reservoir simulation for enhanced gas recovery: an economic evaluation. J. Nat. Gas Sci. Eng. **5**, 42–50 (2012)

A. Iroumé, H. Palacios, Afforestation and changes in forest composition affect runoff in large river basins with pluvial regime and Mediterranean climate, Chile. J. Hydrol. **505**, 113–125 (2013)

S.A. Ishak, H. Hashim, Low carbon measures for cement plant-A review. J. Clean. Prod. **103**, 260–274 (2015)

M.Z. Jacobson, J.E.T. Hoeve, Effects of urban surfaces and white roofs on global and regional climate. J. Clim. **25**, 1028–1044 (2012)
S.H. Jeon, M.S. Kim, Compressor selection methods for multi-stage re-liquefaction system of liquefied CO_2 transport ship for CCS. Appl. Thermal Eng. **82**, 360–367 (2015)
A. Jone, J.M. Haywood, K. Alterskjaer, O. Boucher, J.N.S. Cole, C.L. Curry, P.J. Irvine, D. Ji, B. Kravitz, J.E. Kristjánsson, J.C. Moore, U. Niemeier, A. Robock, H. Schmidt, B. Singh, S. Tilmes, S. Watanabe, J.-H. Toon, The impact of abrupt suspension of solar radiation management (termination effect) in experiment G2 of the Geoengineering Model Intercomparison Project (GeoMIP). J. Geophys. Res. **118**, 17 (2013)
C. Khan, R. Amin, G. Madden, Carbon dioxide injection for enhanced gas recovery and storage (reservoir simulation). Egypt. J. Pet. **22**, 225–240 (2013)
J. Kim, C. Koo, C. Kim, T. Hong, H.S. Park, Integrated CO_2, cost, and schedule management system for building construction projects using the earned value management theory. J. Clean. Prod. **103**, 275–285 (2015)
K. Kravitz, K. Caldeira, O. Boucher, A. Robock, P.J. Rasch, K. Alterskjær, D. Bou Karam, J.N.S. Cole, C.L. Curry, J.M. Haywood, P.J. Irvine, D. Ji, A. Jones, D.J. Lunt, J.E. Kristjánsson, J.C. Moore, U. Niemeier, A. Ridgwell, H. Schmidt, M. Schilz, B. Singh, S. Tilmes, S. Watanabe, J.-H. Yoon, Robust climate response from the Geoengineering Model Intercomparison Project (GeoMIP). J. Geophys. Res. **118**, 3 (2013)
J. Lehmann, J. Gaunt, M. Rondon, Bio-char sequestration in terrestrial ecosystems - a review. Mitig. Adapt. Strateg. Glob. Change, **11**, 403–427 (2006)
J. Lee, T.C. Keener, Y.J. Yang, potential flue ga s impurities in carbon dioxide streams separated from coal-fired power plants. J. Air Waste Manage. Assoc. **59**, 725–732 (2009)
J. Leon, C. Aliaga, G. Boulougouris, M. Hortal, J.L. Marti, Quantifying GHG emissions savings potential in magazine paper production: a case study on supercalendered and light-weight coated papers, 2015
D.Y.C. Leung, G. Caramanna, M.M. Maroto-Valer, An overview of current status of carbon dioxide capture and storage technologies. Renew. Sustain. Energy Rev. **39**, 426–443 (2014)
G. Liu, Z. Yang, B. Chen, J. Zhang, X. Liu, Y. Zhang, M. Su, S. Ulgiati, Scenarios for sewage sludge synergic reduction and reuse in clinker production towards regional eco-industrial development: a comparative energy-based assessment. J. Clean. Prod. **103**, 301–308 (2015)
B. Lin, H. Zhao, Energy efficiency and conservation in China's chemical fiber industry. J. Clean. Prod. **103**, 345–352 (2015)
J. Lu, Y. Chen, J. Ding, W. Wang, High temperature energy storage performances of methane reforming with carbon dioxide in a tubular packed reactor. Appl. Energy **162**, 1473–1482 (2016)
J. Ma, L. Liu, B. Su, B. Xie, Exploring the critical factors and appropriate polices for reducing energy consumption of China's urban civil building sector. J. Clean. Prod. **103**, 446–454 (2015)
M. Mahoutian, Z. Ghouleh, Y. Shao, Carbon dioxide activated ladle slag binder. Constr. Build. Mater. **66**, 214–221 (2014)
R.A. Manfready, Assessing the impacts of desert afforestation on the spread of infections, Agents. Int. J. Environ. Sci. **1**, 901–910 (2011)
Z. Ming, O. Yingjie, S. Hui, CCS technology develo pment in China: Status, problems, and countermeasures-Based on SWOT analysis. Renewable and Sustainable Energy Review **39**, 604–616 (2014)
J.C. Moore, S. Jevrejeva, A. Grinsted, Efficacy of geoengineering to limit 21st century sea-level rise. Proc. Nat. Acad. Sci. **107**, 15699–15703 (2010)
J.C. Moore, A. Rinke, et al, Arctic sea ice and atmospheric circulation under the GeoMIP G1 scenario. J. Geophys. Res. **119**, 567–583 (2014)
J. Morrison, G. Jauffret, J.L. Galvez-Martos, F.P. Glasser, Magnesium-based cements for CO_2 capture and utilization. Cem. Concr. Res. **85**, 183–191 (2016)
S. Muench, Greenhouse gas mitigation potential of electricity from biomass. J. Clean. Prod. **103**, 483–490 (2015)

S. Mushtaq, T.N. Maraseni, K.R. Smith, J. Bundschuh, T. Jackson, Integrated assessment of water-energy-GHG emission tradeoffs in an irrigated lucerne production system in eastern Australia. J. Clean. Prod. **103**, 491–498 (2015)

R.A. Olea, CO_2 retention values in enhanced oil recovery. J. Petroleum Sci. Engineering **129**, 23–28 (2015)

L. Ornstein, I. Aleinov, D. Rind, Irrigated afforestation of the Sahara and Australian outback to end global warming. Clim. Change **97**, 409–437 (2009)

R.W.R. Parker, I.V. Rowe, P.H. Tyedmers, Fuel performance and carbon footprint of the global purse seine tuna fleet. J. Clean. Prod. **103**, 517–524 (2015)

A. Rainville, R. Hawkins, J. Bergerson, Building consensus in life cycle assessment: the potential for a canadian product category rules standard to enhacredibility in greenhouse gas emissions estimates for Alberta's oil sands. J. Clean. Prod. **103**, 525–533 (2015)

J.A. Raven, P.G. Falkowski, Oceanic sinks for atmosphric CO_2. Plant, Cell Environ. **22**, 741–755 (1999)

P. Read, J. Lermi, Bio-energy with carbon storage (BECS): a sequential decision approach to the threat of abrupt climate change. Energy **30**, 2654–2671 (2005)

B. Riano, M.C.G. Gonzalez, Greenhouse gas emissions of an on-farm swine manure treatment plant-comparison with conventional storage in anaerobic tanks. J. Clean. Prod. **103**, 542–548 (2015)

A.J. Ridgwell, J.S. Singarayer, A.M. Hetherington, P.J. Valdes, Tackling regional climate change by leaf albedo bio-geoengineering. Curr. Biol. **19**, 1–5 (2009)

A. Robock, A.B. Marquardt, B. Kravitz, G. Stenchikov, Benefits, risks, and costs of stratospheric geoengineering. Geophys. Res. Lett. **36**, 19 (2009)

J.N. Rosenberg, A. Mathias, K. Korth, M.J. Betenbaugh, G.A. Oyler, Microalgal biomass production and carbon dioxide sequestration from an integrated ethanol biorefinery in Iowa: a technical appraisal and economic feasibility evaluation, 2011

E.S. Rubin, H. Mantripragada, A. Marks, P. Versteeg, J. Kitchin, The outlook for improved carbon capture technology. Prog. Energy Combust. Sci. **38**, 630–671 (2012)

G.P.D. De Silva, P.G. Ranjith, M.S.A. Perera, Geochemical aspects of CO_2 sequestration in deep saline aquifers: a review. Fuel **155**, 128–143 (2015)

S. Sim, I.S. Cole, Y.S. Choi, N. Birbilis, A review of the protection strategies against internal corrosion for the safe transport of supercritical CO_2 via steel pipelines for CCS purposes. Int. J. Greenhouse Gas Control **29**, 185–199 (2014)

J.S. Singarayer, A. Ridgwell, P. Irvine, Assessing the benefits of crop albedo bio-geoengineering. Environ. Res. Lett. **4**, 045110 (2009)

D. Streimikiene, Comparison of carbon dioxide and nuclear waste storage costs in Lithuania. Renew. Sustain. Energy Rev. **16**, 2434–2445 (2012)

H. Taghdisian, M.R. Pishvaie, F. Farhadi, Multi-objective optimization approach for green desigh of methanol plant based on CO_2−efficiency indicator. J. Clean. Prod. **103**, 640–650 (2015)

A. Trabucco, R.J. Zomer, D.A. Bossio, O. van Straaten, L.V. Verchot, Climate change mitigation through afforestation/reforestation: a global analysis of hydrologic impacts with four case studies. Agric. Ecosyst. Environ. **126**, 81–97 (2008)

S. Tilmes, J. Fasullo, J.-F. Lamarque, D.R. Marsh, M. Mills, K. Alterskjær, O. Boucher, J.N.S. Cole, C.L. Curry, J.H. Haywood, P.J. Irvine, D. Ji, A. Jones, D.B. Karam, B. Kravitz, J.E. Kristjánsson, J.C. Moore, H.O. Muri, U. Niemeier, P.J. Rasch, A. Robock, H. Schmidt, M. Schulz, Y. Shuting, B. Singh, S. Watanabe, J.-H. Yoon, The hydrological impact of geo-engineering in the geoengineering model intercomparison project (GeoMIP). J. Geophys. Res. **118**, 2 (2013)

N. Tuana, R. Sriver, T. Svoboda, R. Tonkonojenkov, P. Irvine, J. Haqq-Misra, K. Keller, Towards integrated ethical and scientific analysis of geoengineering: a research agenda. Ethics Policy Env. **15**, 136–157 (2012)

M.A. Velbel, J.R. Price, Solute geochemical mass-balances and mineral weathering rates in small watersheds: methodology, recent advances, and future directions. Appl. Geochem. **22**, 1682–1700 (2007)

Y.S. Vicente, T.C. Drage, M. Ke, J. Poliakoff, M.W. George, Densities of the carbon dioxide+hydrogen, a system of relevance to carbon capture and storage. Int. J. Greenhouse Gas Control **13**, 78–86 (2013)

F. Visser, P. Dargusch, C. Smith, P.R. Grace, Application of the crop carbon progress calculator (CCAP) in a 'farm to ship' cotton production case study in Australia. J. Clean. Prod. **103**, 675–684 (2015)

X. Wang, Z. Duan, L. Wu, D. Yang, Estimation of carbon dioxide emission in highway construction: A case study in southwest region of China. J. Clean. Prod. **103**, 705–714 (2015)

P.D. Warwick, M.S. Blondes, S.T. Brennan, M.D. Corum, M.D. Merrill, Geological survey geologic carbon dioxide storage resource assessment of the United States. Energy Procedia **37**, 5275–5279 (2013)

N. Wei, X. Li, Z. Fang, B. Bai, Q. Li, S. Liu, Y. Jia (2015) Regional resource distribution of onshore carbon geological utilization in China. J. CO_2 Utilization **11**, 20–30

A.C.P. Wickmann, L. Aquilina, C. Weyer, J. Molenat, G. Lischeid, Acidification processes and soil leaching influenced by agricultural practices revealed by strontium isotopic ratios. Geochim. Cosmochim. Acta **73**, 4688–4704 (2009)

P. Williamson, D.W.R. Wallace, C. Law, P.W. Boyd, Y. Collos, P. Croot, K. Denman, U. Riebesell, S. Takedai, C. Vivian, Ocean fertilization for geoengineering: A review of effectiveness, environmental impacts and emerging governance. Process Saf. Environ. Prot. **90**, 475–488 (2012)

L. Wu, X.Q. Mao, A. Zeng, Carbon footprint accounting in support of city water supply infrastructure siting decision making: A case study in Ningbo, China. J. Clean. Prod. **103**, 737–746 (2015)

K. Yang, Y. Jung, M. Cho, S. Tae, Effect of supplementary cementitious materials on reduction of CO_2 emissions from concrete. J. Clean. Prod. **103**, 774–783 (2015)

N. Yang, R. Wang, Sustainable technologies for the reclamation of greenhouse gas CO_2. J. Clean. Prod. **103**, 784–792 (2015)

X. Zhang, L. Luo, M. Skitmore, Household carbon emission research: an analytical review of measurement, influencing factors and mitigation prospects. J. Clean. Prod. **103**, 873–883 (2015)

Z. Zhang, D. Huisingh, Carbon dioxide storage schemes: Technology, assessment, and deployment. J. Clean. Prod. **142**, 1055–1064 (2017)

Z. Zhang, D. Huisingh, Techno-economic assessment of carbon capture and storage facilities coupled to coal-fired power plants. Energy Env. **26**, 1069–1080 (2015)

Z. Zhang, J.C. Moore, Geoengineering: an alternative approach to reducing greenhouse gas effect and mitigating climate change. J. Clean. Prod. **103**, 898–907 (2015)

K. Zhao, Z. Liu, S. Yu, X. Li, H. Huang, B. Li, Analytical energy dissipation in large and medium-sized hydraulic press. J. Clean. Prod. **103**, 908–915 (2015)

B. Zhao, C.W. MacMinn, R. Juanes, Residual trapping, solubility trapping and capillary pinning complement each other to limit CO_2 migration in deep saline aquifers. Energy Procedia **63**, 3833–3839 (2014)

Y. Zhao, Z. Zhang, S. Wang, Y. Zhang, Y. Liu, Linkage analysis of sectoral CO_2 emissions based on the hypothetical extraction method in South Africa. J. Clean. Prod. **103**, 916–924 (2015)

S. Zhou, P.C. Flynn, Geoengineering downwelling ocean currents: a cost assessment. Clim. Change **71**, 203–220 (2005)

R.J. Zomer, A. Trabucco, D.A. Bossio, L.V. Verc, Hot, Climate change mitigation: a spatial analysis of global land suitability for clean development mechanism afforestation and reforestation. Agric. Ecosystems Env. **126**, 67–80 (2008)

Kolibri 360 GmbH
Ferdinand-Jühlke-Straße 7
99095 Erfurt, Deutschland
produktsicherheit@kolibri360.de